Lights and Shadows on Generalizations in Fixed Point Theory

Shahram Rezapour
Department of Mathematics
Azarbaijan Shahid Madani University, Tabriz, Iran

CRC Press
Taylor & Francis Group
Boca Raton London New York

CRC Press is an imprint of the
Taylor & Francis Group, an **informa** business
A SCIENCE PUBLISHERS BOOK

First edition published 2025
by CRC Press
2385 NW Executive Center Drive, Suite 320, Boca Raton FL 33431

and by CRC Press
4 Park Square, Milton Park, Abingdon, Oxon, OX14 4RN

CRC Press is an imprint of Taylor & Francis Group, LLC

Library of Congress Cataloging-in-Publication Data (applied for)

ISBN: 978-1-032-66312-8 (hbk)
ISBN: 978-1-032-69065-0 (pbk)
ISBN: 978-1-032-69067-4 (ebk)

DOI: 10.1201/9781032690674

Typeset in Times New Roman
by Prime Publishing Services

Preface

Fixed point theory is a significant tool within the realm of nonlinear analysis that has found application in diverse fields. The preeminent achievement in this particular theory is the Banach contraction principle, which was established by Stefan Banach in the year 1922. Over the preceding few decades, a multitude of scholars have made valuable contributions to this field from various perspectives. Even though each contraction is a map that exhibits uniform continuity, researchers have endeavored to utilize functions that exhibit discontinuity. This pursuit has consequently led to the introduction of a wide array of contractions that facilitate the utilization of the expansive realm of discontinuous functions.

Researchers have been making various innovations during this time and reviewing the fixed point theory on different mathematical structures like topological and metric spaces, normal structures, partial and cone metrics, b-metrics, etc. However, some of these structures haven't develop much due to the lack of effective applications in other areas.

A wide range of techniques has been used for research, but mistakes and gaps have been made along the way. Most of these mistakes were unintentional. Despite many efforts to create generalizations, it was found that some of them are false. This book aims to familiarize young researchers with mistakes, gaps, and false generalizations. In this book, the goal is not only to show gaps and false generalizations, but on a higher level, with a critical approach derived from famous philosophers of science such as Karl Popper and Rudolf Carnap, we seek to explain the evolution of the fixed point theory.

Acknowledgment

The author of this book would like to express gratitude and acknowledge the efforts of Dr. Mehran Ghaderi (Ph.D. in Mathematics and researcher of Quantum Fractional Differential Inclusions and Philosophy of Science). His cooperation, revision, and scientific-philosophical editing of the book were necessary for its completion. Through hours of dialogues on the subject and editing the text, he showed precision and insight, which played a significant role in the success of this scientific project. Also, the author would like to offer special thanks to the editorial and managerial team from Taylor and Francis for their assistance and excellent cooperative collaboration to produce this valuable scientific work.

Contents

Introduction

In the following, so that the reader can understand the research project intended by the author and the reason for choosing the title, we provide a short introduction of the primary discussions of the philosophy of modern science.

Throughout the last century, philosophers and scientists have discussed various issues that are now classified as part of the philosophy of science. The philosophy of science encompasses fields like physics, mathematics, biology, cosmology, and many other areas. These fields deal with the philosophical implications and consequences of scientific findings. The philosophy of science is built upon the problems that arose with the development of science. It is clear that without science, there would be no philosophy of science.

Science is a magnificent building that has helped humans explain and describe the universe in a precise way. It has also provided amazing technologies for humanity. What was not even conceivable a century ago is now a crucial field of research and study. We live in an exciting time where scientific instruments like dark energy spectrometers and data from the Hubble Space Telescope can identify the nature of dark matter, which will fundamentally change our view of the universe. We are on the verge of great scientific revolutions that may transform our understanding of the universe. Advances in science and technology have accelerated so much that, shortly, our knowledge of the nature of life, science, matter, and the laws of physics may be completely transformed.

The scientific subjects of our time have become so complex and mind-blowing that it's sometimes hard to distinguish them from science fiction. Nowadays, scientists are exploring ideas like the stellar engine[1], which could transport the sun via nuclear fusion at a speed of 50 light years every million years to a safer location within the Milky Way. It's important to note that this story is not limited to cosmology or particle physics. If we include fields like artificial intelligence, medicine, genetics, nanotechnology, and stem cells, we can see that the problems of modern science philosophy are far more complex than those faced by the twentieth-century philosophers of science. Today, philosophers of science are grappling with issues that raise questions about the nature of our existence and consciousness. These questions have the potential to put to rest at least 70,000 years of uncertainty since the cognitive revolution.

Scientists use various concepts such as hypotheses, theories, laws, and principles to express their findings, views, and discoveries. However, understanding the difference between these concepts can be confusing for many, even for students and graduates who have studied for years. One important question to consider is where and how science begins. By understanding that science begins with observation, problems, or hypotheses, we can better understand the evolution and formation of scientific theories, and we can provide better guidance for developing future scientific theories. The view that science begins with observation, originating from English philosopher Francis Bacon, is a long-standing intellectual tradition that continues to be emphasized in authoritative books on physical sciences and biology. However, some philosophers of science, like Karl Popper, have criticized Bacon's teachings, arguing they are meaningless.

From Popper's point of view, mere observation or measurement without the existence of a problem is worthless in science. That science begins with observation immediately brings to mind the question of what exactly to observe. Popper says that observation with an empty mind is not possible because observation requires the existence of a problem, and an empty mind naturally does not have a problem. From this angle, science begins when a common order is challenged, or, in other words, some of our expectations are wrong. He believes that this expectation is exactly what forms the basis of scientific theories.

Regardless of whether we agree or disagree with this interpretation, Bacon's opponents argue that the theory is not useful if it is based solely on empty observation and measurement. Collecting a large amount of data without a specific research program can result in drowning in an ocean of equations. Extracting a theory from a plethora of numbers and digits appears unlikely. A scientist proposes a temporary theory to explain a problem. This theory is then subjected

[1]M. E. Caplan, Stellar engines: Design considerations for maximizing acceleration, *Acta Astronautica*, 165, (2019), 96–104.

to critical discussion within the scientific community, which often leads to the identification of new challenges and problems. As a result, new theories must be developed to explain these emerging issues.

Problem1 $\rightarrow$ Temporary Theory $\rightarrow$ Critical Discussion $\rightarrow$ Problem2

If we look at the history of scientific theories of cosmology, from Thales to Einstein's theory, we can observe that Popper's model aligns with the evolution of cosmological theories. According to Popper's perspective, it is through scientific criticism of each theory that a new theory is developed. This new theory is more coherent, accurate, and consistent with the reality of the world. Kepler's laws, for instance, explain the mechanism of the universe better than previous models. Therefore, the alternative theory is an improved version of the previous theory, which is considered an epistemological step forward by correcting defects and removing errors from the previous theory. Similarly, Newton's dynamics combined Galileo's terrestrial physics with Kepler's astronomical physics.

Popper's contribution to criticizing scientific theories and their historical developments lies in his proposal of falsifiability as an important issue instead of verifiability. According to Popper, the main criterion for distinguishing scientific theories from non-scientific ones is the degree of their falsifiability rather than their verifiability. He believed that this criterion could be used as a method to differentiate between science and non-science. Scientific theories, such as Einstein's relativity, expose themselves to falsification, while non-scientific theories, such as Marxism and Alfred Adler's psychological theory, always seem to be confirmed under any circumstances.

Now, we have explained what we needed from the philosophy of science to express our intention about the title of this book, as well as the goals behind this research project. In fact, with the help of the mentioned material, we intend to prepare the mind of the reader that the contents of this book are not just a collection, or observation of generalizations and theorems in fixed point theory. Based on the research plan that we had in mind, we will show that firstly, we have brought the fixed point theory to the test of falsifiability. Second, the fact that some existing generalizations are not true does not indicate the invalidity and uselessness of the fixed point theory, but the opposite. According to Popper's falsifiability, this is a sign of the scientificity of the mentioned theory, adapting the problems of the philosophy of modern science to the research field of fixed point theory and presenting an improved version of it. Thirdly, we should emphasize the necessity of teaching and learning the philosophy of modern science for researchers in the field of mathematics in order to open new research problems, because science is empty without its existential understanding and its philosophy.

To clarify the mind of the reader, it may be necessary to state that the discussion here is not about the rightness or wrongness of the opinions of the philosophers of science. No one can say that the views of Thomas Kuhn, Karl Popper, Hilary Putnam, Hans Reichenbach, Hempel, Oppenheim, and Jan Hacking about the philosophy of science are wrong. The way to criticize the points of view of these philosophers has always been open, even though there are many criticisms of each of them. What we expect from the reader is, as Lakatos says, to seek to provide rescue hypotheses and to try to solve new problems.

Chapter 1

Orders, cones and graphs

In this chapter we are going to recall some preliminaries briefly.

1.1 A background

The historical trajectory of set theory diverges considerably from that of most other branches of mathematics. In the case of the latter, a protracted process can typically be discerned, wherein ideas gradually evolve until reaching an apex of inspiration, often catalyzed by the concurrent efforts of multiple mathematicians, ultimately yielding a groundbreaking discovery. Perhaps Cantor can be regarded as one of the initial mathematicians to proffer a systematic conceptualization of sets. The concept of infinity had long been a topic of profound contemplation since ancient times, with Zeno of Elea, around 450 BC, making an early and significant contribution through his investigation of infinitude. We abstain from immersing ourselves in the annals of set theory's historical development.

Let us solely contemplate this concept that a "set is a collection of things". A proficient reader ought to endeavor to fully comprehend this concept by posing several inquiries. What precisely defines a collection? What precisely defines things? Who can aid us in accumulating things? A number of additional related inquiries could assist us in gaining a deeper comprehension of the concept. Take a moment to reflect further on infinite sets. This will prove beneficial in due course. Ultimately, two vital associated concepts are member of a set and membership. Is it unequivocally evident that a thing included in a set belongs to said set? For instance, consider the set of tall girls. If we select a girl, can we be certain that she is part of this set?

We liked you know that there are complicate cases. But, we prefer you take it easy because the complicate cases are not our goal. You can consider in your mind only the sets which their members are clear and you have no problem about membership of the members.

Later, mathematicians learned to use structures on sets. In fact, this key note upgraded impact level of mathematics in applying it in another sciences. Thus, it is important we concentrate on basic structures. One of basic questions is related to nature of members of a set. For example, consider (and fix) the set

$$X_* = \{1, 2, Micheal, Apple, Happiness, Airplane, Smile\}.$$

Thus, members of a set are not additive or comparable necessarily. This makes a basic question again. How can we define a structure on a set while we do not have a lot of things actually. The answer is so easy. We have the same set and its subsets. As you know, A is a subset of X whenever each element of A belongs to X. Also, we denote the empty set by $\emptyset$ which has no any element. Note that, the empty set is a subset of any set. Most famous sets in mathematics are the set of natural numbers $\mathbb{N} = \{1, 2, \ldots, n, \ldots\}$, the set of integer numbers $\mathbb{Z} = \{\ldots, -n, \ldots, -2. -1. 0, 1, 2, \ldots, n, \ldots\}$, the set of rational numbers $\mathbb{Q} = \{\frac{k}{m} : m, k \in \mathbb{Z}, m \neq 0\}$ and the set of real numbers $\mathbb{R}$. The set $\mathbb{R} \backslash \mathbb{Q} = \{x \in \mathbb{R} : x \notin \mathbb{Q}\}$ is called the set of irrational numbers. Also, we can make Cartesian product of a set which we shall introduce it later. By using it, we could construct main mathematical structures on a set.

1.2 Orders

First we like to introduce the Cartesian product of a set.

Definition 1.1 Let X and Y be two nonempty sets. The Cartesian product of X and Y is a new set which is defined by $X \times Y = \{(x, y) : x \in X \ y \in Y\}$. The Cartesian product of a set X is defined by $X \times X = \{(x, x') : x, x' \in X\}$.

A well-known Cartesian product set in mathematics is the plane

$$\mathbb{C} = \mathbb{R}^2 = \mathbb{R} \times \mathbb{R} = \{(a, b) : a, b \in \mathbb{R}\}$$

or the set of complex numbers. By continuing this method, for each nutural number n, one can construct the n-dimensional space

$$\mathbb{R}^n = \mathbb{R} \times \cdots \times \mathbb{R} = \{(a_1, \ldots, a_n) : a_1, \ldots, a_n \in \mathbb{R}\}.$$

Example 1 *The Cartesian product of X_* is the following new set*

$$\begin{gathered} X_* \times X_* = \{(1,1),(1,2),(1,Micheal),(1,Apple),(1,Happiness),(1,Airplane), \\ (1,Smile),(2,1),(2,2),(2,Micheal),(2,Apple),(2,Happiness),(2,Airplane), \\ (2,Smile),(Micheal,1),(Micheal,2),(Micheal,Micheal),(Micheal,Apple), \\ (Micheal,Happiness),(Micheal,Airplane),(Micheal,Smile),(Apple,1), \\ (Apple,2),(Apple,Micheal),(Apple,Apple),(Apple,Happiness), \\ (Apple,Airplane),(Apple,Smile),(Happiness,1),(Happiness,2), \\ (Happiness,Micheal),(Happiness,Apple),(Happiness,Happiness), \\ (Happiness,Airplane),(Happiness,Smile),(Airplane,1),(Airplane,2), \\ (Airplane,Micheal),(Airplane,Apple),(Airplane,Happiness), \\ (Airplane,Airplane),(Airplane,Smile),(Smile,1),(Smile,2),(Smile,Micheal), \\ (Smile,Apple),(Smile,Happiness),(Smile,Airplane),(Smile,Smile)\}. \end{gathered}$$

Definition 1.2 Let X and Y be two nonempty sets. A relation from a set X into a set Y is a nonempty subset of $X \times Y$. In case $Y = X$, a relation on a set X is a nonempty subset of $X \times X$. We denote a relation by $\mathcal{R}$. We say that $x\mathcal{R}y$ whenever $(x,y) \in \mathcal{R}$.

Example 2 *For example, we can consider the following relations on X_**

$$\begin{gathered} \mathcal{R}_1 = \{(1,1),(1,2),(1,Micheal),(1,Apple),(1,Happiness),(Apple,Smile)\}, \\ \mathcal{R}_2 = \{(1,1),(2,2),(Micheal,Micheal),(Apple,Apple), \\ (Happiness,Happiness),(Airplane,Airplane),(Smile,Smile)\}, \\ \mathcal{R}_3 = \{(2,Smile),(Smile,2),(2,Micheal),(Micheal,2)\}, \\ \mathcal{R}_4 = \{(1,1),(2,2),(Micheal,Micheal),(Apple,Apple),(2,Apple),(Apple,1) \\ (Happiness,Happiness),(Airplane,Airplane),(Smile,Smile)\}, \\ \mathcal{R}_5 = \{(1,2),(2,Apple),(1,Apple)\}, \\ \mathcal{R}_6 = \{(1,1),(2,2),(Micheal,Micheal),(Apple,Apple), \\ (Happiness,Happiness),(Airplane,Airplane),(Smile,Smile), \\ (2,Airplane),(Airplane,2),(Airplane,Apple),(Apple,Airplane)\}, \end{gathered}$$

$$\mathcal{R}_7 = \{(2, Smile), (Smile, 2), (2,2), (2, Micheal), (Micheal, 2)$$
$$(Smile, Smile), (Micheal, Micheal), (Smile, Micheal), (Micheal, Smile)\},$$
$$\mathcal{R}_8 = \{(1,1), (2,2), (Micheal, Micheal), (Apple, Apple),$$
$$(Happiness, Happiness), (Airplane, Airplane), (Smile, Smile), (2, Airplane)\}$$

and $\mathcal{R}_9 = X_* \times X_*$.

Definition 1.3 We say that a relation $\mathcal{R}$ on a set X is reflexive whenever $(x,x) \in \mathcal{R}$ for all $x \in X$. Also, we say that a relation $\mathcal{R}$ is symmetric whenever $(y,x) \in \mathcal{R}$ whenever $(x,y) \in \mathcal{R}$. Finally, a relation $\mathcal{R}$ is called transitive whenever $(x,y) \in \mathcal{R}$ and $(y,z) \in \mathcal{R}$ implies $(x,z) \in \mathcal{R}$.

The following example shows that these notions are independent.

Example 3 *In last example, the relation $\mathcal{R}_1$ on X_* is not reflexive or symmetric or transitive. The relations $\mathcal{R}_2$ and $\mathcal{R}_9$ on X_* are reflexive and symmetric and transitive. The relation $\mathcal{R}_3$ on X_* is not reflexive or transitive while is symmetric. The relation $\mathcal{R}_4$ on X_* is not transitive or symmetric while is reflexive. The relation $\mathcal{R}_5$ on X_* is not symmetric and reflexive while is transitive. The relation $\mathcal{R}_6$ on X_* is not transitive while is reflexive and symmetric. The relation $\mathcal{R}_7$ on X_* is not reflexive while is transitive and symmetric. The relation $\mathcal{R}_8$ on X_* is not symmetric while is reflexive and transitive.*

Definition 1.4 We say that a relation $\mathcal{R}$ on a set X is antisymmetric whenever $(x,y) \in \mathcal{R}$ and $(y,x) \in \mathcal{R}$ implies $x = y$.

Definition 1.5 Let X be a nonempty set and $\mathcal{R}$ a reflexive, antisymmetric and transitive on X. Then, $(X, \mathcal{R})$ is called a poset or partially ordered set. For simplicity, we call $(X, \mathcal{R})$ an ordered set. Also, we say that $\mathcal{R}$ is an order on X.

Example 4 *Let X be a nonempty set. We denote the set of all subsets of X by $P(X)$. Define the relation $\preceq$ on $P(X)$ by $A \preceq B$ whenever $A \subseteq B$. Note that, $A \subseteq A$ for all $A \in P(X)$. Also, $A \subseteq B$ and $B \subseteq A$ implies $A = B$. Finally, $A \subseteq B$ and $B \subseteq C$ implies $A \subseteq C$. Thus, the relation $\preceq$ on $P(X)$ is reflexive, antisymmetric and transitive. Hence, $(P(X), \preceq)$ is an ordered set.*

Definition 1.6 Let $(X, \mathcal{R})$ be an ordered set. We say that the elements x and y in X are comparable whenever we have $x\mathcal{R}y$ or $y\mathcal{R}x$.

Next example shows that elements of an ordered set are not necessarily comparable.

Example 5 *Consider the ordered set $(P(X_*), \preceq)$. Put $A = \{1, 2, Apple\}$ and*

$$B = \{1, Apple, Airplane\}.$$

Then, A and B are not comparable.

Let X be a nonempty set. An important note about the ordered set $(P(X), \preceq)$ is that for each $A, B \in P(X)$ there exists an element $C \in P(X)$ such that $A \preceq C$ and $B \preceq C$. In fact, $C = A \cup B$ and C is not equal to A or B necessarily. In Example 5, we have $C = \{1, 2, Apple, Airplane\}$.

Definition 1.7 Let $(X, \mathcal{R})$ be an ordered set. We say that $(X, \mathcal{R})$ is totally ordered whenever each two elements of X are comparable, that is, $x\mathcal{R}y$ or $y\mathcal{R}x$ for all $x, y \in X$. If $x\mathcal{R}y$, then we say that $\max\{x, y\} = x$. If $y\mathcal{R}x$, then we say that $\max\{x, y\} = y$.

An important difference between ordered sets and totally ordered sets is that in a totally ordered set $(X_1, \mathcal{R}_1)$ we have $\max\{x, y\} = x$ or $\max\{x, y\} = y$ while this does not true in ordered sets as we checked it in Example 5. The real numbers via the usual order $\leq$ is a totally ordered set.

Let X be a nonempty set and $\mathcal{R}$ an order on X. Hereafter, we denote an ordered set $(X, \mathcal{R})$ by $(X, \leq)$. Also, we denote $x\mathcal{R}y$ by $x \leq y$. If $x \leq y$ and $x \neq y$, then we use the symbol $x < y$.

Let us to rewrite the notion of an ordered set by this new symbol again. We say that X is an ordered set whenever $x \leq x$ for all $x \in X$, $x \leq y$ and $y \leq x$ implies $x = y$, and finally $x \leq y$ and $y \leq z$ implies $x \leq z$.

1.3 Functions

Here, we provide important notion of function.

Definition 1.8 Let X and Y be two nonempty sets and $\mathcal{R}$ a relation from X into Y. We say that $\mathcal{R}$ is a function whenever it is well-defined, that is, $(x, y) \in \mathcal{R}$ and $(x, z) \in \mathcal{R}$ implies $y = z$. We usually denote a function by $\mathcal{F}$.

If $\mathcal{F}$ is a function from a set X into a set Y. If $(x, y) \in \mathcal{F}$, then we use the symbol $y = \mathcal{F}(x)$ instead $(x,y) \in \mathcal{F}$. By using this, well-defined property could be described as $x_1 = x_2$ implies $\mathcal{F}(x_1) = \mathcal{F}(x_1)$. In fact, a function should assign an element of X at most to one element of Y. Hereafter we denote a function from a set X into a set Y by $\mathcal{F} : X \to Y$ or $f : X \to Y$. There is a very important note here. This definition is only a property for relations although there are a lot of applications in different fields for functions. This is not an advantage that a function maps an element of X at most to one element of Y while assigning a subset of Y to each element of X is more interesting. We will study this different type as set-valued mappings (or multifunctions) later.

Definition 1.9 Let X and Y be two nonempty sets and $f : X \to Y$ a function. The domain of f is defined by $Dom(f) = \{x \in X : \textit{there exists } y_0 \in Y \textit{ such that } (x, y_0) \in f\}$ and range of f is defined by $R(f) = \{y \in Y : \textit{there exists } x_0 \in X \textit{ such that } (x_0, y) \in f\}$.

Definition 1.10 Let X and Y be two nonempty sets and $f : X \to Y$ a function. We say that f is a one-to-one function whenever $f(x_1) = f(x_1)$ implies $x_1 = x_2$.

Definition 1.11 Let X and Y be two nonempty sets and $f : X \to Y$ a function. We say that f is onto whenever for each $y \in Y$ there exists $x \in X$ such that $f(x) = y$.

Definition 1.12 Let X, Y and Z be three nonempty sets and $f : X \to Y$ and $g : Y \to Z$ two functions. We say that composition function fog is definable whenever

$$Rang(g) \subseteq Dom(f).$$

In this case, the composition function fog is defined by $fog(x) = f(g(x))$.

Hereafter, when we used composition of functions in this book, we mean that composition of the functions are definable.

Definition 1.13 Let $(X, \leq_1)$ and $(Y, \leq_2)$ be two ordered sets and $f : X \to Y$ a function. We say that f preserves the orders whenever $x_1 \leq_1 x_2$ implies $f(x_1) \leq_2 f(x_2)$.

1.4 Graphs

A graph is a structure amounting to a set of objects in which some pairs of the objects are in some sense "related". The objects correspond to mathematical abstractions called vertices (also called nodes or points) and each of the related pairs of vertices is called an edge (also called link or line). The set of vertices could be a part of a set G which we denote it by $V(G)$. We denote the set of edges by $E(G)$. We denote the graph by $G = (V(G), E(G))$. For example, a graph consisting of three vertices and two edges with $V(G) = \{v_0, v_1, v_2\}$ and $E(G) = \{e_1 = \overrightarrow{v_1 v_0}, e_2 = \overrightarrow{v_2 v_0}\}$ can be shown by

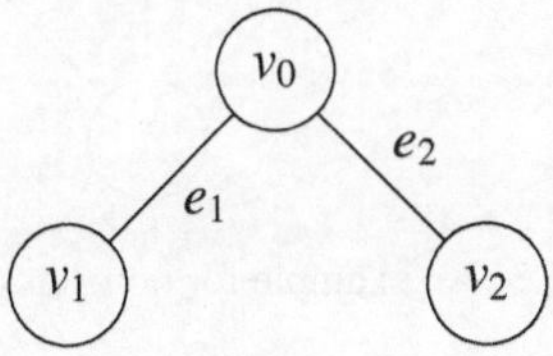

Figure 1.1: A sketch of the star graph G with two edges

In general case a graph can be shown by a figure including nodes and edges. In this book, we consider only undirected graphs. The star-graphs are well-known but we like to work with general graphs. In combinatorics one deals with finite graphs while we will consider not only finite but also infinite graphs in this book.

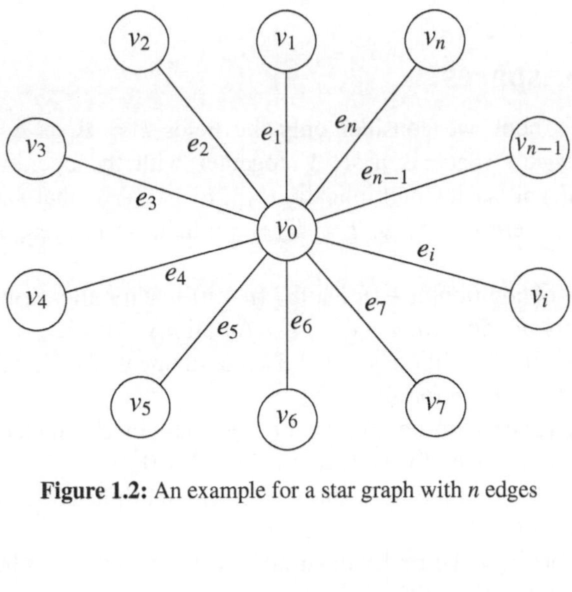

Figure 1.2: An example for a star graph with n edges

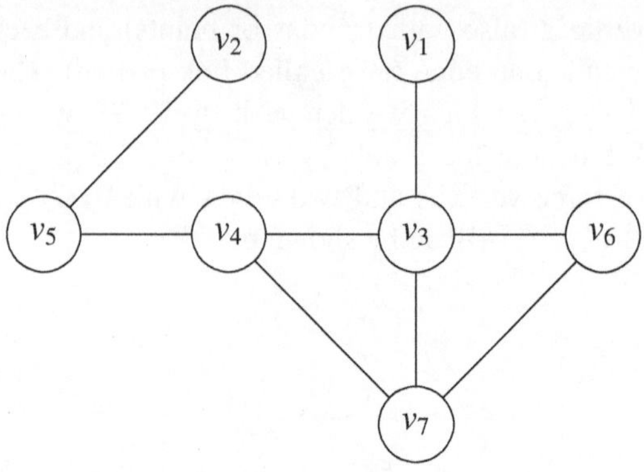

Figure 1.3: An example for non-star graphs

Let G be a graph. We consider the set of vertices of G as the set we deal with, that is, we consider the set $X = V(G)$ as the desired set. Also, each edge between two vertices can be describe as the existence of a relation between the elements. The notion of this relation could be completely independent from the probably existence structure (e.g. a metric) on the set X. In fact, the notion of edge between two vertices x and y in a graph can be completely different from the distance of x and y (if there is a distance).

1.5 Vector spaces

Throughout this book we consider only the fields $F = \mathbb{R}$ or $F = \mathbb{C}$. A real (or complex) vector space is a set V together with the operations addition $+ : V \times V \to V$ and scalar multiplication $+ : F \times V \to V$ that satisfy the following axioms, where $F = \mathbb{R}$ (or $F = \mathbb{C}$) is the field which we consider on the vectors.

1. Associativity of addition: $u + (v + w) = (u + v) + w$ for all $u, v, w \in V$;
2. Commutativity of addition: $u + v = v + u$ for all $u, v \in V$;
3. Identity element of addition: there exists an element $0 \in V$, called the zero vector, such that $v + 0 = v$ for all $v \in V$;
4. Inverse elements of addition: For each $v \in V$, there exists an element $-v \in V$, called the additive inverse of v, such that $v + (-v) = 0$;
5. Compatibility of scalar multiplication with field multiplication: $a(bv) = (ab)v$ for all $v \in V$ and $a, b \in F$;
6. Identity element of scalar multiplication: $1v = v$ for all $v \in V$, where 1 denotes the multiplicative identity in F;
7. Distributivity of scalar multiplication with respect to vector addition: $a(u + v) = au + av$ for all $u, v \in V$ and $a \in F$;

8. Distributivity of scalar multiplication with respect to field addition: $(a+b)v = av+bv$ for all $v \in V$ and $a,b \in F$.

We recall that most well-known real vector spaces are $V_1 = \mathbb{R}$, $V_2 = \mathbb{R}^n$, $V_3 = \mathbb{C}^n$ and $V_4 = \mathcal{M}_{m\times n}(\mathbb{R})$, where $\mathcal{M}_{m\times n}(\mathbb{R})$ is the set of all matrices via m rows and n columns with real entires.

Definition 1.14 Let V be a vector space on the field $F = \mathbb{R}$ (or $F = \mathbb{C}$) and W a subset of V. We say that W is a subspace of V whenever $\lambda x + \mu y \in W$ for all $x, y \in W$ and $\lambda, \mu \in F$.

Definition 1.15 Assume that V is a vector space on the field F, m is a natural number and $\{x_1, \ldots, x_m\}$ is a finite subset of V. The spanned space generated by E is denoted by $< E >$ and is defined by

$$< E >= \{\lambda_1 x_1 + \cdots + \lambda_n x_n : n \in \mathbb{N}, \lambda_1, \ldots, \lambda_n \in F, x_1, \ldots, x_n \in E\}.$$

One can check easily that $< E >$ is a subspace of V.

Definition 1.16 Let V be a vector space on F, m a natural number and $\{x_1, \ldots, x_m\}$ a finite subset of V. We say that $x_1, \ldots, x_m$ are linear independent whenever the relation $\mu_1 x_1 + \cdots + \mu_m x_m = 0$ implies $\mu_1 = \cdots = \mu_m = 0$.

Definition 1.17 Assume that V is a vector space on F and S is a subset of V. We say that S spans V whenever $V =< S >$, that is, for each element of $x \in V$ there are some $\mu_1, \ldots, \mu_m \in F$ and $s_1, \ldots, s_m \in S$ such that $x = \mu_1 s_1 + \cdots + \mu_m s_m$.

Definition 1.18 Let V be a vector space on F. We say that V is finite dimensional whenever there is a natural number n and a finite subset $S = \{x_1, \ldots, x_n\}$ of V such that $x_1, \ldots, x_n$ are linear independent and $V =< S >$. In this case, we write $dimV = n$. Also, we say that $S = \{x_1, \ldots, x_n\}$ is a basis for the vector space V.

Let $F = \mathbb{R}$, $V_1 = \mathbb{R}$, $V_2 = \mathbb{R}^2$, $V_3 = \mathbb{R}^n$, $V_4 = \mathbb{C}^n$, $V_5 = \mathcal{M}_{m\times n}(\mathbb{R})$ and $V_6 = \mathcal{M}_{m\times n}(\mathbb{C})$. Then, one can check that $dimV_1 = 1$, $dimV_2 = 2$, $dimV_3 = n$, $dimV_4 = 2n$, $dimV_5 = mn$ and $dimV_6 = 2mn$. If $F = \mathbb{C}$ and $V_7 = \mathbb{R}^{2k}$, then $dimV_2 = 1$, $dimV_4 = n$, $dimV_6 = mn$ and $dimV_7 = k$. Thus, dimension of a vector space relates to the field directly. Now, try to find some basis for these examples. For example, $S_1 = \{1\}$ is a basis for V_1 and $S_3 = \{(1,0,\ldots,0),\ldots,(0,\ldots,0,1)\}$ via n elements is a basis for V_3 in the case $F = \mathbb{R}$.

Definition 1.19 Let V_1 and V_2 be two vector spaces both on a common field F. We say that the mapping $T : V_1 \to V_2$ is linear transformation whenever $T(\lambda_1 x_1 + \lambda_2 x_2) = \lambda_1 T(x_1) + \lambda_2 T(x_2)$ for all $\lambda_1, \lambda_2 \in F$ and $x_1, x_2 \in V_1$.

We denote the set of all linear transformations from V_1 to V_2 by $L(V_1, V_2)$. If $V_2 = F$, then we denote $L(V_1, F)$ by V_1^*. If $dimV_1 = n$ and $dimV_2 = m$, then one can show that $dimL(V_1, V_2) = mn$. Since $dimF$ on F is equal to 1, $dimV^* = n$ whenever $dimV = n$. Thus, $dim(\mathbb{R}^n)^* = n$.

Let V_1 and V_2 be two vector spaces both on a common field F. We say that $V_1 \equiv V_2$ (or V_1 is isomorphic to V_2) whenever there is a one-to-one and onto linear transformation $T : V_1 \to V_2$. One can prove that for every real finite dimensional V with $dimV = n$, we have $V \equiv \mathbb{R}^n$. This implies that all n-finite dimensional vector spaces are isomorphic.

Let $F = \mathbb{R}$ and $V_1 = V_2 = \mathbb{R}$. Then, $T : V_1 \to V_2$ is a linear map if and only if there is a real constant μ_0 such that $Tx = \mu_0 x$ for all $x \in \mathbb{R}$. In fact, linear maps on $\mathbb{R}$ are the lines in the surface through the origin. This help us to consider the figures of all linear maps on $\mathbb{R}$ in the surface. Now, please try to compare in your mind the set of all linear maps on $\mathbb{R}$ with the set of all nonlinear real maps on $\mathbb{R}$. Which one is the bigger set?

By answering the question, you will find the main reason of interest for researchers to work with nonlinear maps.

It is interesting to consider now the vector spaces $\mathbb{R}^\infty = \{\{x_n\}_{n\geq 1} : x_n \in \mathbb{R}\, for\ all\ n\}$, $\ell^1 = \{\{x_n\}_{n\geq 1} : x_n \in \mathbb{R}\ for\ all\ n\ and\ \sum_{n=1}^{\infty} |x_n| < \infty\}$, the set of all real continuous maps on $[0,1]$, $C_{\mathbb{R}}([0,1])$ and the set of all real differentiable maps on $[0,1]$ via continuous differentiation, $C^1_{\mathbb{R}}([0,1])$. We will provide the notion of continuity in the second chapter.

Definition 1.20 Let V be a vector space on F. We say that V is infinite dimensional whenever there is an infinite subset $S = \{x_1, \ldots, x_n, \ldots\}$ of V such that each n elements of S are linear independent. In this case, we write $dimV = \infty$.

For each natural number m, consider the sequence $e_m = \{x_n^m\}_{n\geq 1} \in \mathbb{R}^\infty$. Now, put $S = \{e_1, \ldots, e_n, \ldots\}$. It is easy to check that each n elements of S are linear independent. Note that, $e_m \in \mathbb{R}^\infty \cap \ell^1$ for all $m \geq 1$. Thus, $dim\mathbb{R}^\infty = dim\ell^1 = \infty$. Consider the set $S_* = \{\psi_0, \psi_1, \psi_2, \ldots, \psi_{2k-1}, \psi_{2k}, \ldots\}$, where $\psi_0(t) = 1$, $\psi_1(t) = \sin t$, $\psi_2(t) = \cos t$, $\psi_{2k-1}(t) = \sin kt$ and $\psi_{2k}(t) = \cos kt$ for all $k \geq 1$ and $t \in [0,1]$. One can check that each n elements of S_* are linear independent. Note that, $S_* \subseteq C_\mathbb{R}([0,1]) \cap C^1_\mathbb{R}([0,1])$. This says us that $dimC_\mathbb{R}([0,1]) = dimC^1_\mathbb{R}([0,1]) = \infty$. Now, consider the vector spaces of real analytical functions on $[0,1]$, that is, the set of all real differentiable maps on $[0,1]$ via continuous differentiation of each order $k \geq 1$, $C^\infty_\mathbb{R}([0,1])$. Since $S_* \subseteq C^\infty_\mathbb{R}([0,1])$, we conclude that $dimC^\infty_\mathbb{R}([0,1]) = \infty$. For each $p \geq 1$, put

$$\ell^p = \{\{x_n\}_{n\geq 1} : x_n \in \mathbb{R} \; for \; all \; n \; and \left(\sum_{n=1}^{\infty} |x_n|^p\right)^{\frac{1}{p}} < \infty\}.$$

Since $e_m \in \ell^p$ for all $m \geq 1$, we get $dim\ell^p = \infty$.

Thus, there are many infinite dimensional spaces which we can deal.

1.6 Cones

During the last years, some researchers generalized some notions in fixed point theory by using orders, cones and graphs. In this section, we provide only the basic notion of a cone. We provide recent mathematical works related to the notions orders, cones and graphs in next chapters.

Definition 1.21 Let V be a vector space on F and P a subset of V. We say that P is a pointed cone (or simply, cone) in V whenever $P \cap (-P) = \{0\}$, $P + P \subseteq P$ and $\lambda P \subseteq P$ for all $\lambda \geq 0$, where $P + P = \{x + y : x, y \in P\}$ and $\lambda P = \{\lambda x : x \in P\}$.

Let $p \geq 1$, $F = \mathbb{R}$, $V_1 = \mathbb{R}$, $V_2 = \mathbb{R}^2$, $V_3 = \mathbb{R}^3$, $V_4 = \mathbb{R}^\infty$, $V5 = \ell^p$, $V_6 = C_\mathbb{R}([0,1])$, $V_7 = C^1_\mathbb{R}([0,1])$ and $V_8 = C^\infty_\mathbb{R}([0,1])$. Put

$$P_1 = \{a \in V_1 : \; a \geq 0\},$$

$$P_2 = \{(x,y) \in V_2 : \; x \geq 0 \; and \; y \geq 0\},$$

$$P_3 = \{(x_1, x_2, x_3) \in V_3 : \; x_1, x_2, x_3 \in (-\infty, 0]\},$$

$$P_4 = \{\{x_n\}_{n\geq 1} \in V_4 : \; x_{2k-1} \geq 0 \; and \; x_{2k} \leq 0 \;\; for \; all \; k \geq 1\},$$

$$P_5 = \{\{x_n\}_{n\geq 1} \in V_5 : \; x_n \geq 0 \; for \; all \; n \geq 1\},$$

$$P_6 = \{f \in V_6 : f(t) \geq 0 \; for \; all \; t \in [0,1]\},$$
$$P_6 = \{f \in V_6 : f(t) \geq 0 \; for \; all \; t \in [0,1]\},$$
$$P_7 = \{f \in V_7 : f(t) \leq 0 \; for \; all \; t \in [0,1]\}$$

and finally

$$P_8 = \{f \in V_8 : f(t) \geq 0 \; for \; all \; t \in [0, \frac{1}{2}) \; and \; f(\frac{1}{2}) = 0 \; and \; f(t) \leq 0 \; for \; all \; t \in (\frac{1}{2}, 1]\}.$$

One can easily check that P_i is a cone in V_i for $i = 1, \ldots, 8$.

A notable question is that what is importance of a cone?

May be most importance property of a cone is creating an order on the vector space. Let us to describe more this issue here.

Let V be a vector space and P a cone in V. Define the order $\preceq$ on V by $x \preceq y$ whenever $y - x \in P$. Since $0 \in P$, $x \preceq x$ for all $x \in V$. Since $P \cap (-P) = \{0\}$, $x \preceq y$ and $y \preceq x$ implies $x = y$. Since $P + P \subseteq P$, $x \preceq y$ and $y \preceq z$ implies $x \preceq z$. Thus, $\preceq$ is an order on V.

For example, consider the vector space V_8 and the cone P_8. Let $f, g \in V_8$ be two analytical functions. Here, $f \preceq g$ mean that $g(t) \geq f(t)$ for all $t \in [0, \frac{1}{2})$, $g(\frac{1}{2}) = f(\frac{1}{2})$ and $g(t) \leq f(t)$ for all $t \in (\frac{1}{2}, 1]$. If we consider the cone

$$P_8^* = \{f \in V_8 : f(t) \geq 0 \; for \; all \; t \in [0,1]\},$$

then $f \preceq g$ mean that $g(t) \geq f(t)$ for all $t \in [0,1]$. Thus, the order generated by a cone on a vector space related completely to the cone and by changing the cone, the order will be changed.

1.7 Generalization in mathematics

Mathematics has important tools such as logic, inference and induction and axioms. One of the most important mathematical tools is the use of generalization which has been a key technique in most of its fields throughout history. Assume that $\mathcal{P}_1$ and $\mathcal{P}_2$ are two statements. We say that $\mathcal{P}_2$ is a generalization for $\mathcal{P}_1$ whenever special case of $\mathcal{P}_2$ is $\mathcal{P}_1$.

Is it possible a generalization could be false?

We say that $\mathcal{P}_2$ is a false generalization for $\mathcal{P}_1$ whenever we can conclude $\mathcal{P}_2$ from $\mathcal{P}_1$, in other word, $\mathcal{P}_1$ is also a generalization for $\mathcal{P}_2$. One of our aims in this book is to show that there some false generalizations and we should be more careful.

Chapter 2

Metric structures

2.1 Metric spaces

2.1.1 A brief history

Metric space, in mathematics, especially topology, is an abstract set with a distance function, called a metric, that specifies a nonnegative distance between any two of its points in such a way that the following properties hold:
(1) the distance from the first point to the second equals zero if and only if the points are the same,
(2) the distance from the first point to the second equals the distance from the second to the first, and
(3) the sum of the distance from the first point to the second and the distance from the second point to a third exceeds or equals the distance from the first to the third.

The last of these properties is called the triangle inequality. The French mathematician Maurice Frechet initiated the study of metric spaces in 1905.

The usual distance function on the real number line is a metric, as is the usual distance function in Euclidean n-dimensional space. There are also more exotic examples of interest to mathematicians. Given any set of points, the discrete metric specifies that the distance from a point to itself equal 0 while the distance between any two distinct points equal 1. The so-called taxicab metric on the Euclidean plane declares the distance from a point (x,y) to a point (z,w) to be $|x-z|+|y-w|$. This "taxicab distance" gives the minimum length of a path from (x, y) to (z, w) constructed from horizontal and vertical line segments. In analysis there are several useful metrics on sets of bounded real-valued continuous or integrable functions.

Thus, a metric generalizes the notion of usual distance to more general settings. Moreover, a metric on a set X determines a collection of open sets, or topology, on X when a subset U of X is declared to be open if and only if for each point p of X there is a positive (possibly very small) distance r such that the set of all points of X of distance less than r from p is completely contained in U. In this way metric spaces provide important examples of topological spaces.

2.1.2 Metric, norm and inner product

The main goal of mathematics is creating mathematical structures on sets. First logical notions in the last century were distance-like structures such as metric on sets and norms and inner products on vector spaces.

Definition 2.1 Let X be a non-empty set and $d : X \times X \to [0,\infty)$ a function such that

(i) $d(x,y) = 0$ if and only if $x = y$,

(ii) $d(x,y) = d(y,x)$ for all $x, y \in X$,

(iii) $d(x,y) + d(y,z) \geq d(x,z)$ for all $x, y, z \in X$ (*triangle inequality*).

In this case, we say that d is a *metric* on X and (X,d) is a *metric space.*

Definition 2.2 Let V be a vector space over F and $\|.\| : V \to [0,\infty)$ a map such that

(i) If $\|\mathbf{u}\| = 0$ if and only if $\mathbf{u} = 0$,

(ii) $\|\lambda\mathbf{u}\| = |\lambda|\|\mathbf{u}\|$,

(iii) $\|\mathbf{u}\| + \|\mathbf{v}\| \geq \|\mathbf{u}+\mathbf{v}\|$

for all $\mathbf{u},\mathbf{v} \in V$ and $\lambda \in F$. In this case, we say that $\| \ \|$ is a *norm* and $(V, \| \ \|)$ is a *normed space.*

In fact each normed space is a metric space. Let $(V, \| \ \|)$ be a normed vector space. Define the function $d_{\|.\|} : X \times X \to [0,\infty)$ by $d(\mathbf{u},\mathbf{v}) = \|\mathbf{u}-\mathbf{v}\|$. Then, it is easy to check that $d_{\|.\|}$ is a metric on V.

Definition 2.3 Let V be a vector space over $\mathbb{R}$ and $\langle .,. \rangle : V \times V \to \mathbb{R}$ a map such that

(i) $\langle \mathbf{u}, \mathbf{u} \rangle \geq 0$,

(ii) If $\langle \mathbf{u}, \mathbf{u} \rangle = 0$, then $\mathbf{u} = 0$,

(iii) $\langle \mathbf{u}, \mathbf{v} \rangle = \langle \mathbf{v}, \mathbf{u} \rangle$,

(iv) $\langle \mathbf{u} + \mathbf{w}, \mathbf{v} \rangle = \langle \mathbf{u}, \mathbf{v} \rangle + \langle \mathbf{w}, \mathbf{v} \rangle$,

(v) $\langle \lambda \mathbf{u}, \mathbf{v} \rangle = \lambda \langle \mathbf{u}, \mathbf{v} \rangle$

for $\mathbf{u}$, $\mathbf{v}$, $\mathbf{w} \in V$, $\lambda \in \mathbb{R}$. Then we say that $\langle .,. \rangle$ is an *inner product* and $(V, \langle .,. \rangle)$ is an *inner product space*.

Every inner product space is a normed space. Let $(V, \langle .,. \rangle)$ be an inner product space. Consider the map $\|\mathbf{u}\| = \langle \mathbf{u}, \mathbf{u} \rangle^{1/2}$ on V. One can check that the Cauchy-Schwartz inequality $\|\mathbf{u}\| \|\mathbf{v}\| \geq |\langle \mathbf{u}, \mathbf{v} \rangle|$ holds for all $\mathbf{u}$, $\mathbf{v} \in V$ and $(V, \| \ \|)$ is a normed vector space. One can easily check that $\langle \mathbf{x}, \mathbf{y} \rangle = \sum_{j=1}^{n} x_j y_j$ is an inner product on $\mathbb{R}^n$ and the parallelogram law $\|\mathbf{u} + \mathbf{v}\|^2 + \|\mathbf{u} - \mathbf{v}\|^2 = 2\|\mathbf{u}\|^2 + 2\|\mathbf{v}\|^2$ holds on inner product spaces. We call the norm $\|\mathbf{x}\|_2 = \left(\sum_{j=1}^{n} x_j^2 \right)^{1/2}$ derived from this inner product the *Euclidean norm* (or sometimes just "the usual norm").

Another notable question: Is there a metric on a given arbitrary set?

The answer is affirmative. However, not all metrics can be derived from norms. Here is a metric that turns out to be more important and less peculiar than it looks at first sight. Let X be a set. Define $d : X \times X \to [0, \infty)$ by

$$d(x,y) = \begin{cases} 0 & \text{if } x = y, \\ 1 & \text{if } x \neq y. \end{cases}$$

Then, d is a metric on X. This metric is called *discrete metric* and (X, d) is called *discrete metric space*. We recommend you to test any putative theorems on metric spaces on both $\mathbb{R}^n$ with the Euclidean metric and $\mathbb{R}^n$ with the discrete metric.

There are different types of metrics. First, we introduce *The counting metric*. If E is a finite set and $\mathcal{E}$ is the collection of subsets of E, we write card C for the number of elements in C and $d(A,B) = \text{card } A \triangle B$. One can check that d is a metric on $\mathcal{E}$. The reader may be inclined to dismiss this metric as uninteresting but it plays an important role (as the *Hamming metric*) in *Codes and Cryptography*. There are two metrics which are included simply to show that metrics do not have to be as simple as the ones above. Define $d : \mathbb{R}^2 \times \mathbb{R}^2 \to [0, \infty)$ by

$$d(\mathbf{u}, \mathbf{v}) = \begin{cases} \|\mathbf{u}\|_2 + \|\mathbf{v}\|_2, & \text{if } \mathbf{u} \neq \mathbf{v}, \\ 0 & \text{if } \mathbf{u} = \mathbf{v}. \end{cases}$$

Then d is a metric and is called the *British Rail express metric* (To get from A to B travel via London). Also, consider the map $d : \mathbb{R}^2 \times \mathbb{R}^2 \to [0, \infty)$ by

$$d(\mathbf{u}, \mathbf{v}) = \begin{cases} \|\mathbf{u} - \mathbf{v}\|_2 & \text{if } \mathbf{u} \text{ and } \mathbf{v} \text{ are linearly dependent,} \\ \|\mathbf{u}\|_2 + \|\mathbf{v}\|_2 & \text{otherwise.} \end{cases}$$

Then d is a metric and is called the *British Rail stopping metric*. (To get from A to B travel via London unless A and B are on the same London route). As you see, there are different metrics on $\mathbb{R}^2$. Here, $\|\mathbf{u}\|_2 = \|(a,b)\|_2 = \sqrt{a^2 + b^2}$. Finally, $d(\mathbf{u}, \mathbf{v}) = \max \|\mathbf{u}\|_2, \|\mathbf{v}\|_2$ and $d(\mathbf{u}, \mathbf{v}) = \|\mathbf{u} - \mathbf{v}\|_2$ are another metrics on $\mathbb{R}^2$.

2.2 Continuity, compactness and open sets for metric spaces

Some definitions and results transfer essentially unchanged from classical analysis on $\mathbb{R}$ to metric spaces. Recall the classical definition of continuity.

Definition 2.4 (Old definition.)

A function $f : \mathbb{R} \to \mathbb{R}$ is called *continuous* if, given $t \in \mathbb{R}$ and $\varepsilon > 0$, we can find a $\delta(t, \varepsilon) > 0$ such that

$$|f(t) - f(s)| < \varepsilon \text{ whenever } |t - s| < \delta(t, \varepsilon).$$

It is not hard to extend this definition to our new, wider context.

Definition 2.5 (New definition.)

Let (X, d) and (Y, ρ) be metric spaces. A function $f : X \to Y$ is called *continuous* if, given $t \in X$ and $\varepsilon > 0$, we can find a $\delta(t, \varepsilon) > 0$ such that

$$\rho(f(t), f(s)) < \varepsilon \text{ whenever } d(t, s) < \delta(t, \varepsilon).$$

It may help you grasp this definition if you read "$\rho(f(t), f(s))$" as "the distance from $f(t)$ to $f(s)$ in Y" and "$d(t, s)$" as "the distance from t to s in X".

Let $\mathbb{R}$ and $\mathbb{R}^2$ have their usual (Euclidean) metric.

(i) Suppose that $f : \mathbb{R} \to \mathbb{R}$ and $g : \mathbb{R} \to \mathbb{R}$ are continuous.Then, the map $(f, g) : \mathbb{R}^2 \to \mathbb{R}^2$ is continuous.

(ii) The map $M : \mathbb{R}^2 \to \mathbb{R}$ given by $M(x, y) = xy$ is continuous.

(iii) The map $m : \mathbb{R}^2 \to \mathbb{R}$ given by $m(x,y) = f(x)g(y)$ is continuous. Now, think about the map $x \mapsto \log\left(2 + \sin\frac{1}{1+x^2}\right)$.

Cantor identified two particular classes of 'well behaved' sets. We start with open sets. For a metric space (X,d), $x \in E$ and $r > 0$, define the neighborhood of x with radius r by $N_r(x) = \{y \in X : d(y,x) < r\}$.

Definition 2.6 Let (X,d) be a metric space, E a subset of X and $x_0 \in E$. We say that x_0 is an *interior point* of E whenever there is $\delta > 0$ such that $N_\delta(x_0) \subseteq E$. We say that E is an *open* set in X whenever every element of E is an interior point. Also, $F \subseteq E$ is called closed whenever F^c is open.

One can easily check that every union of open sets is an open set. Also, finite intersection of open sets is also open. Let (X,d) be a metric space and G a subset of X. We denote the set of interior points of G by $intG$ and define it by

$$intG = \{y \in G : y \text{ is an interior point of } G\}.$$

Here are a couple of simple examples of an open set and a simple example of a set which is not open.

Example 6 *(i) Let (X,d) be a metric space, $r > 0$ and $x \in X$. Then, the neighborhood $N_r(x) = \{y : d(x,y) < r\}$ is open.*

(ii) If we work in $\mathbb{R}^n$ with the Euclidean metric, then the one point set $\{\mathbf{x}\}$ is not open.

(iii) If (X,d) is a discrete metric space, then $\{x\} = N_{\frac{1}{2}}(x)$ and so each subset G of X is open because $G = \cup_{x \in G} N_{\frac{1}{2}}(x)$.

There is a remarkable connection between the notion of open sets and continuity.

Theorem 2.1

Let (X,d) and (Y,ρ) be metric spaces. A function $f : X \to Y$ is continuous if and only if $f^{-1}(U)$ is open in X whenever U is open in Y.

Note that the theorem does not work "in the opposite direction".

Example 7 *Let $X = \mathbb{R}$ and d be the discrete metric. Let $Y = \mathbb{R}$ and ρ be the usual (Euclidean) metric.*

(i) If we define $f : X \to Y$ by $f(x) = x$, then f is continuous but there exist open sets U in X such that $f(U)$ is not open.

(ii) If we define $g : Y \to X$ by $g(y) = y$, then g is not continuous but $g(V)$ is open in X whenever V is open in Y.

Theorem 2.2

If (X,d), (Y,ρ), (Z,σ) are metric spaces and $g : X \to Y$, $f : Y \to Z$ are continuous, then so is the composition fog.

Proof: If U is open in Z, then, by continuity, $f^{-1}(U)$ is open in Y and so, by continuity, $(fg)^{-1}(U) = g^{-1}(f^{-1}(U))$ is open in X. Thus fog is continuous. ♣

Let (X,d) be a metric space, E a subset of X and $t_0 \in X$. We say that t_0 is an *limit point* of E whenever $[N_\varepsilon(t_0)\backslash\{t_0\}] \cap E \neq \emptyset$ for all $\varepsilon > 0$. One can easily check that E is an *closed* set in X if and only if every limit point of E belongs to E.

Consider a sequence $\{x_n\}_{n\geq 1}$ in a metric space (X,d). We say that $\{x_n\}_{n\geq 1}$ converges to $x \in X$ whenever for each $\varepsilon > 0$ there is an natural number N (depending on ε) such that $d(x_n,x) < \varepsilon$ for all $n \geq N$. In this case, we write $x_n \to x$ or $x = \lim_{n\to\infty} x_n$ and say that x is the *limit* of the sequence $\{x_n\}_{n\geq 1}$. One can prove that the limit of a sequence is unique. Consider two metric spaces (X,d) and (Y,ρ). It is not hard to show that a function $f : X \to Y$ is continuous if and only if $f(x_n) \to f(x)$ in Y for all sequence $\{x_n\}_{n\geq 1}$ in X with $x_n \to x$.

Definition 2.7 Let X be a set and τ a collection of subsets of X with the following properties.

(i) $\emptyset, X \in \tau$,

(ii) If $U_\alpha \in \tau$ for all $\alpha \in A$, then $\bigcup_{\alpha\in A} U_\alpha \in \tau$.

(iii) If $U_j \in \tau$ for all $1 \leq j \leq n$, then $\bigcap_{j=1}^{n} U_j \in \tau$.

Then we say that τ is a *topology* on X and that (X,τ) is a *topological space.*

Definition 2.8 Let (X,τ) be a topological space. We say that $\mathcal{B} \subseteq \tau$ is a base for the topology τ whenever

(i) $X = \bigcup_{B\in\mathcal{B}} B$,

(ii) If $B_1, B_2 \in \mathcal{B}$ and $x \in B_1 \cap B_2$, then there is $B_3 \in \mathcal{B}$ such that $B_3 \subseteq B_1 \cap B_2$ and $x \in B_3$.

We say that $\mathcal{S} \subseteq \tau$ is a sub-base whenever the collection of finite intersections of elements from $\mathcal{S}$ forms a basis of τ.

Let (X,d) be a metric space. It is easy to check that the collection of open sets forms a topology. We denote this topology by τ_d and we call it the topology induced by the metric. In fact, $\tau_d = \{E \subseteq X : E\ is\ an\ open\ set\ in\ the\ sense\ of\ metric\ spaces\}$.

Let (X,τ) be a topological metric space and F a subset of X. We say that (X,τ) is metrizable whenever there is a metric d on X such that $\tau_d = \tau$. Assume that $\{A_\alpha\}_{\alpha\in I}$ is a family of subsets of X. We say that $\{A_\alpha\}_{\alpha\in I}$ is an open cover of F whenever $F \subseteq \bigcup_{\alpha\in I} A_\alpha$ and $A_\alpha \in \tau$ for all $\alpha \in I$. Also, we say that F is compact whenever for every open cover of F such $\{A_\alpha\}_{\alpha\in I}$ there are finite elements $A_{\alpha_1},\dots,A_{\alpha_l}$ of the open cover such that $F \subseteq \bigcup_{i=1}^n A_{\alpha_i}$. In metric spaces, a subset F is compact if and only if every bounded sequence $\{x_n\}_{n\geq 1}$ in F has a convergent subsequence to an element of F. All compact subsets of a metric space are bounded and closed but the converse part does not hold necessarily.

Let V be a vector space on F and τ a topology on V. We say that V is a topological vector space whenever $\{x\}$ is closed for all $x \in V$ and the mappings $+ : V \times V \to V$ and $. : F \times V \to V$ defined by $+(x,y) = x+y$ and $.(\lambda,x) = \lambda x$ are continuous maps.

2.3 Cone metrics

2.3.1 A brief history

In the workshop on real and complex singularities which was held in Brazil, August 1992, Rabelo introduced the sentence of cone metrics ([78]). In 1996, Jeffres wrote her ph.d. thesis entitled "Kahler-Einstein cone metrics" in State University of New York at Stony Brook ([56]). Then, she published two papers about cone metrics in the new century ([58] and [57]). In 2005, Schumacher and Trapani published a paper on variation of cone metrics on Riemann surfaces ([95]). In 2006, Parker published a paper about cone metrics on the sphere and Livne's lattices ([79]). But, all of them used cone metrics by using geometric sights. Although there are some works which have used the notion of cone metric spaces (see for example, [69], [103] and [106]), but Huang and Zhang used the notion of cone metric spaces (CMS) systemically in 2007 ([48]).

2.3.2 Some basic questions

For stat of this section, let us ask a question.

Is it necessary that values of a metric be non-negative real numbers? What will happen if we replace the values with vectors?

If we generalize the notion of metric by cone metrics, basic question arises.

Will the notion of cone metrics be an appropriate generalization for metrics? Is it a real generalization in mathematics?

Our goal is to answer similar questions in this book.

2.3.3 Basic topological notions

In this section, we provide some preliminaries.

Let E be a real normed space and P a subset of E. Define the partial order $\leq$ with respect to P by $x \leq y$ if and only if $y - x \in P$. The notation $x < y$ will stand for $x \leq y$ and $x \neq y$, while $x \ll y$ will stand for $y - x \in intP$, where $intP$ denotes the interior of P.

The cone P is called normal if there is a number $M > 0$ such that for all $x, y \in E$

$$0 \leq x \leq y \text{ implies } \|x\| \leq M\|y\|.$$

The least positive number satisfying above is called the normal constant of P ([48]). We shall show that $M \geq 1$.

Consider the vector space $E = \ell^1$, the cone $P = \{\{x_n\}_{n\geq 1} \in \ell^1 : x_n \geq 0 \; for \; all \not\geq 1\}$ and very simple elements $x = \{x_n\}_{n\geq 1} \in \ell^1$ and$y = \{y_n\}_{n\geq 1} \in \ell^1$, where $x_1 = 1$ and $x_k = 0$ for all $k \geq 2$ and $y_1 = 0$ and $y_2 = 1$ and $y_k = 0$ for all $k \geq 3$. Note that, the relations $x \leq y$ or $y \leq x$ do not hold. This is difference of cone metric spaces with the spaces which we deal those usually. Elements of cone metric spaces are not comparable necessarily.

The cone P is called regular if every increasing sequence which is bounded from above is convergent. That is, if $\{x_n\}_{n\geq 1}$ is a sequence such that $x_1 \leq x_2 \leq \cdots \leq y$

for some $y \in E$, then there is $x \in E$ such that $\lim_{n\to\infty} \|x_n - x\| = 0$. Equivalently the cone P is regular if and only if every decreasing sequence which is bounded from below is convergent. It is well known that every regular cone is normal. We shall prove it later.

In [48], the authors considered E as a real normed space while we can consider E as a topological vector space. We always like to consider E as a topological vector space, P is a cone in E with $intP \neq \emptyset$ and $\leq$ is partial ordering with respect to P.

Definition 2.9 Let X be a non-empty set. Suppose the map $d : X \times X \to E$ satisfies
(d_1) $0 \leq d(x,y)$ for all $x,y \in X$ and $d(x,y) = 0$ if and only if $x = y$,
(d_2) $d(x,y) = d(y,x)$ for all $x,y \in X$,
(d_3) $d(x,y) \leq d(x,z) + d(z,y)$ for all $x,y,z \in X$.

Then d is called a cone metric on X, and (X,d) is called a cone metric space.

Note that $0 \leq d(x,y)$ means $d(x,y) \in P$ and so value of $d(x,y)$ is not necessarily a non-negative real number.

Example 8 *Let $E = \mathbb{R}^2$, $P = \{(x,y) \in E : x,y \geq 0\}$, $X = \mathbb{R}$ and $d : X \times X \to E$ defined by $d(x,y) = (|x-y|, \alpha|x-y|)$, where $\alpha \geq 0$ is a constant. Then (X,d) is a cone metric space ([48]).*

Example 9 *Let $E = \ell^1$, $P = \{\{x_n\}_{n\geq 1} \in E : x_n \geq 0, \text{for all } n\}$, (X,ρ) a metric space and $d : X \times X \to E$ defined by $d(x,y) = \{\frac{\rho(x,y)}{2^n}\}_{n\geq 1}$. Then (X,d) is a cone metric space and the normal constant of P is equal to $M = 1$.*

Note that, vale of the cone metric d is a sequence in ℓ^1. We can consider different values for the metric d. This example tells us that the category of cone metric spaces is bigger than the one metric spaces because we can make a cone metric (in different methods) by using a metric space. As you see, the notion of cone metric is a generalization for the metric. What is your opinion?

Definition 2.10 Let (X,d) be a cone metric space, $x \in X$ and $\{x_n\}_{n\geq 1}$ a sequence in X.
(i) $\{x_n\}_{n\geq 1}$ converges to x whenever for every $c \in E$ with $0 \ll c$ there is a natural number N such that $d(x_n,x) \ll c$ for all $n \geq N$. We denote this by $\lim_{n\to\infty} x_n = x$ or $x_n \to x$.
(ii) $\{x_n\}_{n\geq 1}$ is a Cauchy sequence whenever for every $c \in E$ with $0 \ll c$ there is a

natural number N such that $d(x_n, x_m) \ll c$ for all $n, m \geq N$.
(iii) (X, d) is a complete cone metric space if every Cauchy sequence is convergent.

Let us here to list some elementary results of [48].

Let (X, d) be a cone metric space, P a normal cone with normal constant K, $x \in X$ and $\{x_n\}_{n \geq 1}$ a sequence in X. Then
(i) $\{x_n\}_{n \geq 1}$ converges to x if and only if $d(x_n, x) \to 0$.
(ii) Limit point of every sequence is unique.
(iii) Every convergent sequence is Cauchy.
(iv) $\{x_n\}_{n \geq 1}$ is a Cauchy sequence if and only if $d(x_n, x_m) \to 0$ as $n, m \to \infty$.
(v) If $x_n \to x$ and $x'_n \to x'$, then $d(x_n, x'_n) \to d(x, x')$ as $n \to \infty$.

Definition 2.11 Let (X, d) be a cone metric space and $B \subseteq X$.
(i) $b \in B$ is called an interior point of B whenever there is $0 \ll p$ such that $N(b, p) \subseteq B$, where $N(b, p) := \{y \in X : d(y, b) \ll p\}$.
(ii) An element $x \in X$ is called a limit point of B whenever for every $0 \ll e$, $N(x, e) \cap (B \backslash \{x\}) \neq \emptyset$. A subset $B \subseteq X$ is called closed whenever each limit point of B belongs to B.
(iii) A subset $A \subseteq X$ is called open whenever each element of A is an interior point of A.
(iv) A subset $B \subseteq X$ is called bounded whenever there exist $0 \ll c$ and $x_0 \in X$ such that $d(b, x_0) \ll c$ for all $b \in B$.
(v) A subset $B \subseteq X$ is called compact whenever every open cover of B has a finite subcover.

2.4 Different types of cones

You can check that most familiar cones are normal with the normal constant $M = 1$. In this section, we first characterize different types of cones. First, we show that there are some normal cones with normal constant $M > 1$. In this way, we provide the next result.

Proposition 2.1

For each $k > 1$, there is a normal cone with normal constant $M > k$.

Proof: Let $k > 1$ be given. Consider the real vector space

$$E = \{ax+b | a,b \in \mathbb{R}; x \in [1-\frac{1}{k},1]\},$$

with Sup norm and the cone

$$P = \{ax+b \in E | a \leq 0, b \geq 0\}$$

in E. First, we show that P is regular (and so normal). Let $\{a_nx+b_n\}_{n\geq 1}$ be an increasing sequence which is bounded from above, that is there is an element $cx+d \in E$ such that

$$a_1x+b_1 \leq a_2x+b_2 \leq \cdots \leq a_nx+b_n \leq \cdots \leq cx+d,$$

for all $x \in [1-\frac{1}{k},1]$. Then, $\{a_n\}_{n\geq 1}$ and $\{b_n\}_{n\geq 1}$ are two sequences in $\mathbb{R}$ such that

$$b_1 \leq b_2 \leq \cdots \leq d, \quad a_1 \geq a_2 \geq \cdots \geq c.$$

Thus, $\{a_n\}_{n\geq 1}$ and $\{b_n\}_{n\geq 1}$ are convergent. Let $a_n \to a$ and $b_n \to b$. Then, $ax+b \in P$ and $a_nx+b_n \to ax+b$. Therefore, P is regular. Hence, there is $M \geq 1$ such that $0 \leq g \leq f$ implies $\|g\| \leq M\|f\|$, for all $g, f \in E$. Now, we show that $M > k$. First, note that $f(x) = -kx+k \in P$, $g(x) = k \in P$ and $f-g \in P$. So, $0 \leq g \leq f$. Therefore, $k = \|g\| \leq M\|f\| = M$. On the other hand, if we consider $f(x) = -(k+\frac{1}{k})x+k$ and $g(x) = k$, then $f \in P, g \in P$ and $f-g \in P$. Also, $\|g\| = k$ and $\|f\| = 1-\frac{1}{k}+\frac{1}{k^2}$. Thus, $k = \|g\| > k\|f\| = k+\frac{1}{k}-1$. This shows that $M > k$.

♣

Now by providing next key example, we show that there are non-normal cones.

Example 10 *Let $E = C^2_{\mathbb{R}}([0,1])$ with the norm $\|f\| = \|f\|_\infty + \|f'\|_\infty$ and consider the cone $P = \{f \in E : f \geq 0\}$. For each $k \geq 1$, consider the maps $f(x) = x$ and $g(x) = x^{2k}$. Then, $0 \leq g \leq f$, $\|f\| = 2$ and $\|g\| = 2k+1$. Since $k\|f\| < \|g\|$, k is not normal constant of P. Therefore, P is a non-normal cone.*

This example tells us that every new result in cone metric spaces without the condition normality, is a generalization of the old related result in metric spaces and ordered metric spaces. What is your opinion?

2.5 Some topological properties

In this section, we provide some topological properties of cone metric spaces.

Lemma 2.1

Let (X,d) be a cone metric space, $a,b,c \in X$, A a non-empty subset of X and $\alpha > 0$ a real number. Then

(i) If $a \ll b$ and $b \ll c$, then $a \ll c$.
(ii) If $a \leq b$ and $b \ll c$, then $a \ll c$.
(iii) $\alpha intP \subseteq intP$.
(iv) For each $\delta > 0$ and $x \in intP$ there is $0 < \gamma < 1$ such that $\|\gamma x\| < \delta$.
(v) For each $0 \ll c_1$ and $c_2 \in P$, there is an element $0 \ll d$ such that $c_1 \ll d$, $c_2 \ll d$.
(vi) For each $0 \ll c_1$ and $0 \ll c_2$, there is an element $0 \ll e$ such that $e \ll c_1$, $e \ll c_2$.

Proof: Note that E is a topological vector space and $intP$ is an open subset of E. Since the operations $+$ and $.$ are homeomorphism ([5]), $x + intP$ and $\alpha intP$ are open subsets of E. But $intP + intP$ is open, because $intP + intP = \bigcup_{x \in intP} x + intP$. Also, we have $intP + intP \subseteq P$ and $\alpha intP \subseteq P$. Thus, $intP + intP \subseteq intP$ and $\alpha intP \subseteq intP$. This proves (i) and (iii). For (ii), note that $intP + P = \bigcup_{x \in P} x + intP$. Hence, $intP + P \subseteq intP$. This proves (ii). Now, let $\delta > 0$ and $x \in intP$. Choose a natural number $n > 1$ such that $\frac{\delta}{n\|x\|} < 1$. Put $\gamma = \frac{\delta}{n\|x\|}$. Then, $\|\gamma x\| < \delta$. This proves (iv). For (v), choose $\delta > 0$ such that $c_1 + B(\delta, 0) \subseteq intP$, where $B(\delta, 0)$ is the norm neighborhood of 0 in E with radius $\delta > 0$. Since, $B(\delta, 0)$ is absorbing ([5]), we can take $m > 1$ such that $c_2 \in mB(\delta, 0)$. Thus, $-c_2 \in mB(\delta, 0)$ and so $mc_1 - c_2 \in intP$. Now, $d = mc_1$ has the desired properties. For (vi), continue the proof of before part and put $d = \frac{1}{m}c_2$.

♣

Lemma 2.2

Let (X, d) be a cone metric space and A a non-empty subset of X.
(i) Every convergent sequence is bounded.
(ii) If $x \in X$ is limit point of A, then there exists a sequence $\{x_n\}_{n \geq 1}$ in A such that $x_n \to x$.
(iii) For every $0 \ll e$ and $x \in X$, $N(x, e)$ is open.
(iv) $F \subseteq X$ is open if and only if F^c is closed.
(iv) For each $0 \ll c$ and $x \in X$, the set $B(x, c) := \{y \in X : d(x, y) \leq c\}$ is closed.

Proof: (i) Let $\{x_n\}_{n \geq 1}$ be a convergent sequence in X and $0 \ll c$. Choose a natural number N such that $d(x_n, x) \ll c$ for all $n > N$. By Lemma 2.1, there is $0 \ll e_1$ such that $d(x_i, x) \ll e_1$ for $i = 1, 2, \cdots, N$. Again, by Lemma 2.1 part (v), there is $0 \ll e$ such that $e_1 \ll e$ and $c \ll e$. Now, by Lemma 2.1, $d(x_n, x) \ll e$ for all $n \geq 1$. Therefore, $\{x_n\}_{n \geq 1}$ is bounded.
(ii) Let $0 \ll c$ and $0 \ll e$ be given. For each $n \geq 1$, choose

$$x_n \in N(x, \frac{e}{n}) \cap (A \backslash \{x\}).$$

By Lemma 2.1, there is a natural number $N > 1$ such that $e \ll Nc$. Thus, $d(x_n, x) \ll c$ for each $n \geq N$. Therefore, $\{x_n\}_{n \geq 1}$ is convergent.

(iii) Let $0 \ll e$ and $x \in X$ be given and $y \in N(x,e)$. Put $c = e - d(x,y)$ and note that $0 \ll c$. We prove that $N(y,c) \subseteq N(x,e)$. If $z \in N(y,c)$, then $d(z,y) \ll e - d(x,y)$. Since $d(x,z) \leq d(x,y) + d(z,y)$, by using (ii) we have $d(x,z) \ll e$. Hence, $z \in N(x,e)$.
(iv) It is obvious.
(v) If $z \in X$ is a limit point of $B(x,c)$, then $N(z,\frac{c}{n}) \cap B(x,c) \neq \emptyset$, for each $n \geq 1$. Now, if $y \in N(z,\frac{c}{n}) \cap B(x,c)$, then $d(x,y) \ll c$ and $d(y,z) \leq \frac{c}{n}$. Hence, $d(x,y) + d(y,z) \ll c + \frac{c}{n}$. So, $d(x,z) \ll c + \frac{c}{n}$. Since $\lim_{n\to\infty} \|\frac{c}{n}\| = 0$, $d(x,z) \leq c$. Thus, $z \in B(x,c)$.

♣

Lemma 2.3

Let (X,d) be a cone metric space, $0 \ll c$, $0 \ll c'$ and $x,y \in X$.
(i) $N(x,c) \cup N(x,c') \subseteq N(x,c+c')$.
(ii) $N(x,c) \cap N(x,c') \subseteq N(x,\frac{c+c'}{2})$.
(iii) If $N(x,c) \cap N(y,c') \neq \emptyset$, then $y \in N(x,c+c')$.
(iv) $N(x,c) \cup N(y,c') \subseteq N(x,c+c'+d(x,y))$.

Proof: (i) Let $y \in N(x,c)$. Then, $c - d(x,y) \in intP$. Since $c + c' - d(x,y) \in intP$, $y \in N(x,c+c')$.
(ii) Let $y \in N(x,c) \cap N(x,c')$. Then, $c - d(x,y), c' - d(x,y) \in intP$. Since $c + c' - 2d(x,y) \in intP$, $\frac{c+c'}{2} - d(x,y) \in intP$. Thus, $y \in N(x,\frac{c+c'}{2})$.
(iii) Let $z \in N(x,c) \cap N(y,c')$. Then, $c - d(x,y), c' - d(x,y) \in intP$. Since $c + c' - d(x,z) - d(y,z) \in intP$, by Lemma 2.1, $c + c' - d(x,y) \in intP$. Hence, $y \in N(x,c+c')$.
(v) It is clear that $N(x,c) \subseteq N(x,c+c'+d(x,y))$. Let $z \in N(y,c')$. Then, $c' - d(y,z) \in intP$. Since $d(x,z) \ll d(x,y) + d(z,y) + c$ and $d(x,y) + d(z,y) + c \ll c' + c + d(x,y)$, $z \in N(x,c+c'+d(x,y))$.

♣

Lemma 2.4

Let (X,d) be a cone metric space.
(i) For any collection $\{G_\alpha\}_{\alpha\in I}$ of open subsets of X, $\cup_{\alpha\in I} G_\alpha$ is open.
(ii) For any collection $\{F_\alpha\}_{\alpha\in I}$ of closed subsets of X, $\cap_{\alpha\in I} F_\alpha$ is closed.
(iii) For any finite collection $G_1, G_2, \cdots, G_n$ of open sets, $\cap_{i=1}^{n} G_i$ is open.
(iv) For any finite collection $F_1, \cdots, F_n$ of closed sets, $\cup_{i=1}^{n} F_i$ is closed.

Proof: The part (i) is obvious and by Lemma 2.2, the parts (ii) and (iv) are consequences of (i) and (iii). So, it is enough that we prove the part (iii). Let $x \in \cap_{i=1}^{n} G_i$. For each $i = 1,2,\cdots,n$, there is $0 \ll c_1$ such that $N(x,c_i) \subseteq G_i$. Put $N = \cap_{i=1}^{n} N(x,c_i)$. Since $x \in N$, it is sufficient we prove that N is open. If $y \in N$, then $\frac{\sum_{i=1}^{n} c_i}{n} - d(x,y) \in intP$. Let $z \in N(y, \frac{\sum_{i=1}^{n} c_i}{n} - d(x,y))$ and take $r > 0$ such that

$$\{t \in E : \|\frac{\sum_{i=1}^{n} c_i}{n} - d(x,y) - d(z,y) - t\| < r\} \subseteq P.$$

We show that $\cap_{i=1}^{n}\{t \in E : \|c_i - d(x,z) - t\| < r\} \subseteq P$. If

$$\|c_i - d(x,z) - t\| < r$$

for all $1 \leq i \leq n$, then $t - d(x,y) - d(z,y) + d(x,z) + \frac{\sum_{i=1}^{n} c_i}{n} - c_i \in P$. Since $d(x,y) + d(y,z) - d(x,z) \in P, t + \frac{\sum_{i=1}^{n} c_i}{n} - c_i \in P$. Thus,

$$nt = \sum_{i=1}^{n}(t + \frac{\sum_{i=1}^{n} c_i}{n} - c_i) \in P.$$

So, $t \in P$. Thus, $\cap_{i=1}^{n}\{t \in E : \|c_i - d(x,z) - t\| < r\} \subseteq P$. Hence, $z \in \cap_{i=1}^{n} N(x,c_i)$, and so $N(y, \frac{\sum_{i=1}^{n} c_i}{n} - d(x,y)) \subseteq N$. Hence, N is open and this completes the proof.

♣

Lemma 2.5

Let (X,d) be a cone metric space. Then, the family

$$\mathcal{B} = \{N(x,e) : x \in X, 0 \ll e\}$$

is a sub-basis for a topology on X. We denote this cone topology by τ_c.

Proof: It is sufficient we show that if $N(x_1,c_1)$ and $N(x_2,c_2)$ be two elements of $\mathcal{B}$, and $y \in N(x_1,c_1) \cap N(x_2,c_2)$, then there exists $0 \ll e$ such that $N(y,e) \subseteq N(x_1,c_1) \cap N(x_2,c_2)$. For this, choose $0 \ll e_1$ and $0 \ll e_2$ such that $N(y,e_1) \subseteq N(x_1,c_1)$ and $N(y,e_2) \subseteq N(x_2,c_2)$. By using Lemma 2.1, there is $0 \ll e$ such that $e \ll e_1$ and $e \ll e_2$. Thus, $N(y,e) \subseteq N(x_1,c_1) \cap N(x_2,c_2)$.

♣

Proposition 2.2

Let (X,d) be a cone metric space. Then for each distinct elements $x,y \in X$, there is $0 \ll c$ such that $N(x,c) \cap N(y,c) = \emptyset$.

Proof: Suppose that there are distinct elements $x, y \in X$ such that $N(x,c) \cap N(y,c) \neq \emptyset$ for all $0 \ll c$. By Lemma 2.3, $y \in N(x,2c)$ for all $0 \ll c$. Thus, $y \in N(x, \frac{c}{n})$ for all $n \geq 1$. Hence, $\frac{c}{n} - d(x,y) \in P$ for all $n \geq 1$. So, $-d(x,y) \in P$. This implies that $x = y$. This is a contradiction.

♣

Corollary 2.1

Let (X,d) be a cone metric space. Then τ_c, the cone topology on X, is a Hausdorff topology.

Thus, Lemma 2 of [48] is an easy consequence of this corollary without the assumption of normality. Also, it is clear that Lemma 2 of [48] holds without the assumption of normality. Finally, $B \subseteq X$ is an element of τ_c if and only if B is an open set in the sense of usual openness in cone metric spaces.

Theorem 2.3

Let (X,d) be a cone metric space. Then for each distinct elements x and y in X, there is $0 \ll c$ such that $N(x,c) \cap N(y,c) = \emptyset$.

Proof: Suppose that there are distinct elements $x, y \in X$ such that $N(x,c) \cap N(y,c) \neq \emptyset$ for all $0 \ll c$. By Lemma 2.3, $y \in N(x,2c)$ for all $0 \ll c$. Thus, $y \in N(x, \frac{c}{n})$ for all $n \geq 1$. Hence, $\frac{c}{n} - d(x,y) \in P$ for all $n \geq 1$. So, $-d(x,y) \in P$. This implies that $x = y$. This is a contradiction.

♣

2.6 Compactness in cone metric spaces

In this section, we show that compactness properties in cone metric spaces are similar to compactness properties in metric spaces. In this way, we shall prove some results about compactness of metric spaces in cone metric spaces. Also, we shall provide a result with the additional assumption of normality. Here, notions of compactness and sequentially compactness are similar to ones in metric spaces.

Lemma 2.6

Let (X,d) be a sequentially compact cone metric space. If $\{A_\alpha\}_{\alpha \in I}$ is an open cover *of X, then there is $0 \ll c$ such that for each $D \subseteq X$ with $d(x,y) \ll c$ for all $x,y \in D$, there exists $\alpha_o \in I$ such that $D \subseteq A_{\alpha_0}$.*

Proof: Suppose that the statement doesn't hold. Then, for each $0 \ll a$ there exists $B_a \subseteq X$ such that $d(x,y) \ll e$ for all $x,y \in B_a$ but B_a doesn't include in nothing A_α. Let $0 \ll e$ be given. For each $n \geq 1$, choose $B_n \subseteq X$ such that $d(x,y) \ll \frac{e}{n}$ for all $x,y \in B_n$ but B_n doesn't include in nothing A_α. Choose $x_n \in B_n$ for all $n \geq 1$. Now, we claim that $\{x_n\}_{n\geq 1}$ has not convergent subsequence in X. Let $\{x_{n_k}\}_{k\geq 1}$ be a convergent subsequence of $\{x_n\}_{n\geq 1}$ to the point $x \in X$. Take $\alpha_o \in I$ such that $x \in A_{\alpha_0}$. Since A_{α_0} is open, there is $0 \ll f$ such that $N(x,f) \subseteq A_{\alpha_0}$. Now, choose a natural number K such that $\frac{e}{n_k} \ll \frac{f}{2}$ and $d(x_{n_k},x) \ll \frac{f}{2}$ for all $k \geq K$. Then, $d(x,t) \leq d(x,x_{n_k}) + d(x_{n_k},t) \ll f$ for all $t \in B_{n_k}$ $(k \geq K)$. Thus, $B_{n_k} \subseteq N(x,f) \subseteq A_{\alpha_0}$. This shows that $\{x_n\}_{n\geq 1}$ has not convergent subsequence in X. Therefore, this contradiction completes the proof.

♣

Lemma 2.7

Let (X,d) be a sequentially compact cone metric space. Then, X is totally bounded, that is for every $0 \ll c$ there exist $x_1,\cdots,x_m \in X$ such that $X = \cup_{i=1}^{m} N(x_i,c)$.

Proof: Suppose that the statement doesn't hold. Then, there is $0 \ll c$ such that $X \neq \bigcup_{i=1}^{m} N(x_i,c)$ for all $x_1,\cdots,x_m \in X$. Thus, for every $x_1 \in X$ there exists a sequence $\{x_n\}_{n\geq 1}$ such that $x_k \in X \setminus \cup_{i=1}^{k-1} N(x_i,c)$ for all $k \geq 2$. But this implies that $\{x_n\}_{n\geq 1}$ has not convergent subsequence in X. This contradiction completes the proof.

♣

Theorem 2.4

Let (X,d) be a cone metric. Then, the following statements are equivalent.
(a) X is compact.
(b) Every infinite subset of X has a limit point in X.
(c) X is sequentially compact.

Proof: $(a) \Rightarrow (b)$. Let $A \subseteq X$ be an infinite subset without any limit point in X. Then, for every $p \in X$ there exists $0 \ll e_p$ such that $N(p,e_p) \cap A \subseteq \{p\}$. Thus, $\{N(p,e_p)\}_{p\in X}$ is an open cover of X which has not finite subcover.
$(b) \Rightarrow (c)$. It is obvious.
$(c) \Rightarrow (a)$. Let $\{A_\alpha\}_{\alpha\in I}$ be an open cover of X. By Lemma 2.6, there is $0 \ll c$

such that for each $D \subseteq X$ with $d(x,y) \ll c$ for all $x,y \in D$, there exists $\alpha_o \in I$ such that $D \subseteq A_{\alpha_0}$. By Lemma 2.7, there exist $x_1,\cdots,x_m \in X$ such that $X = \cup_{i=1}^m N(x_i,\frac{c}{2})$. Thus, there exist $\alpha_1,\cdots,\alpha_m \in X$ such that $X = \cup_{i=1}^m A_{\alpha_i}$. Thus, X is compact.

♣

Lemma 2.8

Let (X,d) be a complete cone metric space, P a normal cone with normal constant K and $\{A_n\}_{n\geq 1}$ a sequence of non-empty, nested, closed and bounded subsets of X such that $\lim_{n\to\infty}\sup_{x,y\in A_n}\|d(x,y)\| = 0$. Then, $\cap_{n=1}^\infty A_n$ consists only one point.

Proof: Choose $x_n \in A_n$ for all $n \geq 1$. By our assumption, $\lim_{m,n\to\infty}\|d(x_n,x_m)\| = 0$. By using Lemmas 1 and 4 in [48], $\{x_n\}_{n\geq 1}$ is a Cauchy sequence. Since (X,d) is complete, there is $x \in X$ such that $x_n \to x$. Note that $x \in A_n$ for all $n \geq 1$, because $\{A_n\}_{n\geq 1}$ is a sequence of nested and closed sets. Thus, $x \in \cap_{n=1}^\infty A_n$. Let $x,y \in \cap_{n=1}^\infty A_n$. Since

$$\lim_{n\to\infty}\sup_{x,y\in A_n}\|d(x,y)\| = 0,$$

$\|d(x,y)\| = 0$. Hence, $x = y$. Therefore, $\cap_{n=1}^\infty A_n$ consists only one point.

♣

Theorem 2.5

Let (X,d) be a complete cone metric space and P a normal cone with normal constant K. Then, X is compact if and only if (X,d) is complete and totally bounded.

Proof: It is trivial that every compact cone metric space is complete and totally bounded. For the converse part, it is sufficient we prove that every infinite subset of X has a limit point in X. Let A be an infinite subset of X and $0 \ll c$. Choose a finite subset $F \subseteq X$ such that $X = \bigcup_{x\in F} B(x,c)$. There is $x_1 \in F$ such that $A \cap B(x_1,c)$ is infinite set. Let $x_1,\cdots,x_n$ be chosen. Choose $x_{n+1} \in X$ such that

$$E_n = \overline{A} \cap B(x_1,c) \cap B(x_2,\frac{c}{2}) \cap \cdots \cap B(x_{n+1},\frac{c}{n+1})$$

is infinite set. Thus, $\{E_n\}_{n\geq 1}$ is a sequence of non-empty, nested, closed and bounded subsets of X. Also, $\lim_{n\to\infty}\sup_{x,y\in E_n}\|d(x,y)\| = 0$. By Lemma 2.8, $\cap_{n=1}^\infty E_n$ consists only one point. If $\cap_{n=1}^\infty E_n = \{x\}$ for some $x \in X$, then $d(x_n,x) \ll \frac{c}{n}$. This implies that $x_n \to x$ and so x is a limit point of A.

♣

We do not provide more notions such as weak compactness, weak continuity, etc. The readers should read some specialized concepts from other books.

2.7 Partial metric spaces

In 1992, Matthews introduced partial metric spaces ([72]). In a partial metric space, the distance of a point in the self may not be zero. A partial metric is a function $p : X \times X \to [0,\infty)$ satisfying the following conditions

(a) $p(x,y) = p(y,x)$,

(b) If $p(x,x) = p(x,y) = p(y,y)$, then $x = y$,

(c) $p(x,x) \leq p(x,y)$,

(d) $p(x,z) + p(y,y) \leq p(x,y) + p(y,z)$,

for all $x,y,z \in X$. Then (X,p) is called a partial metric space. If p is a partial metric p on X, then $p^s : X \times X \to [0,\infty)$ defined by $p^s(x,y) = 2p(x,y) - p(x,x) - p(y,y)$ is a metric on X (see [72]).

Definition 2.12 ([12, 72]) Let (X,p) be a partial metric space.
(i) A sequence $\{x_n\}_{\not\geq 1}$ converges to $x \in X$ whenever $\lim_{n\to\infty} p(x_n,x) = p(x,x)$,
(ii) $\{x_n\}_{\not\geq 1}$ is called a Cauchy sequence if $\lim_{n,m\to\infty} p(x_n,x_m)$ exists (and finite),
(iii) (X,p) is said to be complete if every Cauchy sequence $\{x_n\}$ in X converges, with respect to τ_p, to a point $x \in X$, that is, $\lim_{n,m\to\infty} p(x_n,x_m) = p(x,x)$.

Definition 2.13 ([50]) A sequence $\{x_n\}_{n\geq 1}$ in a partial metric space (X,p) is called 0-Cauchy if $\lim\limits_{n,m\to\infty} p(x_n,x_m) = 0$. We say that (X,p) is 0−complete if each 0-Cauchy sequence in X converges to a point $x \in X$ such that $p(x,x) = 0$.

Note that, each 0-Cauchy sequence in (X,p) is Cauchy in (X,p^s) and every complete partial metric space is 0-complete.

Lemma 2.9

([12]) Let (X,p) be a partial metric space. Then
(i) $\{x_n\}_{n\geq 1}$ is a Cauchy sequence in (X,p) if and only if $\{x_n\}_{n\geq 1}$ is a Cauchy sequence in the metric space (X,p^s),
(ii) (X,p) is complete if and only if the metric space (X,p^s) is complete. In fact, $\lim\limits_{n\to\infty} p^s(x_n,x) = 0$ if and only if $p(x,x) = \lim\limits_{n\to\infty} p(x_n,x) = \lim\limits_{n,m\to\infty} p(x_n,x_m)$.

It seems that main idea of working with partial metrics is taken from the maximum function $p(x,y) = \max\{|x|,|y|\}$ on the real numbers. We will find out later that most generalizations on this structure are false.

2.8 b-metric spaces

In 1998, Czerwik introduced the concept of b-metric spaces ([33]).

Definition 2.14 Let X be a non-empty set and $d : X \times X \to [0,\infty)$ a function such that

(i) $d(x,y) = 0$ if and only if $x = y$,

(ii) $d(x,y) = d(y,x)$ for all $x, y \in X$,

(iii) There exists $K \geq 1$ satisfying $d(x,z) \leq K(d(x,y) + d(y,z))$ for all $x, y, z \in X$ (*K-relaxed triangle inequality*).

In this case, we say that d is a *b-metric* on X and (X,d) is a *b-metric space*.

Lemma 2.10

([33]) Let (X,d) be a b-metric space. Then for each $x_0, x_1, \ldots, x_n \in X$ we have $d(x_0,x_n) \leq \sum_{j=0}^{n-2} K^{j+1} d(x_j,x_{j+1}) + K^{n-1} d(x_{n-1},x_n)$.

Consider the function $f : \mathbb{N} \to \mathbb{N} \cup \{0\}$ defined by $f(n) = -[-\log_2 n]$. Then, we have $d(x_0,x_n) \leq K^{f(n)} \sum_{j=0}^{n-2} d(x_j,x_{j+1})$ for all n ([100]). Also, $f(n) \in \{f(n), f(n)+1\}$, $f(2n) = f(n)+1$ for all $n \geq 1$ and f is increasing ([100]). Now, consider the function $g : \mathbb{N} \cup \{0\} \to [0,\infty)$ defined by $g(0) = 0$ and $g(n) = (2n - 2^{f(n)})K^{f(n)} + (2^{f(n)} - n)K^{f(n)-1}$ for all $n \geq 1$ ([100]). Then, we have $g(n) = K(g([\frac{n}{2}])) + g(n - ([\frac{n}{2}]))$,

$$g(n) - g(n-1) \geq g(n-1) - g(n-2) > 0$$

and $g(n) \leq K((g(k)) + g(n-k))$ for all $n > 2$ and $k < n$ ([100]). By using the functions f and g, we can make some non-trivial b-metric space ([100]).

Definition 2.15 ([100]) Let (X,d) be a b-metric space, A a subset of X, $x_0 \in X$ and $\{x_n\}_{\not\geq 1}$ a sequence in X. We say that

(i) $\{x_n\}_{\not\geq 1}$ converges to x_0 whenever $\lim_{n\to\infty} d(x_n,x_0) = 0$,

(ii) $\{x_n\}_{\not\geq 1}$ a Cauchy sequence whenever $\limsup_{n\to\infty} \sup\{d(x_n,x_m) : m > n\} = 0$,

(iii) A is said to be closed whenever for each convergent sequence in A, its limit belongs to A,

(iv) A is said to be bounded whenever $\sup\{d(x,y) : x,y \in A\} < \infty$.

Let (X,d) be a b-metric space and $\{x_n\}_{\not\geq 1}$ a sequence in X. Assume that there exists $0 \leq r < 1$ such that $d(x_{n+1}, x_{n+2}) \leq rd(x_n, x_{n+1})$ for all $n \geq 1$. Then $\{x_n\}_{\not\geq 1}$ is a Cauchy sequence ([100]).

By using a little change in the notion of b-metric spaces, Khamsi introduced metric type spaces, but the structure and techniques are the same ([66]).

When mathematical researchers wants to work on a new space, they should consider whether this space is a real generalization of the previous spaces or not. Again, we ask our main question.

Are b-metric spaces a real generalization of metric spaces?

Chapter 3

Some modern fixed point results

3.1 Introduction

Our aim in this chapter is to recall some classic fixed point theory. In 1922, Stefan Banach published his famous fixed point result which we call the Banach contraction principle. Not only can one find a lot of works which use Lipschitz or contraction conditions but also in hidden used techniques in mathematical softwares such Mathematica and Matlab (and in this reason we find some illogical errors in our computer calculations specially in numerical solutions of singular differential equations). After the development of Nonlinear Analysis, during last decades many researchers tried to work with contractive type mappings that are nonlinear and discontinuous. This has contributed significantly to the bizarre development of the fixed point theory in last decades. We first state some classic fixed point results in the next two sections and then we deal with some recent ones.

3.2 Banach contraction principal

One can find different generalizations of the Banach contraction principle (see for examples, [98], [61] and [101]). Here, we provide the Banach contraction principle.

Theorem 3.1

Assume that (X,d) is a complete metric space and the mapping $T : X \to X$ satisfies the contractive condition $d(Tx,Ty) \leq kd(x,y)$ for all $x,y \in X$, where $k \in [0,1)$ is a *fix constant. Then, T has a unique fixed point in X. For each $x_0 \in X$, the iterative sequence $\{T^n x_0\}_{n\geq 1}$ converges to the fixed point.*

Proof: For each $x_0 \in X$ and $n \geq 1$, set $x_1 = Tx_0$ and $x_{n+1} = Tx_n = T^{n+1}x_0$. Then,

$$\begin{aligned} d(x_{n+1},x_n) &= d(Tx_n,Tx_{n-1}) \leq kd(x_n,x_{n-1}) \\ &\leq k^2 d(x_{n-1},x_{n-2}) \leq \cdots \leq k^n d(x_1,x_0). \end{aligned}$$

So for $n > m$,

$$\begin{aligned} d(x_n,x_m) &\leq d(x_n,x_{n-1}) + d(x_{n-1},x_{n-2}) + \cdots + d(x_{m+1},x_m) \\ &\leq (k^{n-1} + k^{n-2} + \cdots + k^m) d(x_1,x_0) \leq \frac{k^m}{1-k} d(x_1,x_0). \end{aligned}$$

Let $\varepsilon > 0$ be given. We can choose a natural number N_1 such that $\frac{k^m}{1-k}d(x_1,x_0) < \frac{\varepsilon}{2}$ for all $m \geq N_1$. Thus, $d(x_n,x_m) \leq \frac{k^m}{1-k}d(x_1,x_0) < \frac{\varepsilon}{2}$ for all $n > m$. Therefore, $\{x_n\}_{n\geq 1}$ is a Cauchy sequence in (X,d). Since (X,d) is a complete metric space, there exists $x^* \in X$ such that $x_n \to x^*$. Choose a natural number N_2 such that $d(x_n,x^*) < \frac{\varepsilon}{2}$ for all $n \geq N_2$. Hence,

$$\begin{aligned} d(Tx^*,x^*) &\leq d(Tx_n,Tx^*) + d(Tx_n,x^*) \leq kd(x_n,x^*) + d(x_{n+1},x^*) \\ &\leq d(x_n,x^*) + d(x_{n+1},x^*) < \frac{\varepsilon}{2} + \frac{\varepsilon}{2} = \varepsilon \end{aligned}$$

for all $n \geq N_2$. Thus, $d(Tx^*,x^*) = 0$ and so $Tx^* = x^*$, that is, x^* is a fixed point of T. One can easily check T has unique fixed point.

♣

3.3 Kannan, Chatterjea and Zamfirescu contractions

In 1968, Kannan introduced a new class of contractive type mapping which we call the Kannan contractive mappings ([64]).

Theorem 3.2

Let (X,d) be a complete metric space and the mapping $T : X \to X$ satisfy the contractive condition $d(Tx,Ty) \leq k(d(Tx,x) + d(Ty,y))$ for all $x,y \in X$, where $k \in [0, \frac{1}{2})$ is a fix constant. Then, T has a unique fixed point in X. For each $x_0 \in X$, the iterative sequence $\{T^n x_0\}_{n\geq 1}$ converges to the fixed point.

Proof: For each $x_0 \in X$ and $n \geq 1$, set $x_1 = Tx_0$ and $x_{n+1} = Tx_n = T^{n+1}x_0$. Then,

$$d(x_{n+1},x_n) = d(Tx_n,Tx_{n-1}) \leq k(d(Tx_n,x_n) + d(Tx_{n-1},x_{n-1}))$$

$$= k(d(x_{n+1},x_n)+d(x_n,x_{n-1}))$$

and so $d(x_{n+1},x_n) \le \frac{k}{1-k}d(x_n,x_{n-1}) = hd(x_n,x_{n-1})$, where $h = \frac{k}{1-k}$. For $n > m$,

$$d(x_n,x_m) \le d(x_n,x_{n-1})+d(x_{n-1},x_{n-2})+\cdots+d(x_{m+1},x_m)$$

$$\le (h^{n-1}+h^{n-2}+\cdots+h^m)d(x_1,x_0) \le \frac{h^m}{1-h}d(x_1,x_0).$$

Let $\varepsilon > 0$ be given. Choose a natural number N_1 such that $\frac{h^m}{1-h}d(x_1,x_0) < \varepsilon$ for all $m \ge N_1$. Thus, $d(x_n,x_m) < \varepsilon$ for $n > m$. Therefore, $\{x_n\}_{n\ge1}$ is a Cauchy sequence in (X,d). Since (X,d) is a complete cone metric space, there exists $x^* \in X$ such that $x_n \to x^*$. Choose a natural number N_2 such that $d(x_{n+1},x_n) < \frac{\varepsilon(1-k)}{2k}$ and $d(x_{n+1},x^*) < \frac{\varepsilon(1-k)}{2}$ for all $n \ge N_2$. Hence, for $n \ge N_2$ we have

$$d(Tx^*,x^*) \le d(Tx_n,Tx^*)+d(Tx_n,x^*) \le k(d(Tx_n,x_n)+d(Tx^*,x^*))+d(x_{n+1},x^*).$$

Thus, $d(Tx^*,x^*) \le \frac{1}{1-k}(kd(x_{n+1},x_n)+d(x_{n+1},x^*)) < \frac{\varepsilon}{2}+\frac{\varepsilon}{2} = \varepsilon$. Thus, $d(Tx^*,x^*) = 0$ and so $Tx^* = x^*$. Now, if y^* is another fixed point of T, then

$$d(x^*,y^*) = d(Tx^*,Ty^*) \le k(d(Tx^*,x^*)+d(Ty^*,y^*)) = 0.$$

Hence, $x^* = y^*$. Thus, the fixed point of T is unique.

♣

In 1972, Chatterjea by making a little change, introduced a different contractive type mappings ([27]).

Theorem 3.3

Let (X,d) be a complete metric space and the mapping $T : X \to X$ satisfy the contractive condition $d(Tx,Ty) \le k(d(Tx,y)+d(x,Ty))$ for all $x,y \in X$, where $k \in [0,\frac{1}{2})$ is a fix constant. Then, T has a unique fixed point in X. For each $x_0 \in X$, the iterative sequence $\{T^n x_0\}_{n\ge1}$ converges to the fixed point.

Proof: For each $x_0 \in X$ and $n \ge 1$, set $x_1 = Tx_0$ and $x_{n+1} = Tx_n = T^{n+1}x_0$. Then,

$$d(x_{n+1},x_n) = d(Tx_n,Tx_{n-1}) \le k(d(Tx_n,x_{n-1})+d(Tx_{n-1},x_n))$$

$$\le k(d(x_{n+1},x_n)+d(x_n,x_{n-1}))$$

and so $d(x_{n+1},x_n) \le \frac{k}{1-k}d(x_n,x_{n-1}) = hd(x_n,x_{n-1})$, where $h = \frac{k}{1-k}$. For $n > m$,

$$d(x_n,x_m) \le d(x_n,x_{n-1})+d(x_{n-1},x_{n-2})+\cdots+d(x_{m+1},x_m)$$

$$\le (h^{n-1}+h^{n-2}+\cdots+h^m)d(x_1,x_0) \le \frac{h^m}{1-h}d(x_1,x_0).$$

Let $\varepsilon > 0$ be given. Choose a natural number N_1 such that $\frac{h^m}{1-h}d(x_1,x_0) < \varepsilon$ for all $m \geq N_1$. Thus, $d(x_n,x_m) < \varepsilon$ for $n > m$. Therefore, $\{x_n\}_{n\geq 1}$ is a Cauchy sequence in (X,d). Since (X,d) is a complete cone metric space, there exists $x^* \in X$ such that $x_n \to x^*$. Choose a natural number N_2 such that $d(x_n,x^*) < \frac{\varepsilon(1-k)}{3}$ for all $n \geq N_2$. Hence, for $n \geq N_2$ we have

$$d(Tx^*,x^*) \leq d(Tx_n,Tx^*) + d(Tx_n,x^*)$$

$$\leq k(d(Tx^*,x_n) + d(Tx_n,x^*)) + d(x_{n+1},x^*)$$
$$\leq k(d(Tx^*,x^*) + d(x_n,x^*) + d(x_{n+1},x^*)) + d(x_{n+1},x^*).$$

Thus, $d(Tx^*,x^*) \leq \frac{1}{1-k}(kd(x_n,x^*) + d(x_{n+1},x^*)) + d(x_{n+1},x^*) < \frac{\varepsilon}{3} + \frac{\varepsilon}{3} + \frac{\varepsilon}{3} = \varepsilon$and so we get $d(Tx^*,x^*) = 0$. Hence, $Tx^* = x^*$. Now, if y^* is another fixed point of T, then

$$d(x^*,y^*) = d(Tx^*,Ty^*) \leq k(d(Tx^*,y^*) + d(Ty^*,x^*)) = 2kd(x^*,y^*).$$

Hence, $d(x^*,y^*) = 0$ and so $x^* = y^*$. Thus, the fixed point of T is unique.

♣

One can find distinct similar results in the literature. Here, we provide the next one.

Theorem 3.4

Let (X,d) be a complete metric space and the mapping $T : X \to X$ satisfy the contractive condition $d(Tx,Ty) \leq kd(x,y) + ld(y,Tx)$ for all $x,y \in X$, where $k,l \in [0,1)$ are fix constants. Then, T has a fixed point in X. Also, the fixed point of T is unique whenever $k+l \neq 1$.

Proof: For each $x_0 \in X$ and $n \geq 1$, set $x_1 = Tx_0$ and $x_{n+1} = Tx_n = T^{n+1}x_0$. Then,

$$d(x_{n+1},x_n) = d(Tx_n,Tx_{n-1}) \leq k(d(x_n,x_{n-1}) + d(Tx_{n-1},x_n))$$
$$= kd(x_n,x_{n-1}) \leq k^n d(x_1,x_0).$$

Thus for $n > m$, we have

$$d(x_n,x_m) \leq d(x_n,x_{n-1}) + d(x_{n-1},x_{n-2}) + \cdots + d(x_{m+1},x_m)$$

$$\leq (k^{n-1} + k^{n-2} + \cdots + k^m)d(x_1,x_0) \leq \frac{k^m}{1-k}d(x_1,x_0).$$

Let $\varepsilon > 0$ be given. Choose a natural number N_1 such that $\frac{k^m}{1-k}d(x_1,x_0) < \varepsilon$ for all $m \geq N_1$. Thus, $d(x_n,x_m) < \varepsilon$ for $n > m$. Therefore, $\{x_n\}_{n\geq 1}$ is a Cauchy sequence in (X,d). Since (X,d) is a complete cone metric space, there exists $x^* \in X$ such that $x_n \to x^*$. Choose a natural number N_2 such that $d(x_n,x^*) < \frac{\varepsilon}{3}$ for all $n \geq N_2$. Hence, for $n > N_2$ we have

$$d(Tx^*,x^*) \leq d(x_n,Tx^*) + d(x_n,x^*) = d(Tx_{n-1},Tx^*) + d(x_n,x^*)$$

$$\leq kd(x_{n-1},x^*) + ld(Tx_{n-1},x^*) + d(x_n,x^*)$$

$$\leq d(x_{n-1},x^*) + d(x_n,x^*) + d(x_n,x^*)$$

and so $d(Tx^*,x^*) < \frac{\varepsilon}{3} + \frac{\varepsilon}{3} + \frac{\varepsilon}{3} = \varepsilon$. Thus, $d(Tx^*,x^*) = 0$ and so $Tx^* = x^*$. Now, if y^* is another fixed point of T and $k+l \neq 1$, then

$$d(x^*,y^*) = d(Tx^*,Ty^*) \leq kd(x^*,y^*) + l(Ty^*,x^*)) = (k+l)d(x^*,y^*).$$

Hence, $d(x^*,y^*) = 0$ and so $x^* = y^*$. Thus, the fixed point of T is unique whenever $k+l \neq 1$.

♣

In 1972, Zamfirescu introduced another contractive type mappings which we call now Zamfirescu contractions ([107]). We omit proof of the next result because its proof is similar to the last results.

Theorem 3.5

Let (X,d) be a complete metric space and the mapping $T: X \to X$ satisfy the contractive condition

$$d(Tx,Ty) \leq k\max\{d(x,y), \frac{1}{2}[d(x,Ty)+d(y,Tx)], \frac{1}{2}[d(x,Tx)+d(y,Ty)]\}$$

for all $x,y \in X$, where $k \in [0,1)$ is a fix constant. Then, T has a fixed point in X.

One can find different types of contractive type mappings in literature such Bianchini, Hardy-Rogers and Ciric contractions. Most researchers call Ciric type mappings as quasi-contractions. There a many different types of contractive mappings. For example in section 6, we will reprove the Caristi fixed point theorem by using graphs.

Why should we consider complete metric spaces in fixed point results? What will be happened if we delete it?

You will see the answer in the approximate fixed point part. The answer is this: Nothing. You will obtain approximate fixed points instead fixed points in most results.

3.4 The Property (P)

Let (X,d) be a metric space and $T : X \to X$ a selfmap. Denote as usual, by $F(T)$ the set of fixed points of T. We say that a selfmap T has the property (P) whenever $F(T) = F(T^n)$ for all $n \geq 1$. Note that, $F(T) \subseteq F(T^n)$ for all $n \geq 1$, that is, $Tx_0 = x_0$ implies $T^n x_0 = x_0$ for all n. Thus, the interesting part of the property (P) is that we could conclude $Tx_0 = x_0$ whenever we have $T^m x_0 = x_0$ for some natural number m.

Now, the main question is which mappings have the property (P)?

There are some simple mappings which have the property (P). For more details about the property (P) read [42] and [62]. We like to show you that some complicate nonlinear and discontinuous mappings also have the property (P).

We say that $(X,d,\leq)$ is an ordered metric space whenever $\leq$ is an order on X and (X,d) is a metric space. There is no relation between d and the order. If $(X,\leq)$ is a partially ordered set, then we write $X_{\leq} = \{(x,y) \in X \times X : x \leq y \text{ or } y \leq x\}$. Also, we say that a selfmap $T : X \to X$ is orbitally continuous at x whenever for each sequence $\{n(i)\}_{i\geq 1}$ with $T^{n(i)}x \to a$ for some $a \in X$, we have $T^{n(i)+1}x \to Ta$. We define also the orbit of T at x by $O(x,\infty) := \{x, Tx, T^2x, \cdots, T^n x, \cdots\}$ and we say that T has the strongly comparable property whenever $(T^{n-1}y, T^n y) \in X_{\leq}$ for all $n \geq 1$ and $m \geq 2$, where $y \in F(T^m)$, $F(T^m)$ is the set of fixed points of T^m and $T^m = To\ldots oT$ (m-times).

Theorem 3.6

Let $(X,d,\leq)$ be a complete ordered metric space, $\lambda \in (0,1)$ and T a selfmap on X satisfy the condition

$$\min\{d^2(Tx,Ty), d(x,y)d(Tx,Ty), d^2(y,Ty)\}$$
$$-\min\{d^2(x,Tx), d(y,Ty)d(x,Ty), d^2(y,Tx)\} \leq \lambda\, d(x,Tx)d(y,Ty)$$

for all $x,y \in X_{\leq}$. If there exists $x_0 \in X$ such that $(T^{n-1}x_0, T^n x_0) \in X_{\leq}$ for all $n \geq 1$ and T is orbitally continuous at x_0, then T has a fixed point. Moreover, if T has the strongly comparable property, then T has the property (P).

Proof: Define $x_{n+1} = Tx_n = T^{n+1}x_0$ for all $n \geq 0$. If $x_{n_0} = x_{n_0-1}$ for some natural number n_0, then $x_n = x_{n_0}$ for all $n \geq n_0$ and x_{n_0} is a fixed point of T. Suppose that $x_n \neq x_{n-1}$ for all $n \geq 1$. Now for each $n \geq 1$, by using the assumption, we can put $x = x_{n-1}$ and $y = x_n$ in the condition. Thus, we obtain

$$\min\{d^2(x_n,x_{n+1}), d(x_{n-1},x_n)d(x_n,x_{n+1})\} \leq \lambda\, d(x_{n-1},x_n)d(x_n,x_{n+1}).$$

Since $\lambda < 1$, $\min\{d^2(x_n,x_{n+1}),d(x_{n-1},x_n)d(x_n,x_{n+1})\} = d^2(x_n,x_{n+1})$. Hence,

$$d(x_n,x_{n+1}) \leq \lambda d(x_{n-1},x_n).$$

By continuing this process we obtain $d(x_n,x_{n+1}) \leq \lambda^n d(x_0,x_1)$ for all $n \geq 1$. Thus, for each natural number k we have

$$d(x_n,x_{n+k}) \leq \sum_{i=n}^{n+k-1} d(x_i,x_{i+1}) \leq \sum_{i=n}^{n+k-1} \lambda^i d(x_0,x_1) \leq \frac{\lambda^n}{1-\lambda} d(x_0,x_1).$$

Therefore, $\{x_n\}$ is a Cauchy sequence. Since (X,d) is a complete metric space, there exists $u \in X$ such that $x_n \to u$. Since T is orbitally continuous, $x_{n+1} = Tx_n \to Tu$. This implies that $Tu = u$. Now, we prove that T has the property (P). Let $n \geq 2$ be given and $v \in F(T^n)$. Since T has the strongly comparable property, we can put $x = T^{n-1}v$ and $y = T^n v$ in the condition. Thus, we obtain

$$\min\{d^2(T^n v,T^{n+1}v),d(T^{n-1}v,T^n v)d(T^n v,T^{n+1}v)\} \leq \lambda\, d(T^{n-1}v,T^n v)d(T^n v,T^{n+1}v).$$

Thus, $\min\{d^2(v,Tv),d(T^{n-1}v,v)d(v,Tv)\} \leq \lambda\, d(T^{n-1}v,v)d(v,Tv)$ and so we get two cases.

Case I. $d^2(v,Tv) \leq \lambda\, d(T^{n-1}v,v)d(v,Tv)$.

We claim that $d(v,Tv) = 0$. If $d(v,Tv) > 0$, then

$$d(v,Tv) = d(T^n v,T^{n+1}v) \leq \lambda\, d(T^{n-1}v,T^n v).$$

Again, by putting $x = T^{n-2}v$ and $y = T^{n-1}v$ in the condition, we obtain

$$\min\{d^2(T^{n-1}v,T^n v),d(T^{n-2}v,T^{n-1}v)d(T^{n-1}v,T^n v)\} \leq \lambda\, d(T^{n-2}v,T^{n-1}v)d(T^{n-1}v,T^n v)$$

Again, we get two cases. Let $d^2(T^{n-1}v,T^n v) \leq \lambda\, d(T^{n-2}v,T^{n-1}v)d(T^{n-1}v,T^n v)$. If $d(T^{n-1}v,T^n v) = 0$, then $T^{n-1}v = v$ and so $v = T^n v = Tv$. If $d(T^{n-1}v,T^n v) > 0$, then $d(T^{n-1}v,T^n v) \leq \lambda\, d(T^{n-2}v,T^{n-1}v)$. Now, let

$$d(T^{n-2}v,T^{n-1}v)d(T^{n-1}v,T^n v) \leq \lambda\, d(T^{n-2}v,T^{n-1}v)d(T^{n-1}v,T^n v).$$

In this case we should have $d(T^{n-2}v,T^{n-1}v) = 0$ or $d(T^{n-1}v,T^n v) = 0$ (and so $v = Tv$), because if $d(T^{n-2}v,T^{n-1}v) > 0$ and $d(T^{n-1}v,T^n v) > 0$, then we get $\lambda \geq 1$ which is a contradiction. By continuing this process, we obtain

$$d(v,Tv) = d(T^n v,T^{n+1}v) \leq \lambda\, d(T^{n-1}v,T^n v) \leq \lambda^2\, d(T^{n-2}v,T^{n-1}v) \cdots \leq \lambda^n\, d(v,Tv)$$

which leads us to $\lambda \geq 1$ which is a contradiction. Therefore, in this case we have $d(v,Tv) = 0$ and so $Tv = v$.

Case II. $d(T^{n-1}v,v)d(v,Tv) \leq \lambda\, d(T^{n-1}v,v)d(v,Tv)$.

In this case we should have $d(T^{n-1}v,v) = 0$ or $d(v,Tv) = 0$ (and so $v = Tv$). In fact, if $d(T^{n-1}v,v) > 0$ and $d(v,Tv) > 0$, then $\lambda \geq 1$ which is a contradiction. Thus, we consequence that $F(T^n) \subseteq F(T)$. Therefore, T has the property (P).

♣

The following example shows that there are nonlinear and discontinuous mappings that satisfy the condition of Theorem 3.6.

Example 11 *Let $X = [0,\infty)$, $d(x,y) = |x-y|$ and T be a selfmap on X defined by $Tx = 0$ whenever $0 \leq x \leq 10$, $Tx = x - 10$ whenever $10 \leq x \leq 11$ and $Tx = 1.1$ whenever $x > 11$. Then, by putting $\lambda = \frac{1}{2}$, T satisfies the condition of Theorem 3.6.*

Theorem 3.7

Let $(X,d,\leq)$ be a complete ordered metric space, $\lambda \in (0,1)$, $a \geq 0$, T an orbitally continuous selfmap on X satisfy the condition

$$\min\{d(Tx,Ty),d(x,y),d(x,Tx),d(y,Ty),\frac{d(x,Tx)[1+d(y,Ty)]}{1+d(x,y)},$$

$$\frac{d(y,Ty)[1+d(x,Tx)]}{1+d(x,y)},\frac{\min\{d^2(Tx,Ty),d^2(x,Tx),d^2(y,Ty)\}}{d(x,y)}\}$$

$$-a\min\{\frac{d^2(x,Tx)[1+d^2(y,Ty)]}{1+d^2(x,y)},\frac{d^2(y,Ty)[1+d^2(x,Tx)]}{1+d^2(x,y)},d(x,Tx),d(y,Tx)\}$$

$$\leq \lambda\ \max\{d(x,y),d(x,Tx)\}$$

for all $x,y \in X_{\leq}$. If there exists $x_0 \in X$ such that $(T^{n-1}x_0, T^n x_0) \in X_{\leq}$ for all $n \geq 1$ and T is orbitally continuous at x_0, then T has a fixed point. Moreover, if T has the strongly comparable property, then T has the property (P).

Proof: Define $x_{n+1} = Tx_n = T^{n+1}x_0$ for all $n \geq 0$. If $x_{n_0} = x_{n_0-1}$ for some natural number n_0, then $x_n = x_{n_0}$ for all $n \geq n_0$ and x_{n_0} is a fixed point of T. Suppose that $x_n \neq x_{n-1}$ for all $n \geq 1$. Now, by using the assumption, we can put $x = x_0$ and $y = x_1 = Tx_0$ in the condition. Then, we obtain

$$\min\{d(Tx_0,T^2x_0),d(x_0,Tx_0),\frac{d(x_0,Tx_0)[1+d(Tx_0,T^2x_0)]}{1+d(x_0,Tx_0)},$$

$$\frac{\min\{d^2(Tx_0,T^2x_0),d^2(x_0,Tx_0)\}}{d(x_0,Tx_0)}\} \leq \lambda\ d(x_0,Tx_0).$$

If $d(x_0,Tx_0) \leq d(Tx_0,T^2x_0)$, then we deduce that $d(x_0,Tx_0) \leq \lambda\, d(x_0,Tx_0)$ and so $\lambda \geq 1$ which is a contradiction. This implies that $d(Tx_0,T^2x_0) \leq d(x_0,Tx_0)$. Thus,

$$\min\{d(Tx_0,T^2x_0), d(x_0,Tx_0), \frac{d(x_0,Tx_0)[1+d(Tx_0,T^2x_0)]}{1+d(x_0,Tx_0)}, \frac{d^2(Tx_0,T^2x_0)}{d(x_0,Tx_0)}\} \leq \lambda\, d(x_0,Tx_0).$$

But, it is easy to see that

$$\min\{d(Tx_0,T^2x_0), d(x_0,Tx_0), \frac{d(x_0,Tx_0)[1+d(Tx_0,T^2x_0)]}{1+d(x_0,Tx_0)}, \frac{d^2(Tx_0,T^2x_0)}{d(x_0,Tx_0)}\} = \frac{d^2(Tx_0,T^2x_0)}{d(x_0,Tx_0)}$$

and so $\frac{d^2(Tx_0,T^2x_0)}{d(x_0,Tx_0)} \leq \lambda\, d(x_0,Tx_0)$. Hence, $d(Tx_0,T^2x_0) \leq \sqrt{\lambda}\, d(x_0,Tx_0)$. Now, by continuing this process, for each $n \geq 1$ we obtain

$$d(T^n x_0, T^{n+1}x_0) \leq \sqrt{\lambda}\, d(T^{n-1}x_0, T^n x_0) \leq \cdots \leq \sqrt{\lambda^n}\, d(x_0,Tx_0).$$

Thus, for each natural number k we have

$$d(x_n, x_{n+k}) = d(T^n x_0, T^{n+k}x_0) \leq \sum_{i=n}^{n+k-1} \sqrt{\lambda^i}\, d(x_0,Tx_0) \leq \frac{\sqrt{\lambda^n}}{1-\sqrt{\lambda}} d(x_0,Tx_0).$$

This implies that $\{x_n\}$ is a Cauchy sequence. Since (X,d) is a complete metric space, there exists $u \in X$ such that $x_n \to u$. Since T is orbitally continuous, $Tx_n \to Tu$. Hence, u is a fixed point of T. Now, we prove that T has the property (P). Let $n \geq 2$ be given and $v \in F(T^n)$. If $d(v,Tv) = 0$, then $Tv = v$. Suppose that, $d(v,Tv) > 0$. A similar argument in the first part of this proof shows that

$$\frac{d^2(Tv,T^2v)}{d(v,Tv)} = \min\{d(Tv,T^2v), d(v,Tv), \frac{d(v,Tv)[1+d(Tv,T^2v)]}{1+d(v,Tv)}, \frac{d^2(Tv,T^2v)}{d(v,Tv)}\}$$

$$= \min\{d(Tv,T^2v), d(v,Tv), \frac{d(v,Tv)[1+d(Tv,T^2v)]}{1+d(v,Tv)}, \frac{\min\{d^2(Tv,T^2v), d^2(v,Tv)\}}{d(v,Tv)}\} \leq \lambda\, d(v,Tv).$$

Hence, $d(Tv,T^2v) \leq \sqrt{\lambda}\, d(v,Tv)$. Since T has the strongly comparable property, by using a similar argument in Theorem 3.6, we obtain

$$d(v,Tv) = d(T^n v, T^{n+1} v) \leq \sqrt{\lambda}\, d(T^{n-1}v, T^n v) \leq \cdots \leq \sqrt{\lambda^n}\, d(v,Tv).$$

This implies that $Tv = v$ because $\lambda < 1$. Thus, $F(T^n) \subseteq F(T)$. Therefore, T has the property (P).

♣

The following example shows that there are nonlinear and discontinuous mappings that satisfy the condition of Theorem 3.7.

Example 12 *Let $X = [0,\infty)$, $d(x,y) = |x-y|$ and T be a selfmap on X defined by $Tx = 0$ whenever $0 \leq x \leq 20$, $Tx = x-20$ whenever $20 \leq x \leq 21$ and $Tx = 1.1$ whenever $x > 21$. Then, by putting $\lambda = \frac{1}{2}$ and $a = 1$, T satisfies the condition of Theorem 3.7.*

The following result generalizes Proposition 1.1 of [59].

Theorem 3.8

Let $(X,d,\leq)$ be a complete ordered metric space, $b \in [0,1)$, $c \geq 0$, m a nonnegative integer and T a selfmap on X satisfy the condition

$$d^2(T^{m+1}x, T^{m+2}y) \leq b\, d(T^m x, T^{m+1}x) d(T^{m+1}y, T^{m+2}y)$$

$$+c\, d(T^m x, T^{m+2}y) d(T^{m+1}y, T^{m+1}x)$$

for all $x,y \in X_{\leq}$. Suppose that there exists $x_0 \in X$ such that $(T^{n-1}x_0, T^n x_0) \in X_{\leq}$ for all $n \geq 1$. If T is orbitally continuous at x_0 or $m = 0$, then T has a fixed point. Moreover, T has a unique fixed point whenever $c < 1$. If T has the strongly comparable property, then T has the property (P).

Proof: Define $x_1 = T^{m+1}x_0$ and $x_{n+1} = Tx_n$ for all $n \geq 1$. Then

$$d^2(x_1,x_2) = d^2(T^{m+1}x_0, T^{m+2}x_0) \leq b\, d(T^m x_0, T^{m+1}x_0) d(T^{m+1}x_0, T^{m+2}x_0)$$

$$+c\, d(T^m x_0, T^{m+2}x_0) d(T^{m+1}x_0, T^{m+1}x_0)$$

$$= b\, d(T^m x_0, T^{m+1}x_0) d(T^{m+1}x_0, T^{m+2}x_0) = b\, d(T^m x_0, T^{m+1}x_0) d(x_1,x_2).$$

If $d(x_1,x_2) = 0$, then $Tx_1 = x_2 = x_1$ and so T has a fixed point. If $d(x_1,x_2) > 0$, then $d(x_1,x_2) \leq bd(T^m x_0, x_1)$. Similarly, we have

$$d^2(x_2,x_3) = d^2(T^{m+2}x_0, T^{m+3}x_0) \leq b\, d(T^{m+1}x_0, T^{m+2}x_0) d(T^{m+2}x_0, T^{m+3}x_0)$$

$$+c\, d(T^{m+1}x_0, T^{m+3}x_0) d(T^{m+2}x_0, T^{m+2}x_0)$$

$$= b\, d(T^{m+1}x_0, T^{m+2}x_0)d(T^{m+2}x_0, T^{m+3}x_0) = b\, d(x_1,x_2)d(x_2,x_3).$$

If $d(x_2,x_3) = 0$, then $Tx_2 = x_3 = x_2$ and so T has a fixed point. If $d(x_2,x_3) > 0$, then $d(x_2,x_3) \le bd(x_1,x_2)$ and so $d(x_2,x_3) \le b^2 d(T^m x_0,x_1)$. By continuing this process we get that $d(x_n,x_{n+1}) \le b^n d(T^m x_0,x_1)$ for all $n \ge 1$. This implies that $\{x_n\}$ is a Cauchy sequence. Since (X,d) is a complete metric space, there exists $u \in X$ such that $x_n \to u$. If T is orbitally continuous, then $Tx_n \to Tu$. Hence, $Tu = u$.

If $m = 0$, then for each $n \ge 2$ we have

$$d^2(Tu,T^n x_0) \le b\, d(u,Tu)d(Tx_{n-2},T^2x_{n-2}) + c\, d(u,T^2x_{n-2})d(Tx_{n-2},Tu).$$

Since $x_n \to u$, we have $d^2(Tu,u) \le c\, d(u,u)d(u,Tu) = 0$ and so $Tu = u$. Now, we show that T has a unique fixed point whenever $c < 1$. Let u and v be fixed points of T. Then, we have

$$d^2(u,v) = d^2(T^{m+1}u,T^{m+2}v) \le b\, d(T^m u,T^{m+1}u)d(T^{m+1}v,T^{m+2}v)$$
$$+c\, d(T^m u,T^{m+2}v)d(T^{m+1}v,T^{m+1}u) = c\, d^2(u,v).$$

Hence, $d(u,v) = 0$ because $c < 1$. Thus, $u = v$ and so T has a unique fixed point.

Finally, we prove that T has the property (P) whenever T has the strongly comparable property. Let $n \ge 2$ be given and $v \in F(T^n)$. We consider the following cases.

Case I. $m = 0$.

In this case, we have

$$d^2(v,Tv) = d^2(T(T^{n-1}v),T^2(T^{n-1}v)) \le b\, d(T^{n-1}v,T^n v)d(T^n v,T^{n+1}v)$$
$$+c\, d(T^{n-1}v,T^{n+1}v)d(T^n v,T^n v) = b\, d(T^{n-1}v,v)d(v,Tv).$$

If $d(v,Tv) = 0$, then $Tv = v$. If $d(v,Tv) > 0$, then

$$d(T^n v,T^{n+1}v) \le bd(T^{n-1}v,T^n v).$$

By continuing the process and using a similar argument in Theorem 3.6, we obtain

$$d(v,Tv) = d(T^n v,T^{n+1}v) \le bd(T^{n-1}v,T^n v)$$
$$\le b^2 d(T^{n-2}v,T^{n-1}v)\cdots \le b^n d(v,Tv).$$

Since $b < 1$, $Tv = v$.

Case II. $m \geq 1$ and $n > m$.

In this case, we have

$$\begin{aligned} d^2(v,Tv) &= d^2(T^{m+1}(T^{n-m-1}v), T^{m+2}(T^{n-m-1}v)) \\ &\leq b\, d(T^m(T^{n-m-1}v), T^{m+1}(T^{n-m-1}v))d(T^{m+1}(T^{n-m-1}v), T^{m+2}(T^{n-m-1}v)) \\ &+ c\, d(T^m(T^{n-m-1}v), T^{m+2}(T^{n-m-1}v))d(T^{m+1}(T^{n-m-1}v), T^{m+1}(T^{n-m-1}v)) \\ &= b\, d(T^{n-1}v, T^n v)d(T^n v, T^{n+1}v) + c\, d(T^{n-1}v, T^{n+1}v)d(T^n v, T^n v) \\ &= b\, d(T^{n-1}v, v)d(v, Tv). \end{aligned}$$

If $d(v,Tv) = 0$, then $Tv = v$. If $d(v,Tv) > 0$, then

$$d(T^n v, T^{n+1}v) \leq bd(T^{n-1}v, T^n v).$$

By continuing the process and using a similar argument in Theorem 3.6, we obtain

$$\begin{aligned} d(v,Tv) = d(T^n v, T^{n+1}v) &\leq bd(T^{n-1}v, T^n v) \\ &\leq b^2 d(T^{n-2}v, T^{n-1}v) \cdots \leq b^n d(v, Tv). \end{aligned}$$

Since $b < 1$, $Tv = v$.

Case III. $m \geq 1$ and $n \leq m$.

In this case, choose a natural number r and an integer number $0 \leq s < n$ such that $m+1 = rn+s$. Then, we have $T^m(T^{n-s}v) = T^{n-1}v$, $T^{m+1}(T^{n-s}v) = v$ and so

$$\begin{aligned} d^2(v,Tv) &= d^2(T^{m+1}(T^{n-s}v), T^{m+2}(T^{n-s}v)) \\ &\leq b\, d(T^m(T^{n-s}v), T^{m+1}(T^{n-s}v))d(T^{m+1}(T^{n-s}v), T^{m+2}(T^{n-s}v)) \\ &+ c\, d(T^m(T^{n-s}v), T^{m+2}(T^{n-s}v))d(T^{m+1}(T^{n-s}v), T^{m+1}(T^{n-s}v)) \\ &= b\, d(T^{n-1}v, v)d(v, Tv). \end{aligned}$$

If $d(v,Tv) = 0$, then $Tv = v$. If $d(v,Tv) > 0$, then

$$d(T^n v, T^{n+1}v) \leq bd(T^{n-1}v, T^n v).$$

By continuing the process and using a similar argument in Theorem 3.6, we obtain

$$\begin{aligned} d(v,Tv) = d(T^n v, T^{n+1}v) &\leq bd(T^{n-1}v, T^n v) \\ &\leq b^2 d(T^{n-2}v, T^{n-1}v) \cdots \leq b^n d(v, Tv). \end{aligned}$$

Since $b < 1$, $Tv = v$. Thus, $F(T^n) \subseteq F(T)$. Thus, T has the property (P).

♣

The following example shows that there are nonlinear and discontinuous mappings that satisfy the condition of Theorem 3.8.

Example 13 *Let* $X = [0,\infty)$, $d(x,y) = |x-y|$ *and* T *be a selfmap on* X *defined by* $Tx = 0$ *whenever* $0 \leq x \leq 100$, $Tx = x - 100$ *whenever* $100 \leq x \leq 100.1$ *and* $Tx = 0.15$ *whenever* $x > 100.1$. *Then, by putting* $m = 0$, $b = \frac{1}{4}$ *and* $c = \frac{1}{2}$, T *satisfies the condition of Theorem 3.8.*

The following examples show that there are some mappings satisfying the condition of Theorem 3.8 while do not satisfy the condition of Proposition 1.1 of [59].

Example 14 *Let* $X = \{1,6,8\}$, $d(x,y) = |x-y|$, $\leq = \{(1,1),(6,6),(8,8)\}$ *and* T *be a selfmap on* X *defined by* $T1 = 6$, $T6 = 8$ *and* $T8 = 8$. *Then, by putting* $m = 0$, $x_0 = 8$, $b = \frac{1}{2}$ *and* $c = 0.9$, *it is easy to see that* T *satisfies the condition of Theorem 3.8 while for* $x = 1$ *and* $y = 8$ *we have* $49 = d^2(1,8) > bd(1,6)d(6,8) + cd(1,8)d(8,6) = 17.6$. *This implies that* T *does not satisfy the condition of Proposition 1.1 of [59].*

Example 15 *Let* $X = \{1,2,3,4,5\}$,

$$\leq = \{(1,1),(2,2),(3,3),(4,4),(5,5),(1,3)\},$$

$d(x,y) = |x-y|$ *and* T *be a selfmap on* X *defined by* $T1 = 4$, $T2 = 3$, $T3 = 3$, $T4 = 4$ *and* $T5 = 5$. *Then, by putting* $m = 0$, $x_0 = 4$, $b = \frac{1}{2}$ *and* $c = 0.6$, *it is easy to see that* T *satisfies the condition of Theorem 3.8 while for* $x = 2$ *and* $y = 5$ *we have*

$$4 = d^2(3,5) > bd(2,3)d(5,5) + cd(2,5)d(5,3) = 3.6.$$

This implies that T *does not satisfy the condition of Proposition 1.1 of [59].*

In 2011, Alghamdi, Alnafei, Radonevic and Shahzad ([11]) has studied convex contraction and two-sided convex contraction mappings, which were introduced by Istratescu [53]. These notions were for continuous mappings. We change them here as following.

Definition 3.1 Let (X,d) be a metric space and T a selfmap on X. Then T is said to be a convex contraction if there exist $a,b \in (0,1)$ with $a+b < 1$ such that

$$d(T^2x,T^2y) \leq a\, d(Tx,Ty) + b\, d(x,y)$$

for all $x,y \in X$. Also, T is said to be a convex contraction of order 2 if there exist $a_1,a_2,b_1,b_2 \in (0,1)$ with $a_1+a_2+b_1+b_2 < 1$ such that

$$d(T^2x,T^2y) \leq a_1\, d(x,Tx) + a_2\, d(Tx,T^2x) + b_1\, d(y,Ty) + b_2\, d(Ty,T^2y)$$

for all $x,y \in X$.

It is known that there are convex contractions of order 2 which are not contraction on metric spaces ([11, 53]). Our next result generalizes a result of Istratescu [53] (see also ([11]).

Theorem 3.9

Let $(X,d,\leq)$ be a complete ordered metric space, $a,b \in (0,1)$ with $a+b<1$ and T an orbitally continuous selfmap on X satisfy the condition

$$d(T^2x,T^2y) \leq a\,d(Tx,Ty)+b\,d(x,y)$$

for all $x,y \in X_{\leq}$. If there exists $x_0 \in X$ such that $(T^{n-1}x_0, T^n x_0) \in X_{\leq}$ for all $n \geq 1$, then T has a unique fixed point. Also, $F(T)=F(T^2)$.

Proof: Define $x_n = T^n x_0$ for all $n \geq 1$, $v = d(Tx_0,T^2x_0)+d(x_0,Tx_0)$ and $\lambda = a+b$. Then by using a similar technique in proof of Theorem 3 of [11], it is easy to see that $d(T^{m+1}x_0,T^m x_0) \leq 2\lambda^l v$, where $m=2l$ or $m=2l+1$. Also, $d(T^m x_0,T^n x_0) \leq \frac{4\lambda^l}{1-\lambda}v$ for all $n>m$ and so $\{x_n\}$ is a Cauchy sequence. Since (X,d) is complete, there exists $u \in X$ such that $x_n \to u$. Since T is orbitally continuous, $Tx_n \to Tu$. Hence, we get $Tu=u$. Now, we show that T has a unique fixed point. Let y and z be fixed points of T. Then $d(y,z)=d(T^2y,T^2z) \leq a\,d(Ty,Tz)+b\,d(y,z)=(a+b)d(y,z)$. Since $a+b<1$, we consequence that $y=z$. Now, we prove that $F(T)=F(T^2)$. Let $y \in F(T^2)$. Then, we have

$$d(Ty,y)=d(T^2Ty,T^2y) \leq a\,d(T^2y,Ty)+b\,d(Ty,y)=(a+b)d(y,Ty).$$

Since $a+b<1$, we get $Ty=y$. This completes the proof.

♣

Similar to the last examples, we can show that there are nonlinear and discontinuous mappings that satisfy the condition of Theorem 3.9. The following examples show that there are some mappings satisfying the condition of Theorem 3.9 while they do not satisfy the condition of Theorem 1.2 of [53] (see also [11]). In 2011, Haghi, Rezapour and Shahzad proved that some fixed point generalizations are not real generalizations ([45]). But, the following examples show that our results are real generalizations.

Example 16 *Let $X=\{1,3,5\}$, $d(x,y)=|x-y|$, $\leq=\{(1,1),(3,3),(5,5)\}$ and T be a selfmap on X defined by $T1=3$, $T3=1$ and $T5=5$. Then, by putting $x_0=5$, $a=\frac{1}{2}$ and $b=\frac{1}{4}$, it is easy to see that T satisfies the condition of Theorem 3.9 while for $x=1$ and $y=3$ we have $2=d(T^21,T^23)>ad(T1,T3)+bd(1,3)=1.5$. This implies that T does not satisfy the condition of Theorem 1.2 of [53].*

Example 17 *Let $X=\{1,2,3,4,5\}$,*

$$\leq=\{(1,1),(2,2),(3,3),(4,4),(5,5),(1,3)\},$$

$d(x,y) = |x-y|$ and T be a selfmap on X defined by

$$T1 = 4,\ T2 = 3,\ T3 = 3,\ T4 = 4 \text{ and } T5 = 5.$$

Then, by putting $a = \frac{1}{10}$, $x_0 = 4$ and $b = \frac{1}{2}$, it is easy to see that T satisfies the condition of Theorem 3.9 while for $x = 2$ and $y = 5$ we have

$$2 = d(3,5) > ad(3,5) + bd(2,5) = 1.7.$$

This implies that T does not satisfy the condition of Theorem 1.2 of [53].

Let $(X,d,\leq)$ be an ordered metric space, $\varphi : [0,\infty) \to [0,\infty)$ an increasing map such that $\varphi^n(t) \to 0$ for all $t \geq 0$ and T a selfmap on X. We say that T is an ordered φ-contraction whenever $d(Tx,Ty) \leq \varphi(d(x,y))$ for all $x,y \in X_{\leq}$. Note that this notion extends the notion of φ-contractions (see [76]). Since $\varphi(t) < t$ for all $t > 0$, each ordered φ-contraction is a convex contraction. As an application for Theorem 3.9, we emphasize that by using appropriate assumptions, the result holds for ordered φ-contractions. Next result generalizes the Theorem 2.3 of [53] (see also [11]).

Theorem 3.10

Let $(X,d,\leq)$ be a complete ordered metric space, $a_1,a_2,b_1,b_2 \in (0,1)$ with $a_1 + a_2 + b_1 + b_2 < 1$ and T an orbitally continuous selfmap on X satisfy the condition $d(T^2x,T^2y) \leq a_1\, d(x,Tx) + a_2\, d(Tx,T^2x) + b_1\, d(y,Ty) + b_2\, d(Ty,T^2y)$ for all $x,y \in X_{\leq}$. If there exists $x_0 \in X$ such that $(T^{n-1}x_0, T^n x_0) \in X_{\leq}$ for all $n \geq 1$, then T has a unique fixed point. Also, $F(T) = F(T^2)$.

Proof: Define $x_n = T^n x_0$ for all $n \geq 1$, $v = d(Tx_0,T^2x_0) + d(x_0,Tx_0)$. Also, put $\lambda = a_1 + a_2 + b_1$ and $\beta = 1 - b_2$. First, note that

$$\begin{aligned} d(T^3x_0,T^2x_0) &\leq a_1\, d(x_0,Tx_0) + a_2\, d(Tx_0,T^2x_0) + b_1\, d(Tx_0,T^2x_0) + b_2\, d(T^3x_0,T^2x_0) \\ &\leq a_1 v + (a_2+b_1)v + b_2\, d(T^3x_0,T^2x_0). \end{aligned}$$

Hence, $d(T^3x_0,T^2x_0) \leq (\frac{\lambda}{\beta})v$. Now, by using the assumption, we can put $x = Tx_0$ and $y = T^2x_0$ in the condition. Thus, we obtain

$$\begin{aligned} d(T^3x_0,T^4x_0) &\leq a_1\, d(Tx_0,T^2x_0) + a_2\, d(T^2x_0,T^3x_0) \\ &\quad + b_1\, d(T^2x_0,T^3x_0) + b_2\, d(T^3x_0,T^4x_0) \\ &\leq a_1 v + (a_2+b_1)\frac{a_1+a_2+b_1}{1-b_2}v + b_2\, d(T^3x_0,T^4x_0). \end{aligned}$$

Hence, $d(T^3x_0,T^4x_0) \leq (\frac{\lambda}{\beta})v$. Similarly, we have

$$d(T^5x_0,T^4x_0) \leq a_1\, d(T^2x_0,T^3x_0) + a_2\, d(T^3x_0,T^4x_0)$$

$$+b_1\, d(T^3x_0, T^4x_0) + b_2\, d(T^5x_0, T^4x_0)$$
$$\leq a_1 \frac{a_1+a_2+b_1}{1-b_2} v + (a_2+b_1)\frac{a_1+a_2+b_1}{1-b_2} v + b_2\, d(T^5x_0, T^4x_0).$$

Hence, $d(T^5x_0, T^4x_0) \leq (\frac{\lambda}{\beta})^2 v$. Also, by using the assumption and putting $x = T^3x_0$ and $y = T^4x_0$ in the condition, we obtain

$$d(T^5x_0, T^6x_0) \leq a_1\, d(T^3x_0, T^4x_0) + a_2\, d(T^4x_0, T^5x_0)$$
$$+b_1\, d(T^4x_0, T^5x_0) + b_2\, d(T^5x_0, T^6x_0)$$
$$\leq a_1 \frac{\lambda}{\beta} v + (a_2+b_1)(\frac{\lambda}{\beta})^2 v + b_2\, d(T^5x_0, T^6x_0)$$
$$= (\frac{\lambda}{\beta})^2 [a_1(\frac{\beta}{\lambda})v + (a_2+b_1)v + (\frac{\beta}{\lambda})^2 b_2\, d(T^5x_0, T^6x_0)]$$
$$\leq (\frac{\lambda}{\beta})^2 [a_1(\frac{\beta}{\lambda})v + (\frac{\beta}{\lambda})(a_2+b_1)v + (\frac{\beta}{\lambda})^2 b_2\, d(T^5x_0, T^6x_0)]$$
$$= (\frac{\lambda}{\beta})^2 [(\frac{\beta}{\lambda})(a_1+a_2+b_1)v + (\frac{\beta}{\lambda})^2 b_2\, d(T^5x_0, T^6x_0)]$$
$$\leq (\frac{1}{\beta})^2 [(\frac{\beta}{\lambda})(a_1+a_2+b_1)^3 v + (\beta)^2 b_2\, d(T^5x_0, T^6x_0)]$$

which implies $(\frac{\lambda}{\beta})\beta^3 d(T^5x_0, T^6x_0) \leq (1-b_2)^3 d(T^5x_0, T^6x_0) \leq (\lambda)^3 v$. Hence,

$$d(T^5x_0, T^6x_0) \leq (\frac{\lambda}{\beta})^2 v.$$

By continuing this process and using a similar technique in proof of Theorem 4 of [11], it is easy to see that $\{x_n\}$ is a Cauchy sequence. Since (X,d) is a complete metric space, there exists $u \in X$ such that $x_n \to u$. Since T is orbitally continuous, $Tx_n \to Tu$ and so $Tu = u$. Now, we show that T has a unique fixed point. Let y and z be fixed points of T. Then,

$$d(y,z) = d(T^2y, T^2z) \leq a_1\, d(y,Ty) + a_2\, d(Ty, T^2y) + b_1\, d(z,Tz) + b_2\, d(Tz, T^2z)$$

and so $y = z$. Now, we prove that $F(T) = F(T^2)$. Let $y \in F(T^2)$. Then, we have

$$d(Ty,y) = d(T^2Ty, T^2y) \leq a_1 d(Ty, T^2y) + a_2 d(T^2y, T^3y) + b_1 d(y,Ty) + b_2 d(Ty, T^2y)$$
$$= (a_1+a_2+b_1+b_2) d(y,Ty).$$

Since $a+b<1$, we get $Ty = y$. This completes the proof.

♣

Again, we can show that there are nonlinear and discontinuous mappings satisfying the condition of Theorem 3.10. The following examples show that there are some mappings satisfying the condition of Theorem 3.10 while not satisfying the condition of Theorem 2.3 of [53].

Example 18 *Let* $X = \{1,3,5\}$, $d(x,y) = |x-y|$, $\leq = \{(1,1),(3,3),(5,5)\}$ *and* T *be a selfmap on* X *defined by* $T1 = 3$, $T3 = 1$ *and* $T5 = 5$. *Then, by putting* $x_0 = 5$, $a_1 = a_2 = b_1 = b_2 = \frac{1}{8}$, *it is easy to see that* T *satisfies the condition of Theorem 3.10 while for* $x = 1$ *and* $y = 3$ *we have*

$$2 = d(T^2 1, T^2 3) > a_1 d(1,T1) + a_2 d(T1,T^2 1) + b_1 d(3,T3) + b_2 d(T3,T^2 3) = 0.8.$$

This implies that T *does not satisfy the condition of Theorem 2.3 of [53].*

Example 19 *Let* $X = \{1,2,3,4,5\}$,

$$\leq = \{(1,1),(2,2),(3,3),(4,4),(5,5),(1,3)\},$$

$d(x,y) = |x-y|$ *and* T *be a selfmap on* X *defined by* $T1 = 4$, $T2 = 3$, $T3 = 3$, $T4 = 4$ *and* $T5 = 5$. *Then, by putting* $x_0 = 4$, $a_1 = \frac{1}{2}$, $a_2 = b_1 = b_2 = \frac{1}{10}$ *it is easy to see that* T *satisfies the condition of Theorem 3.10 while for* $x = 2$ *and* $y = 5$ *we have*

$$2 = d(3,5) > a_1 d(2,3) + a_2 d(3,3) + b_1 d(5,5) + b_2 d(5,5) = 0.5.$$

This implies that T *does not satisfy the condition of Theorem 2.3 of [53].*

Let (X,d) be a metric space and T a selfmap on X satisfy the contractive condition $d(Tx,Ty) \leq \lambda(d(x,Tx) + d(y,Ty))$ for all $x,y \in X$, where $\lambda \in [0,\frac{1}{2})$ is a constant (see [85]). Note that T is a convex contraction of order 2. As an application, we can emphasize that Theorem 3.10 is a strong generalization of Theorem 2.6 in [85].

3.5 Generalized α-ψ-contractions

We think that in the last two decades most important contractive type mappings to be introduced were the Suzuki Type contractions (by Tomonari Suzuki, [98] and [101]) and α-ψ-contractions (by Samet et al., [94]). Between these new types of contractions, the α-ψ-contractions found appropriate application in the solving of singular fractional differentiable equations and it was natural that researchers focused on it more with relation to the notion introduced by Suzuki. In this section, we concentrate on generalized α-ψ-contractions.

Denote by Ψ the family of increasing functions $\psi : [0,\infty) \to [0,\infty)$ such that

$$\sum_{n=1}^{\infty} \psi^n(t) < \infty$$

for all $t > 0$, where ψ^n is the nth iterate of ψ. If there exist $\psi_1 \in \Psi$ $t_1 > 0$ such that $\psi(t_1) \geq t_1$, then $\psi_1^n(t_1) \geq t_1$ for all n because ψ_1 is increasing. Hence, $\lim_{n\to\infty} \psi_1^n(t_1) \geq t_1 > 0$ for all n and so $\lim_{n\to\infty} \psi_1^n(t_1) \neq 0$ while $\sum_{n=1}^{\infty} \psi_1^n(t) < \infty$ and so we should have $\lim_{n\to\infty} \psi_1^n(t_1) = 0$. This contradiction shows that $\psi(t) < t$ for all $t > 0$ and $\psi \in \Psi$.

A mapping $\varphi : [0,\infty) \to [0,\infty)$ is called a sublinear whenever

$$\varphi(t+s) \leq \varphi(t) + \varphi(s)$$

for all $t,s \in [0,\infty)$. We show that there are a lot of sublinear mappings in Ψ. Let $\lambda < 1$ and $g : [0,\infty) \to [0,\lambda)$ be a decreasing function. Consider the map $\varphi(t) = \int\limits_0^{\frac{t}{4}} g(x)dx$. Then, φ is increasing because $g > 0$. Also, $\varphi(t) < \min\{\frac{t}{4}, \lambda\}$. Hence, $\sum_{n=1}^{\infty} \varphi^n(t) < \infty$ for all $t \geq 0$ and so $\varphi \in \Psi$. Since g is decreasing, we have

$$\begin{aligned}\varphi(u+v) &= \int_0^{\frac{u+v}{4}} g(x)dx = \int_0^{\frac{u}{4}} g(x)dx + \int_{\frac{u}{4}}^{\frac{u+v}{4}} g(x)dx \\ &\leq \int_0^{\frac{u}{4}} g(x)dx + \int_0^{\frac{v}{4}} g(x)dx = \varphi(u) + \varphi(v)\end{aligned}$$

for all $u,v \geq 0$.

Let (X,d) be a metric space, $\psi \in \Psi$, T a selfmap on X and $\alpha : X \times X \to [0,\infty)$ a function. We say that the mapping T is a α-ψ-contraction whenever

$$\alpha(x,y)d(Tx,Ty) \leq \psi(d(x,y))$$

for all $x,y \in X$ ([94]). Do you see here any graph? In fact, we can consider that the coefficient mapping α has role of a graph in these type relations. If we consider that X is our space and there is a graph G on this space and $V(G) = X$, by considering $\alpha(x,y) = 1$ whenever $(x,y) \in E(G)$, that is there is an edge between x and y, and $\alpha(x,y) = 0$ otherwise, the relations $\alpha(x,y)d(Tx,Ty) \leq \psi(d(x,y))$ for all $x,y \in X$ reduces to $d(Tx,Ty) \leq \psi(d(x,y))$ for all $x,y \in X$ with $(x,y) \in E(G)$. This means that fixed point theory on graphs is a special cases of the relations using a coefficient mapping α. It is not bad idea if we recall the coefficient mapping α as an invisible graph. Young researchers can focus more on applying invisible graphs in their works. We expect that applications of invisible graphs will be appearing in different modern areas of engineering such as grids, new electronic theories in economy, complicate lattices and softwares, artificial intelligence, etc.

Also, T is called α-admissible whenever $\alpha(x,y) \geq 1$ implies $\alpha(Tx,Ty) \geq 1$ ([94]). Also, we say that X has the property (B) respect to α if $\{x_n\}$ is a sequence in X such that $\alpha(x_n,x_{n+1}) \geq 1$ for all $n \geq 1$ and $x_n \to x$, then $\alpha(x_n,x) \geq 1$

for all $n \geq 1$ ([94]). Suppose that (X,d) is a complete metric space, T is a α-admissible α-ψ-contractive mapping on X and there exists $x_0 \in X$ such that $\alpha(x_0,Tx_0) \geq 1$. If T is continuous or X has the property (B) respect to α, then T has a fixed point ([94]; Theorems 2.1 and 2.2). Finally, we say that X has the property (H) whenever for each $x,y \in X$ there exists $z \in X$ such that $\alpha(x,z) \geq 1$ and $\alpha(y,z) \geq 1$. If X has the property (H) in the Theorems 2.1 and 2.2, then X has a unique fixed point ([94]; Theorem 2.3). It is considerable that the results of Samet, Vetro and Vetro generalize similar ordered results in the literature (see the results of third section in [94]). Now, we introduce a new notion. We say that T is a generalized α-ψ-contractive mapping whenever $\alpha(x,y)d(Tx,Ty) \leq \psi(M(x,y))$ for all $x,y \in X$, where $M(x,y) = \max\{d(x,y),d(x,Tx),d(y,Ty),\frac{1}{2}\max\{d(x,Ty),d(y,Tx)\}\}$. Throughout this section, we suppose that $\psi \in \Psi$ is sublinear and $\alpha : X \times X \to [0,\infty)$ is a function.

Theorem 3.11

Let (X,d) be a complete metric space and T a continuous generalized α-ψ-contractive and α-admissible selfmap on X. If there exists $x_0 \in X$ such that $\alpha(x_0,Tx_0) \geq 1$, then T has a fixed point.

Proof: Take $x_0 \in X$ such that $\alpha(x_0,Tx_0) \geq 1$. Define the sequence $\{x_n\}_{n\geq 0}$ in X by $x_{n+1} = Tx_n$ for all $n \geq 0$. If $x_n = x_{n+1}$ for some $n \geq 0$, then $x^* = x_n$ is a fixed point for T. Assume that $x_n \neq x_{n+1}$ for all $n \geq 0$. Since T is α-admissible, we get $\alpha(x_n,x_{n+1}) \geq 1$ for all $n \geq 1$. But, we have

$$d(x_1,x_2) = d(Tx_0,Tx_1) \leq \alpha(x_0,x_1)d(Tx_0,Tx_1) \leq \psi(M(x_0,x_1)),$$

where

$$\begin{aligned} M(x_0,x_1) &= \max\{d(x_0,x_1),d(x_0,Tx_0),d(x_1,Tx_1),\frac{1}{2}\max\{d(x_0,Tx_1),d(x_1,Tx_0)\}\} \\ &= \max\{d(x_0,x_1),d(x_1,x_2),\frac{1}{2}d(x_0,x_2)\}. \end{aligned}$$

Note that, $M(x_0,x_1) \neq d(x_1,x_2)$ because if $M(x_0,x_1) = d(x_1,x_2)$, then we have $d(x_1,x_2) \leq \psi(d(x_1,x_2)) < d(x_1,x_2)$ which is a contradiction. Thus,

$$M(x_0,x_1) = \max\{d(x_0,x_1),\frac{1}{2}d(x_0,x_2)\}.$$

If $M(x_0,x_1) = d(x_0,x_1)$, then $d(x_1,x_2) \leq \psi(d(x_0,x_1))$ and if $M(x_0,x_1) = \frac{1}{2}d(x_0,x_2)$, then

$$\begin{aligned} d(x_1,x_2) \leq \psi(\frac{d(x_0,x_2)}{2}) &\leq \frac{\psi(d(x_0,x_1)) + \psi(d(x_1,x_2))}{2} \\ &< \frac{1}{2}\psi(d(x_0,x_1)) + \frac{1}{2}d(x_1,x_2) \end{aligned}$$

because ψ is sublinear. Hence, $d(x_1,x_2) \leq \psi(d(x_0,x_1))$. Now by using induction, we obtain $d(x_n,x_{n+1}) \leq \psi^n(d(x_0,x_1))$ for all n. Fix $\varepsilon > 0$ and choose $n(\varepsilon) \geq 1$ such that $\sum_{n\geq n(\varepsilon)} \psi^n(d(x_0,x_1)) < \varepsilon$. Let $m > n > n(\varepsilon)$. By using the triangular inequality, we obtain $d(x_n,x_m) \leq \sum_{k=n}^{m-1} d(x_k,x_{k+1}) \leq \sum_{k=n}^{m-1} \psi^k(d(x_0,x_1)) \leq \sum_{n\geq n(\varepsilon)} \psi^n(d(x_0,x_1)) < \varepsilon$. Thus, $\{x_n\}_{n\geq 0}$ is a Cauchy sequence. Hence, there exists $x^* \in X$ such that $x_n \to x^*$. Thus, $x_{n+1} = Tx_n \to Tx^*$ and so x^* is a fixed point of T.

♣

Example 20 *Let $X = [0,+\infty)$ and $d(x,y) = |x-y|$ for all $x,y \in \mathbb{R}$. Define the selfmap T on X by $Tx = x+8$ whenever $0 \leq x \leq 1$ and $Tx = 9$ whenever $x > 1$, and*

$$\alpha(x,y) = \begin{cases} 2 & x,y \in [0,1] \text{ or } x,y \in [8n, 8n+1], \text{ for some } n \geq 1, \\ 1 & x \in [0,1] \text{ and } y \in [8,9], \\ 0 & \text{else.} \end{cases}$$

If we define $\psi(t) = \frac{t}{2}$ for all $t \geq 0$, then it is easy to check that T is a generalized α-ψ-contractive mapping. In fact, for each $x,y \in [0,1]$ with $x \leq y$ we have

$$2 \times |x-y| = \alpha(x,y)d(Tx,Ty) \leq \psi(M(x,y)) = \frac{1}{2} \times 8 = 4.$$

For $x \in [0,1]$ and $y \in (1, \frac{81}{19})$ we have

$$0 \times |x-9| = \alpha(x,y)d(Tx,Ty) \leq \psi(M(x,y)) = \frac{1}{2} \times |y-9|.$$

For $x \in [0,1]$ and $y \geq \frac{81}{19}$ we have

$$0 \times |x-9| = \alpha(x,y)d(Tx,Ty) \leq \psi(M(x,y)) = \frac{1}{2}|x-y|.$$

Let $x,y \in [8(n-1), 8n]$ for some $n \geq 1$ and $x \leq y$. Then

$$2 \times 0 = \alpha(x,y)d(Tx,Ty) \leq \psi(M(x,y)) = \frac{1}{2} \times |x-9|.$$

If $x \in [0,1]$ and $y \in [8,9]$, then

$$1 \times |x-1| = \alpha(x,y)d(Tx,Ty) \leq \psi(M(x,y)) = \frac{1}{2} \max\{|x-y|, 8\}.$$

Also, for $x_0 = 0$ *we have* $\alpha(0,T0) = \alpha(0,8) = 1$. *Obviously* T *is continuous and so it remains to show that* T *is* α*-admissible. If* $x,y \in [0,1]$ *or* $x,y \in [8n,8n+1]$ *for some* $n \geq 1$, *then* $\alpha(x,y) = \alpha(Tx,Ty) = 2$. *If* $x \in [0,1]$ *and* $y \in [8,9]$, *then* $\alpha(x,y) = 1$ *and* $\alpha(Tx,Ty) = 2$. *Hence,* T *is* α*-admissible. Now, note that* T *has the fixed point* $x_0 = 9$. *Finally, note that* $\alpha(x,y)d(Tx,Ty) \nleq \psi(d(x,y))$ *for all* $x,y \in [0,1]$. *Thus, the last result is a generalization of Theorem 2.1 in [94].*

3.6 Fixed points results on graphs

In this section our aim is to familiar young researchers to some fixed point results on graphs.

Let (X,d) be a metric space and $\Delta = X \times X$. Consider a directed graph G such that the set $V(G)$ of its vertices coincides with X and the set $E(G)$ of its edges contains all loops, that is, $\Delta \subset E(G)$. We assume that G has no parallel edges. By G^{-1} we denote the conversion of a graph G, that is, the graph obtained from G by reversing the direction of the edges. Also, $\widetilde{G}$ denotes the undirected graph obtained from$G$ by ignoring direction of the edges. We say that a mapping $f : X \to X$ preserves edges of G whenever $(x,y) \in E(G)$ implies $(fx,fy) \in E(G)$ for all $x,y \in X$. Also, a mapping $f : X \to X$ is called a G-contraction [G-nonexpansive] whenever f preserves edges of G and there is $\alpha \in (0,1)$ such that $d(fx,fy) \leq \alpha d(x,y)$ [$d(fx,fy) \leq d(x,y)$] for all $(x,y) \in E(G)$.

Example 21 *Each Banach contraction is a* G_0*-contraction, where the graph* G_0 *is defined by* $E(G_0) = X \times X$.

Example 22 *Let* $\preceq$ *be a partial order on X. Define the graph* G_1 *by*

$$E(G_1) = \{(x,y) \in X \times X : x \preceq y\}.$$

The preserving edges of G_1 *means* f *is nondecreasing with respect to this order.*

Definition 3.2 Let x and y be two vertices in a graph G. A path in G from x to y of length n is a sequence $\{x_i\}_{i=0}^n$ of $n+1$ distinct vertices such that $x_0 = x$, $x_n = y$ and $(x_i,x_{i+1}) \in E(G)$ for all $i = 0,1,...,n-1$.

We denote by $r(x,y)$, the sum of edges distance between x and y, that is,

$$r(x,y) = \sum_{i=1}^{n} d(x_{i-1},x_i).$$

A graph G is connected if there is a path between any two vertices. G is weakly connected if $\widetilde{G}$ is connected. We denote by $[x]_G$ the set of all vertices in G that there is a path between x and those.

Throughout this section, (X,d) is a complete metric space. We say that G is a (C)-graph whenever for each sequence $\{x_n\}_{n\geq 0}$ in X with $x_n \to x$ and $(x_n, x_{n+1}) \in E(G)$ for all $n \geq 0$, there is a subsequence $\{x_{n_k}\}_{k\geq 0}$ such that $(x_{n_k}, x) \in E(G)$ for all $k \geq 0$. We say that G is a (P)-graph whenever $\{x_n\}_{n\geq 0}$ is a convergent sequence to a point x and $x_n \in [x]_G$ for all $n \geq 0$, we have $r(x_n, x) \to 0$. Now by providing the following examples, we show that the notions are independent on infinite graphs. Of course, it is normal we deal infinite graphs because we have the notion of limit here.

Example 23 *Let $X = \{\frac{1}{n} : n \geq 1\} \cup \mathbb{N} \cup \{0\}$ via the Euclidean metric. Define the undirected graph G_2 by $V(G_2) = X$ and $E(G_2) = \{(\frac{1}{n}, n) : n \geq 1\} \cup \{(n, 0) : n \geq 1\}$. Let $x_n = \frac{1}{n}$ for all $n \geq 1$. Then $x_n \to 0$, $x_n \in [0]_{G2}$ while $r(x_n, 0) = 2n - \frac{1}{n} \nrightarrow 0$. Hence G_2 is not a (P)-graph. But, it is easy to see that G_2 is a (C)-graph.*

Example 24 *Let $X = \{\frac{1}{n} : n \geq 1\} \cup \{\frac{1}{\sqrt{2+n}} : n \geq 1\} \cup \{0\}$ via the Euclidean metric. Define the undirected graph G_3 by $V(G_3) = X$ and*

$$E(G_3) = \{(\frac{1}{n}, \frac{1}{n+1}) : n \geq 1\} \cup \{(\frac{1}{n}, \frac{1}{\sqrt{2+n}}) : n \geq 1\} \cup \{(\frac{1}{\sqrt{2+n}}, 0) : n \geq 1\}.$$

Let $x_n = \frac{1}{n}$ for all $n \geq 1$. Then, $x_n \to 0$ and $(x_n, x_{n+1}) \in E(G_3)$ for all $n \geq 1$ while there is no subsequence $\{x_{n_k}\}_{k\geq 1}$ such that $(x_{n_k}, x) \in E(G_3)$ for all $k \geq 1$. Thus, G_3 is not a (C)-graph. But, it is easy to see that G_3 is a (P)-graph.

Example 25 *Let $(X, d, \preceq)$ be an ordered metric space. Define the graph G by $E(G) = \{(x,y) \in X \times X : x \preceq y \text{ or } x \preceq y\}$. It is easy to see that G is a (P)-graph.*

Theorem 3.12

Let G be a weakly connected (C)-graph or (P)-graph. Suppose that T is a G-contraction selfmap on X with constant α and $\{\varepsilon_i\}_{i\geq 0}$ is a sequence in $(0, \infty)$ satisfying $\sum_{i=0}^{\infty} \varepsilon_i < \infty$. Assume that $\{T_i\}_{i\geq 0}$ is a sequence of self-maps on X such that T_i preserves edges of G, $d(T_i x, Tx) < \varepsilon_i$ and $(T_i x, T_{i+1} y) \in E(G)$ for all $x \in G$, $i \geq 0$ and $(x,y) \in E(G)$. Let $x_0 \in X$ and $x_{i+1} = T_i x_i$ for all $i \geq 0$. Then $\{x_i\}_{i\geq 0}$ converges to a fixed point of T.

Proof: Since $[x_0]_{\widetilde{G}} = X$, $T_0 x_0 \in [x_0]_{\widetilde{G}}$ and so there is a path $\{\gamma_i\}_{i=0}^{N}$ from x_0 to x_1. Now, we show that there is a path from x_i to x_{i+1} for all i. Put $x = x_0$ and $y = \gamma_1$ and take $i = 0$. By using the assumption, we get $(T_0 x_0, T_1 \gamma_1) \in E(G)$. Again, put $x = T_0 x_0$ and $y = T_1 \gamma_1$ and take $i = 1$. Then, by using the assumption, we get $(T_1T_0x_0, T_2T_1\gamma_1) \in E(G)$. By continuing this process and using induction, we conclude that $(T_{i-1}\cdots T_1T_0x_0, T_i\cdots T_2T_1\gamma_1) \in E(G)$ for all $i \geq 1$. Thus, $(x_i, T_i\cdots T_2T_1\gamma_1) \in E(G)$ for all i. Now, it is easy to checked that

$$x_i, T_iT_{i-1}...T_1\gamma_1, T_iT_{i-1}...T_1\gamma_2, ..., T_iT_{i-1}...T_1\gamma_{N-1}, x_{i+1}$$

is a path from x_i to x_{i+1} for all i. Since T is a $\widetilde{G}$-contraction, for each $i \geq 1$ we have

$$\begin{aligned}
r(x_i,x_{i+1}) &= d(x_i, T_{i-1}T_{i-2}...T_0\gamma_1) + d(T_{i-1}T_{i-2}...T_0\gamma_1, T_iT_{i-1}...T_0\gamma_1) \\
&+ d(T_iT_{i-1}...T_1\gamma_1, T_iT_{i-1}...T_1\gamma_2) + ... + d(T_iT_{i-1}...T_1\gamma_{n-1}, x_{i+1}) \\
&\leq 2\varepsilon_{i-1} + \alpha d(x_{i-1}, T_{i-2}...T_0\gamma_1) + \varepsilon_{i-1} + \varepsilon_i + \alpha d(T_{i-2}...T_0\gamma_1, T_{i-1}...T_0\gamma_1) \\
&+ 2\varepsilon_i + \alpha d(T_{i-1}...T_1\gamma_1, T_{i-1}...T_1\gamma_2) + ... + 2\varepsilon_i + d(T_{i-1}...T_1\gamma_{n-1}, x_i) \\
&= 3\varepsilon_{i-1} + (2N-1)\varepsilon_i + \alpha r(x_{i-1}, x_i).
\end{aligned}$$

Thus, $r(x_1,x_2) \leq \alpha r(x_0,x_1) + 3\varepsilon_0 + (2N-1)\varepsilon_1$ and so

$$\begin{aligned}
r(x_2,x_3) &\leq 3\varepsilon_1 + (2N-1)\varepsilon_2 + \alpha r(x_1,x_2) \\
&\leq \alpha^2 r(x_0,x_1) + \alpha(3\varepsilon_0 + (2N-1)\varepsilon_1) + 3\varepsilon_1 + (2N-1)\varepsilon_2.
\end{aligned}$$

Now, by induction we can prove that for each $n \geq 1$ we have

$$r(x_n,x_{n+1}) \leq \alpha^n r(x_0,x_1) + \sum_{i=0}^{n-1} \alpha^i (3\varepsilon_{n-i-1} + (2N-1)\varepsilon_{n-i}).$$

Hence,

$$\begin{aligned}
d(x_{n+1},x_{n+2}) &\leq \varepsilon_n + \varepsilon_{n+1} + \alpha d(x_n,x_{n+1}) \\
&\leq \varepsilon_n + \varepsilon_{n+1} + \alpha r(x_n,x_{n+1}) \\
&\leq \varepsilon_n + \varepsilon_{n+1} + \alpha^{n+1} r(x_0,x_1) + \sum_{i=0}^{n-1} \alpha^{i+1} (3\varepsilon_{n-i-1} + (2N-1)\varepsilon_{n-i}).
\end{aligned}$$

Hence, we obtain

$$\begin{aligned}
\sum_{n=0}^{\infty} d(x_{n+1},x_{n+2}) &\leq \sum_{n=0}^{\infty} (\varepsilon_n + \varepsilon_{n+1} + \alpha^{n+1} r(x_0,x_1) + \sum_{i=0}^{n-1} \alpha^{i+1} (3\varepsilon_{n-i-1} + (2N-1)\varepsilon_{n-i}) \\
&= \sum_{n=0}^{\infty} (\varepsilon_n + \varepsilon_{n+1} + \alpha^{n+1} r(x_0,x_1) + \sum_{i=1}^{n} \alpha^{n-i+1} (3\varepsilon_{i-1} + (2N-1)\varepsilon_i) \\
&\leq \sum_{n=0}^{\infty} \varepsilon_n + \sum_{n=0}^{\infty} \varepsilon_{n+1} + \sum_{n=0}^{\infty} \alpha^{n+1} r(x_0,x_1) + \sum_{i=1}^{\infty} (\sum_{j=1}^{\infty} \alpha^j (3\varepsilon_{i-1} + (2N-1)\varepsilon_i)
\end{aligned}$$

$$\leq \sum_{n=0}^{\infty} \varepsilon_n + \sum_{n=0}^{\infty} \varepsilon_{n+1} + \sum_{n=0}^{\infty} \alpha^{n+1} [r(x_0, x_1) + \sum_{n=1}^{\infty} (3\varepsilon_{n-1} + (2N-1)\varepsilon_n)] < \infty.$$

Thus, $\{x_n\}$ is a Cauchy sequence and so there exists $x_* \in X$ such that $x_n \to x_*$. If G is a (P)-graph, then we have

$$\begin{aligned} d(x_*, Tx_*) &\leq d(x_*, x_{n+1}) + d(x_{n+1}, Tx_*) \\ &\leq d(x_*, x_{n+1}) + d(x_{n+1}, Tx_n) + d(Tx_n, Tx_*) \\ &\leq d(x_*, x_{n+1}) + \varepsilon_n + \alpha r(x_n, x_*). \end{aligned}$$

In this case, we get $Tx_* = x_*$ because $x_n \in [x_*]_{\widetilde{G}}$ for all n. Now, suppose that G is a (C)-graph. Since $(x_n, x_{n+1}) \in E(G)$ for all n, there is a subsequence $\{x_{n_k}\}_{k \geq 0}$ such that $(x_{n_k}, x_*) \in E(G)$ for all $k \geq 0$. Thus,

$$\begin{aligned} d(x_*, Tx_*) &\leq d(x_*, x_{n_{k+1}}) + d(x_{n_{k+1}}, Tx_*) = d(x_*, x_{n_{k+1}}) + d(T_{n_k} x_{n_k}, Tx_*) \\ &\leq d(x_*, x_{n_k+1}) + d(T_{n_k} x_{n_k}, Tx_{n_k}) + d(Tx_{n_k}, Tx_*) \\ &\leq d(x_*, x_{n_k+1}) + \varepsilon_{n_k} + \alpha d(x_{n_k}, x_*). \end{aligned}$$

In this case, we conclude that $Tx_* = x_*$.

♣

The following result shows that Theorem 3.12 is a generalization of Theorem 2.1 in [82].

Corollary 3.1

Let $(X, \preceq, d)$ be an ordered complete metric space such that every pair of elements of X has an upper and a lower bound. Suppose that $f : X \to X$ is a continuous and monotone mapping and there exists $\alpha \in (0,1)$ such that

$$d(fx, fy) \leq \alpha d(x, y)$$

for all $x, y \in X$ with $x \preceq y$. If there exists $x_0 \in X$ such that $x_0 \preceq fx_0$ or $fx_0 \preceq x_0$, then f is a Picard operator.

Proof: Define the graph G by $E(G) = \{(x,y) \in X \times X : x \preceq y \text{ or } x \preceq y\}$. Since each pair of elements of X has an upper and a lower bound, $\widetilde{G}$ is weakly connected. Thus, by using example 25, G is a weakly connected (P)-graph. Let $\{\varepsilon_i\}_{i \geq 0}$ be an arbitrary sequence in $(0, \infty)$ satisfying $\sum_{i=0}^{\infty} \varepsilon_i < \infty$ and $T_i = f$ for all $i \geq 0$. Now, it is easy to check that by using Theorem 3.12, f is a Picard operator.

♣

In 1967, Kelisky and Rivlin defined the Bernstein operator B_n on the space $C[0,1]$ by $(B_n\varphi)(t) = \sum_{k=0}^{n} \varphi(\frac{k}{n}) \binom{n}{k} t^k(1-t)^{n-k}$ for all $\varphi \in C[0,1]$, $t \in [0,1]$ and $n \geq 1$ ([65]). They proved that each Bernstein operator B_n is a weakly Picard operator. Moreover,

$$\lim_{j\to\infty} (B_n^j\varphi)(t) = \varphi(0) + (\varphi(1) - \varphi(0))t$$

for all $\varphi \in C[0,1]$, $t \in [0,1]$ and $n \geq 1$, where $\{B_n^j\}_{j\geq 1}$ is the sequence of the iterates of B_n. In 2004, a simple proof of the Kelisky-Rivlin theorem was given by Rus with the help of some trick with the Contraction Principle ([91]). In 2008, Jachymski extended the Kelisky-Rivlin theorem (Theorem 4.1 in [54]). Now, by using the idea of Reich which we used in Theorem 3.12, we give another version of the result of Jachymski.

Theorem 3.13

Let X be a Banach space, X_0 a closed subspace of X, T a linear operator on X such that $\|T|_{X_0}\| < 1$ and $(I-T)(X) \subseteq X_0$. Suppose that $\{\varepsilon_i\}_{i\geq 0}$ is a sequence in $(0,\infty)$ satisfying $\sum_{i=0}^{\infty} \varepsilon_i < \infty$ and $\{T_i\}_{i\geq 0}$ is a sequence of selfmaps on X such that T_i preserves edges of G, $d(T_ix, Tx) < \varepsilon_i$ and $(T_ix, T_{i+1}y) \in E(G)$ for all $x \in G$, $i \geq 0$ and $(x,y) \in E(G)$. Let $x_0 \in X$ and $x_{i+1} = T_ix_i$ for all $i \geq 0$. Then $\{x_i\}_{i\geq 0}$ converges to a fixed point of T.

Proof: Define the graph G by $E(G) = \{(x,y) \in X \times X : x - y \in X_0\}$. First, we show that T is a G-contraction. If $(x,y) \in E(G)$, then

$$Tx - Ty = (y - Ty) - (x - Tx) + (x - y) \in X_0.$$

Thus, $(Tx, Ty) \in E(G)$. Moreover by putting $\alpha = \|T|_{X_0}\|$, for each $(x,y) \in E(G)$ we have $\|Tx - Ty\| = \|T(x-y)\| \leq \|T|_{X_0}\|\|x-y\| = \alpha\|x-y\|$. Hence, T is a G-contraction. Now, we show that G is a (C)-graph. Let $\{x_n\}_{n\geq 0}$ be a sequence in X with $x_n \to x$ and $(x_n, x_{n+1}) \in E(G)$ for all $n \geq 0$. It is easy to see that $(x_n, x_m) \in E(G)$ for all n and $m \geq n$. For each n, define the set A_n by

$$A_n = \{y \in X : (x_n, y) \in E(G)\}.$$

Thus, $x_m \in A_n$ for all n and $m \geq n$. But, note that $A_n = x_n + X_0$. Since X_0 is closed, A_n so is. Hence, $(x_n, x) \in E(G)$ for all n. Therefore, G is a (C)-graph. Now by using Theorem 3.12, we get $\{x_i\}_{i\geq 0}$ converges to a fixed point of T.

♣

Now, we provide our next result.

Theorem 3.14

Let G be a weakly connected graph such that for each $x,y \in X$ we have $d(x,y) = r(x,y)$, $T : X \to X$ a G-nonexpansive mapping such that T preserves edges of G, $\{\varepsilon_i\}_{i=0}^{\infty}$ a sequence in $(0,\infty)$ with $\sum_{i=0}^{\infty} \varepsilon_i < \infty$ and F a nonempty subset of X. Assume that $\{T_i\}_{i\geq 0}$ is a sequence of selfmaps on X such that $d(T_i x, Tx) < \varepsilon_i$ and $(T_i x, Ty) \in E(G)$ for all $x \in G$, $i \geq 0$ and $(x,y) \in E(G)$. Let $x_0, t_0 \in X$. If the sequence $\{x_{i+1} = Tx_i\}_{i=0}^{\infty}$ satisfies $\lim_{i\to\infty} d(x_i, F) = 0$, then $d(t_i, F) \to 0$, where $\{t_{i+1} = T_i t_i\}_{i=0}^{\infty}$.

Proof: Let $\delta > 0$ be given. Choose a natural number p such that $\sum_{i=p}^{\infty} \varepsilon_i < \frac{\delta}{2}$. Since $[t_{p+1}]_{\widetilde{G}} = X$, $x_{p+1} = Tx_p \in [t_{p+1}]_{\widetilde{G}}$. Thus, there is a path between t_{p+1} and x_{p+1}. Let $k \geq p$. If $\{\gamma_i\}_{i=0}^{N}$ be a path from t_{p+1} to x_{p+1}, then it is easy to see that for each $k \geq p+1$, $t_k, T^{k-(p+1)}\gamma_1, T^{k-(p+1)}\gamma_2, \ldots, T^{k-(p+1)}\gamma_{N-1}, x_k$ is a path between t_k and x_k. But $d(T_k\lambda_j, T^{k-(p+1)}\lambda_{j+1}) \leq \varepsilon_k + d(\lambda_j, T^{k-(p+1)-1}\lambda_{j+1})$ for all $(\lambda_j, \lambda_{j+1}) \in E(G)$ and $k \geq p+1$. Thus,

$$r(t_k,x_k) = d(t_k, T^{k-(p+1)}\gamma_1) + d(T^{k-(p+1)}\gamma_1, T^{k-(p+1)}\gamma_2) + \cdots + d(T^{k-(p+1)}\gamma_{N-1}, x_k)$$

$$\leq \varepsilon_{k-1} + d(t_{k-1}, T^{k-(p+1)-1}\gamma_1) + d(T^{k-(p+1)-1}\gamma_1, T^{k-(p+1)-1}\gamma_2)$$
$$+ \cdots + d(T^{k-(p+1)-1}\gamma_{N-1}, x_{k-1}) = \varepsilon_{k-1} + r(t_{k-1}, x_{k-1}).$$

Now by using induction, we can show that $r(t_k,x_k) \leq \sum_{i=p}^{k} \varepsilon_i - \varepsilon_k$. Choose $q > p$ such that $d(x_k,F) < \frac{\delta}{2}$ for all $k \geq q$. Then, for each $k \geq q$ we have

$$d(t_k,F) \leq d(t_k,x_k) + d(x_k,F) \leq \sum_{i=k}^{\infty} \varepsilon_i + \frac{\delta}{2} < \frac{\delta}{2} + \frac{\delta}{2} = \delta.$$

This shows that $d(t_i,F) \to 0$.

♣

Now, we provide some fixed point results on subgraphs of directed graphs.

Let (X,d) be a metric space and G a directed graph G such that $V(G) = X$ and the set $E(G)$ of its edges contains all loops. We denote the conversion of a graph G by G^{-1}, that is, the graph obtained from G by reversing the direction of the edges. A mapping $f : X \to X$ preserves edges of G whenever $(x,y) \in E(G)$ implies $(fx, fy) \in E(G)$ for all $x,y \in X$ ([54]). Since G is a directed graph, direct of the edge (x,y) is inverse direct of (y,x), that is, $(x,y) \neq (y,x)$. Let G be the directed graph. A finite path of length n in G from x to y is a sequence $\{x_i\}_{i=0}^{n}$ of distinct vertices such that $x_0 = x$, $x_n = y$ and $(x_i, x_{i+1}) \in E(G)$ for $i = 0, 1, ..., n-1$ ([10]). In fixed point theory we like to deal with infinite graphs (see [10]). In this reason, we consider infinite paths. In fact, $B \subseteq E(G)$ is an infinite path whenever between its any two vertices there is a finite path. Throughout this chapter, a path could be finite or infinite and the vertices of the path are pairwise distinct. Also, we consider cycles as finite paths. We denote by $[x]_G$ the set of all vertices in G that there is a (finite or infinite) path from those to x.

Let G' be a subgraph of the directed graph G and $x \in G'$. We emphasize that $[x]_G$ denotes the set of all vertices in G that there is a path from those to x via the edges in G. Also, we remind here that $V(G') \subseteq V(G)$ and $E(G') \subseteq E(G)$. Let G' be a subgraph of the directed graph G. we say that $b \in G$ is an upper bound for G' whenever $g' \in [b]_G$ for all $g' \in G'$. Also, we say that $c \in G$ is a supremum of G' whenever $c \in [b]_G$ for all upper bounds b. In fact, c is a least upper bound in a sense.

Example 26 *Let G be the directed graph via the vertices $V(G) = \{a, b, c, d\}$ and the edges $E(G) = \{(a,b), (b,c), (c,d), (d,a)\}$. Suppose that G' is a subgraph of G denied by $V(G') = \{a, b, c\}$ and $E(G') = \{(a,b), (b,c)\}$, then c, d are upper bounds of G'. Thus, an upper bound is not unique in a subgraph necessarily.*

Example 27 *Let G be the directed graph via the vertices $V(G) = \{0, 2, \frac{1}{n} : n \geq 1\}$ and the edges $E(G) = \{(\frac{1}{n}, \frac{1}{n+1}) : n \geq 1\} \bigcup \{(\frac{1}{n}, 0)\} \bigcup (0, 2) \bigcup \{(\frac{1}{n}, 2) : n \geq 1\}$. If $V(G') = \{0, \frac{1}{n} : n \geq 1\}$ and $E(G') = \{(\frac{1}{n}, \frac{1}{n+1})\}$, then 0 and 2 are supremum of G'. Thus, a supremum is not unique in a subgraph necessarily.*

Let G be a directed graph and $x_0 \in G$. We say that x_0 is an end point whenever there is no $x \in G$ such that $(x_0, x) \in G$ and $x \neq x_0$. There are many directed graph via end points. In the following result we give a class of directed graphs which have end points. The proof of this result is straightforward.

Lemma 3.1

Let G be the directed graph, $X = V(G)$, $\varphi : X \to \mathbb{R}$ a function, $E(G) = \{(x,y) : d(x,y) \leq \varphi(x) - \varphi(y)\}$ and d a metric on X. If there exists $x_0 \in X$ such that $\varphi(x_0) = \inf_{x \in X} \varphi(x)$, then x_0 is an end point of G.

Let G be the directed graph and $\mathcal{M}$ the set of all paths in G. Then, $\subseteq$ is a partial order on $\mathcal{M}$. By using the Hausdorff's maximum principle, $\mathcal{M}$ has a maximal element. This means that G has a maximal path. We use this subject in our results.

Theorem 3.15

Let G be a directed graph such that every path in G has an upper bound. Then G has an end point or a cycle.

Proof: Suppose that G has no cycle. Let B be the maximal path in G and u an upper bond of B. If u is not end point, there exists $x \in G$ such that $x \neq u$ and $(u,x) \in E(G)$. Thus, $B \bigcup \{x\}$ is a path in G and $B \subset B \bigcup \{x\}$. This contradiction shows that u is an end point of G.

♣

Let G be a directed graph and T a selfmap on G. We say that T is a self-path-map whenever $x \in [Tx]_G$ for all $x \in G$.

Theorem 3.16

Let G be a directed graph. Then, G has an end point if and only if each self-path-map on G has a fixed point.

Proof: Suppose that G has an end point x_0 and T is a self-path-map. We prove that x_0 is a fixed point of T. Since $x_0 \in [Tx_0]_G$, there is a (finite or infinite) path $\{\lambda_i\}_{i \geq 0}$ between x_0 and Tx_0. Since x_0 is the end point of G and $\lambda_0 = x_0$, we have $x_0 = \lambda_1$. By continuing this process, it is easy to see that $x_0 = \lambda_i$ for all i. Thus, $x_0 = Tx_0$. Now, assume that G is a directed graph and each self-path-map on G has a fixed point but G has no any end point. Then for each $x \in G$ there exists $y \in G$ such that $y \neq x$ and $(x,y) \in E(G)$. By using the selection principle, we can define a selfmap T on G by $Tx = y$. Note that T is a self-path-map which has no fixed point.

♣

Example 28 *Let G be the directed graph via the vertices* $V(G) = \{0, \frac{1}{n} : n \geq 1\}$ *and the edges* $E(G) = \{(\frac{1}{n}, \frac{1}{n+1}) : n \geq 1\} \bigcup \{(\frac{1}{n}, 0)\} \bigcup (0,1)$. *Define the selfmap T on G by* $T_0 = 1$ *and* $T\frac{1}{n} = \frac{1}{n+1}$ *for all* $n \geq 1$. *Then* $x \in [Tx]_G$ *for all* $x \in G$, *T has no fixed point and G has no end point.*

Theorem 3.17

Let G be a directed graph such that every path in G has a supremum, T a selfmap on G such that $Tx \in [Ty]_G$ *for all* $x \in [y]_G$, $G' = \{x \in G : x \in [Tx]_G\} \neq \emptyset$ *and* G' *has no cycle. Then T has a fixed point in* G'.

Proof: Suppose that B is a path in G and b is the supremum of B in G. Since $c \in [b]_G$ for all $c \in B$, $Tc \in [Tb]_G$ and so $c \in [Tb]_G$. It follows that Tb is an upper bound for B. Since b is supremum, $b \in Tb$. Thus, $b \in G'$. By using Theorem 3.15, G' has an end point. Since $x \in [Tx]$ for all $x \in G'$, $Tx \in [T^2x]_G$ and so T is a self-path-map on G'. Now by using Theorem 3.16, T has a fixed point in G'.

♣

Now, we show that a version of Knaster-Tarski fixed point theorem is a consequence of last result.

Theorem 3.18

Let $(X, \preceq)$ be a partially ordered set such that each chain in X has a supremum and T a monotone selfmap on X. Assume that there exists $a \in X$ such that $a \preceq Ta$. Then T has a fixed point.

Proof: Define the graph G by $V(G) = X$ and $E(G) = \{(x,y) : x \preceq y \text{ and } x \neq y\}$. Then $Tx \in [Ty]_G$ for all $x \in [y]_G$. Since $G' = \{x \in G : x \in [Tx]_G\} \neq \emptyset$ and G' has no cycle, by using Theorem 3.17, T has a fixed point.

♣

Let X be a set and $\varphi : X \to (-\infty, \infty)$ a map. Suppose that G is the directed graph defined by $V(G) = X$ and $E(G) = \{(x,y) : d(x,y) \leq \varphi(x) - \varphi(y)\}$. We say that φ is lower semi-continuous whenever $\varphi(x) \leq \varphi(x_n)$ for all sequence $\{x_n\}$ in X with $x_n \to x$.

Lemma 3.2

Let X be a complete metric space and $\varphi : X \to (-\infty, \infty)$ a bounded from below map. Suppose that G is the directed graph defined by $V(G) = X$ and $E(G) = \{(x,y) : d(x,y) \leq \varphi(x) - \varphi(y)\}$. If φ is lower semi-continuous, then G has an end point.

Proof: First we prove that G has no cycle. If G has a cycle, then there exists a path $\{\lambda_i\}_{i=1}^n$ in G such that $\lambda_1 = \lambda_n$. It is easy to check that $d(\lambda_1, \lambda_i) \leq \varphi(\lambda_1) - \varphi(\lambda_i)$ and $d(\lambda_i, \lambda_n) \leq \varphi(\lambda_i) - \varphi(\lambda_n)$ for all $i = 2, 3, \ldots, n-1$ and so $\lambda_i = \lambda_1$ for $i \geq 2$. This contradiction shows that G has no cycle. Now we prove that each path in G has an upper bound. Let $\{x_\alpha\}_{\alpha \in \Omega}$ be a path in G. Then $\{\varphi(x_\alpha)\}_\Omega$ is a decreasing net of real numbers. Since φ is bounded from below, there is an increasing sequence $\{\alpha_n\}_{n \geq 1}$ in Ω such that $\lim_{n \to \infty} \varphi(x_{\alpha_n}) = \inf_{\alpha \in \Omega} \varphi(x_\alpha)$. One can easily show that $\{x_{\alpha_n}\}_{n \geq 1}$ is a Cauchy sequence and so converges to some $x \in X$. Since φ is lower semi-continuous, $x_{\alpha_n} \in [x]_G$ for all $n \geq 1$. Thus, x is an upper bound for $\{x_{\alpha_n}\}_{n \geq 1}$.

Now we show that x is an upper bound for $\{x_\alpha\}_{\alpha\in\Omega}$. If there exists $\beta \in \Omega$ such that $x_{\alpha_n} \in [x_\beta]_G$ for all $n \geq 1$, then $\varphi(x_\beta) \leq \varphi(x_{\alpha_n})$ for all $n \geq 1$ which implies $\varphi(x_\beta) = \inf_{\alpha\in\Omega}\varphi(x_\alpha)$. Since $d(x_{\alpha_n}, x_\beta) \leq \varphi(x_{\alpha_n}) - \varphi(x_\beta)$, we get $x_{\alpha_n} \to x_\beta$ which implies $x_\beta = x$. Hence, $\varphi(x) = \inf_{\alpha\in\Omega}\varphi(x_\alpha)$. Now we claim that $x_\alpha \in [x]_G$ and so x is an upper bound for $\{x_{\alpha_n}\}_{n\geq 1}$. In fact if there is $\alpha \in \Omega$ such that $x \in [x_\alpha]_G$, then $d(x, x_\alpha) \leq \varphi(x) - \varphi(x_\alpha) \leq \varphi(x_\alpha) - \varphi(x_\alpha) = 0$ and so $x = x_\alpha$. Since $\{x_\alpha\}_{\alpha\in\Omega}$ is a path in G, if the last case does not hold, then for each $\alpha \in \Omega$ there exists $n \geq 1$ such that $x_\alpha \in [x_{\alpha_n}]_G$. Hence, $x_\alpha \in [x]_G$ for all $\alpha \in \Omega$. Thus, x is an upper bound for $\{x_\alpha\}_{\alpha\in\Omega}$. Now by using Theorem 3.15, G has an end point.

♣

Now, we can consequent the Caristi fixed point theorem.

Theorem 3.19

Let X be a complete metric space, $\varphi : X \to (-\infty, \infty)$ a bounded from below and lower semi-continuous map and $T : X \to X$ a selfmap satisfying $d(x, Tx) \leq \varphi(x) - \varphi(Tx)$ for all $x \in X$. Then T has a fixed point.

Proof: Suppose that G is the directed graph via the vertices $V(G) = X$ and the edges $E(G) = \{(x,y) : d(x,y) \leq \varphi(x) - \varphi(y)\}$. By using Lemma 3.2, G has an end point. It is easy to see that T is a self-path map on G. Now by using Theorem 3.16, T has a fixed point.

♣

3.7 Best proximity points for cyclic mappings

In 2003, Kirk et al proved a generalization of Banach contraction principle (see [68]).

Theorem 3.20

Let A and B be nonempty closed subsets of a complete metric space (X,d). Suppose that $T : A \cup B \to A \cup B$ be a map satisfying $T(A) \subseteq B$, $T(B) \subseteq A$ and there exists $k \in (0,1)$ such that $d(Tx, Ty) \leq kd(x,y)$ for all $x \in A$ and $y \in B$. Then, T has a unique fixed point in $A \cap B$.

Let A and B be nonempty closed subsets of a metric space (X,d) and $\varphi : [0,\infty) \to [0,\infty)$ a strictly increasing map. We say that $T : A \cup B \to A \cup B$

is a cyclic φ-contraction map whenever $T(A) \subseteq B$, $T(B) \subseteq A$ and

$$d(Tx,Ty) \leq d(x,y) - \varphi(d(x,y)) + \varphi(d(A,B))$$

for all $x \in A$ and $y \in B$ ([9]). Also, $x \in A \cup B$ is called a best proximity point if $d(x,Tx) = d(A,B)$. As a special case, when $\varphi(t) = (1-\alpha)t$, in which $\alpha \in (0,1)$ is a constant, T is called cyclic contraction.

In 2005, Petrusel proved some results about periodic points of cyclic contraction maps ([81]). Then, Eldered and Veeramani proved some results about best proximity points of cyclic contraction maps in 2006 ([38]). They raised a question about the existence of a best proximity point for a cyclic contraction map in a reflexive Banach space. In 2009, Al-Thagafi and Shahzad gave a positive answer to the question ([9]). In fact, they solved the problem for cyclic φ-contraction maps. But, they proved some results about best proximity point of weakly continuous cyclic contraction maps satisfying the proximal property on reflexive (and strictly convex) Banach spaces ([9]; Theorems 9, 10, 11 and 12). In this way, they raised the following question in ([9]).

Question. It is interesting to ask whether Theorems 9 and 10 [resp. Theorems 11 and 12] hold for cyclic φ-contraction maps where the space is only reflexive [resp. reflexive and strictly convex] Banach space.

We are going to provide a positive answer to the question. For obtaining the answer, note that the key result in last section of [9] is proposition 3.3 of [38]. Therefore, it is sufficient that we prove the proposition 3.3 of [38] for cyclic φ-contraction maps.

First, we give the following extension of Proposition 3.3 in [38] for cyclic φ-contraction maps, where φ is unbounded.

Theorem 3.21

Let $\varphi : [0,\infty) \to [0,\infty)$ be a strictly increasing unbounded map. Also, let A and B be nonempty subsets of a metric space (X,d), $T : A \cup B \to A \cup B$ a cyclic φ-contraction map, $x_0 \in A \cup B$ and $x_{n+1} = Tx_n$ for all $n \geq 0$. Then, the sequences $\{x_{2n}\}$ and $\{x_{2n+1}\}$ are bounded.

Proof: Suppose that $x_0 \in A$ (the proof when $x_0 \in B$ is similar). By using Theorem 3 in [9], $d(x_{2n}, x_{2n+1}) \to d(A,B)$. Hence, it is sufficient we prove that $\{x_{2n+1}\}$ is bounded. Since φ is unbounded, there exists $M > 0$ such that

$$\varphi(M) > d(x_0, Tx_0) + \varphi(d(A,B)).$$

If $\{x_{2n+1}\}$ is not bounded, then there exists a natural number n_0 such that

$$d(T^2x_0, T^{2n_0+1}x_0) > M \quad \text{and} \quad d(T^2x_0, T^{2n_0-1}x_0) \leq M.$$

Then, we have

$$\begin{aligned}
M &< d(T^2x_0, T^{2n_0+1}x_0) \leq d(Tx_0, T^{2n_0}x_0) - \varphi(d(Tx_0, T^{2n_0}x_0)) + \varphi(d(A,B)) \\
&\leq d(x_0, T^{2n_0-1}x_0) - [\varphi(d(Tx_0, T^{2n_0}x_0)) + \varphi(d(x_0, T^{2n_0-1}x_0))] + 2\varphi(d(A,B)) \\
&\leq d(x_0, T^2x_0) + d(T^2x_0, T^{2n_0-1}x_0) - [\varphi(d(Tx_0, T^{2n_0}x_0)) + \varphi(d(x_0, T^{2n_0-1}x_0))] \\
&\quad + 2\varphi(d(A,B)) \\
&\leq d(x_0, Tx_0) + d(Tx_0, T^2x_0) + M - [\varphi(d(Tx_0, T^{2n_0}x_0)) + \varphi(d(x_0, T^{2n_0-1}x_0))] \\
&\quad + 2\varphi(d(A,B)).
\end{aligned}$$

Since $d(Tx, Ty) \leq d(x,y)$ for all $x \in A$ and $y \in B$, we obtain

$$M < 2d(x_0, Tx_0) + M - [\varphi(d(Tx_0, T^{2n_0}x_0)) + \varphi(d(x_0, T^{2n_0-1}x_0))] + 2\varphi(d(A,B)).$$

Since $d(Tx_0, T^{2n_0}x_0) \leq d(x_0, T^{2n_0-1}x_0)$, we have

$$\varphi(d(Tx_0, T^{2n_0}x_0)) \leq \varphi(d(x_0, T^{2n_0-1}x_0)).$$

Thus, we obtain $\varphi(d(Tx_0, T^{2n_0}x_0)) < d(x_0, Tx_0) + \varphi(d(A,B))$. Since

$$M < d(T^2x_0, T^{2n_0+1}x_0) \leq d(Tx_0, T^{2n_0}x_0),$$

$\varphi(M) < \varphi(d(Tx_0, T^{2n_0}x_0))$. Hence, $\varphi(M) < d(x_0, Tx_0) + \varphi(d(A,B))$. This contradiction completes the proof.

♣

The following example shows that $\varphi : [0,\infty) \to [0,\infty)$ could be a bounded strictly increasing map. Thus, unboundedness of φ is a necessary condition in Theorem 3.21 which we used it in the proof.

Example 29 *Consider the map* $\varphi : [0,\infty) \to [0,\infty)$ *by* $\varphi(x) = \frac{x}{1+x}$. *Then* φ *is a bounded strictly increasing map because* $\varphi'(x) = \frac{1}{(x+1)^2} > 0$ *for all* $x \in (0,\infty)$.

Since the proof of the last result was classic, we presented it separately. Here, we provide our key result via a special proof which is a general case of Theorem 3.21.

Theorem 3.22

Let $\varphi : [0,\infty) \to [0,\infty)$ be a strictly increasing map. Also, let A and B be nonempty subsets of a metric space (X,d), $T : A \cup B \to A \cup B$ a cyclic φ-contraction map, $x_0 \in A \cup B$ and $x_{n+1} = Tx_n$ for all $n \geq 0$. Then, the sequences $\{x_{2n}\}$ and $\{x_{2n+1}\}$ are bounded.

Proof: Suppose that $x_0 \in A$ (the proof when $x_0 \in B$ is similar). By using Theorem 3 in [9], $d(x_{2n}, x_{2n+1}) \to d(A,B)$. Hence, either $\{x_{2n+1}\}$ and $\{x_{2n}\}$ are bounded or both are unbounded. Suppose that both sequences are unbounded. Fix $n_1 \in \mathbb{N}$ and define

$$e_{n,k} = d(T^{2n}x_0, T^{2(n_1+k)+1}x_0),$$

for all $n,k \geq 1$. Since $\{x_{2n+1}\}$ is unbounded, $\limsup_{k\to\infty} e_{n,k} = \infty$ for all $n \geq 1$. Thus, we can choose a strictly increasing subsequence $\{e_{1,k_i^1}\}_{i\geq 1}$ of the sequence $\{e_{1,k}\}_{k\geq 1}$. Since $d(T^2x_0, T^{2(n_1+k_i^1)+1}x_0) \leq d(T^2x_0, T^4x_0) + d(T^4x_0, T^{2(n_1+k_i^1)+1}x_0)$, we have

$$\limsup_{i\to\infty} e_{2,k_i^1} = \infty.$$

Again, we can choose a strictly increasing subsequence $\{e_{2,k_i^2}\}_{i\geq 1}$ of the sequence $\{e_{2,k_i^1}\}_{i\geq 1}$ such that $\limsup_{i\to\infty} e_{2,k_i^2} = \infty$. By continuing this process, for each natural number n, we can choose a strictly increasing subsequence $\{e_{n,k_i^n}\}_{i\geq 1}$ of the sequence $\{e_{n,k_i^{n-1}}\}_{i\geq 1}$ such that $\limsup_{i\to\infty} e_{n,k_i^n} = \infty$. By the construction, if we consider the sequence $\{k_i^i\}_{i\geq 1}$, then $\lim_{i\to\infty} k_i^i = \infty$, $\{e_{n,k_i^i}\}_{i\geq 1}$ is a strictly increasing subsequence of $\{e_{n,k_i^n}\}_{i\geq 1}$ and $\limsup_{i\to\infty} e_{n,k_i^i} = \infty$ for all $n \geq 1$. Now, define $n_2 = n_1 + k_2^2 - k_1^1$. Also, by induction define the sequence $\{n_m\}_{m\geq 1}$ by $n_m = n_1 + k_m^m - k_1^1$. Note that, the sequence $\{n_m\}_{m\geq 1}$ is strictly increasing and $\limsup_{m\to\infty} n_m = \infty$. Since T is a cyclic φ-contraction map, $\{d(T^{2n_m}x_0, T^{2(n_m+k_1^1)+1}x_0)\}_{m\geq 1}$ is a decreasing sequence. Hence by the construction of the sequence $\{n_m\}_{m\geq 1}$, $\{d(T^{2n_m}x_0, T^{2(n_1+k_m^m)+1}x_0)\}_{m\geq 1}$ is a decreasing sequence. Let $m \geq 1$ be given. Since $e_{n_m,k_1^1} \leq e_{n_m,k_m^m}$, we have

$$d(T^{2n_m}x_0, T^{2(n_1+k_1^1)+1}x_0) \leq d(T^{2n_m}x_0, T^{2(n_1+k_m^m)+1}x_0).$$

Thus,

$$d(T^{2n_m}x_0, T^{2(n_1+k_1^1)+1}x_0) \leq d(T^{2n_1}x_0, T^{2(n_1+k_1^1)+1}x_0)$$

for all $m \geq 1$. Hence, we have

$$d(T^{2(n_1+k_m^m)+1}x_0, T^{2(n_1+k_1^1)+1}x_0)$$

$$\leq d(T^{2n_m}x_0, T^{2(n_1+k_1^1)+1}x_0) + d(T^{2n_m}x_0, T^{2(n_1+k_m^m)+1}x_0)$$

$$= d(T^{2n_m}x_0, T^{2(n_1+k_1^1)+1}x_0) + d(T^{2n_m}x_0, T^{2(n_m+k_1^1)+1}x_0)$$
$$\leq d(T^{2n_1}x_0, T^{2(n_1+k_1^1)+1}x_0) + d(T^{2n_m}x_0, T^{2(n_m+k_1^1)+1}x_0),$$

for all $m \geq 1$. Since $d(Tx,Ty) \leq d(x,y)$ for all $x \in A$ and $y \in B$, we obtain

$$d(T^{2(n_1+k_m^m)+1}x_0, T^{2(n_1+k_1^1)+1}x_0)$$
$$\leq d(T^{2n_1}x_0, T^{2(n_1+k_1^1)+1}x_0) + d(T^{2n_m-1}x_0, T^{2(n_m+k_1^1)}x_0)$$
$$\leq d(T^{2n_1}x_0, T^{2(n_1+k_1^1)+1}x_0) + d(x_0, T^{2k_1^1+1}x_0),$$

for all $m \geq 1$. Consequently

$$d(T^{2(n_1+k_m^m)+1}x_0, T^{2(n_1+k_1^1)}x_0)$$
$$\leq d(T^{2(n_1+k_m^m)+1}x_0, T^{2(n_1+k_1^1)+1}x_0) + d(T^{2(n_1+k_1^1)+1}x_0, T^{2(n_1+k_1^1)}x_0)$$
$$\leq d(T^{2n_1}x_0, T^{2(n_1+k_1^1)+1}x_0) + d(x_0, T^{2k_1^1+1}x_0) + d(T^{2(n_1+k_1^1)+1}x_0, T^{2(n_1+k_1^1)}x_0),$$

for all $m \geq 1$. This implies that

$$e_{(n_1+k_1^1),k_m^m} \leq \mu$$

for all $m \geq 1$, where

$$\mu = d(T^{2n_1}x_0, T^{2(n_1+k_1^1)+1}x_0) + d(x_0, T^{2k_1^1+1}x_0) + d(T^{2(n_1+k_1^1)+1}x_0, T^{2(n_1+k_1^1)}x_0)$$

is a constant. But, $\limsup_{i\to\infty} e_{n,k_i^i} = \infty$ for all $n \geq 1$. This contradiction completes the proof.

♣

Now by using this key result, we provide our main results which gives a positive answer to the question.

Theorem 3.23

Let $\varphi : [0,\infty) \to [0,\infty)$ be a strictly increasing map. Also, let A and B be nonempty weakly closed subsets of a reflexive Banach space and $T : A \cup B \to A \cup B$ a cyclic φ-contraction map. Then there exists $(x,y) \in A \times B$ such that

$$\|x - y\| = d(A,B).$$

Proof: If $d(A,B) = 0$, the result follows from Theorem 1 in [9]. Thus, we assume that $d(A,B) > 0$. For $x_0 \in A$, define $x_{n+1} = Tx_n$ for all $n \geq 1$. By Theorem 3.22, the sequences $\{x_{2n}\}$ and $\{x_{2n+1}\}$ are bounded. Since X is reflexive and A is weakly closed, the sequence $\{x_{2n}\}$ has a subsequence $\{x_{2n_k}\}$ with $x_{2n_k} \xrightarrow{w} x \in A$. As $\{x_{2n_k+1}\}$ is bounded and B is weakly closed, we can say, without loss of generality, that $x_{2n_k+1} \xrightarrow{w} y \in B$ as $k \to \infty$. Since $x_{2n_k} - x_{2n_k+1} \xrightarrow{w} x - y \neq 0$ as $k \to \infty$, there exists a bounded linear functional $f : X \to [0, \infty)$ such that $\|f\| = 1$ and $f(x-y) = \|x-y\|$. For each $k \geq 1$, we have

$$|f(x_{2n_k} - x_{2n_k+1})| \leq \|f\| \|x_{2n_k} - x_{2n_k+1}\| = \|x_{2n_k} - x_{2n_k+1}\|.$$

Since $\lim_{k\to\infty} f(x_{2n_k} - x_{2n_k+1}) = f(x-y) = \|x-y\|$, by using Theorem 3 in [9] we get

$$\|x-y\| = \lim_{k\to\infty} |f(x_{2n_k} - x_{2n_k+1})| \leq \lim_{k\to\infty} \|x_{2n_k} - x_{2n_k+1}\| = \|x_{2n_k} - x_{2n_k+1}\| = d(A,B).$$

Thus, $\|x-y\| = d(A,B)$.

♣

Definition 3.3 (see [9]) Let A and B be nonempty subsets of a normed space X, $T : A \cup B \to A \cup B$, $T(A) \subseteq B$ and $T(B) \subseteq A$. We say that T satisfies the proximal property whenever $x_n \xrightarrow{w} x \in A \cup B$ and $\|x_n - Tx_n\| \to d(A,B) \Rightarrow \|x - Tx\| = d(A,B)$.

Theorem 3.24

Let $\varphi : [0,\infty) \to [0,\infty)$ be a strictly increasing map. Also, let A and B be nonempty subsets of a reflexive Banach space X such that A is weakly closed and $T : A \cup B \to A \cup B$ a cyclic φ-contraction map. Then, there exists $x \in A$ such that $\|x - Ty\| = d(A,B)$ provided one of the following conditions is satisfied:

(a) T is weakly continuous on A.
(b) T satisfies the proximal property.

Proof: If $d(A,B) = 0$, the result follows from Theorem 1 in [9]. So, we assume that $d(A,B) > 0$. For $x_0 \in A$, define $x_{n+1} = Tx_n$ for all $n \geq 1$. By Theorem 3.22, the sequence $\{x_{2n}\}$ is bounded. Since X is reflexive and A is weakly closed, the sequence $\{x_{2n}\}$ has a subsequence $\{x_{2n_k}\}$ with $x_{2n_k} \xrightarrow{w} x \in A$ as $k \to \infty$.

(a) Since T is weakly continuous on A and $T(A) \subseteq B$, we have $x_{2n_k+1} \xrightarrow{w} Tx \in B$ as $k \to \infty$. Thus $x_{2n_k} - x_{2n_k+1} \xrightarrow{w} x - Tx \neq 0$ as $k \to \infty$. The rest of the proof is similar to that of Theorem 3.23.

(b) By using Theorem 3 in [9], we have

$$\|x_{2n_k} - Tx_{2n_k}\| = \|x_{2n_k} - x_{2n_k+1}\| \to d(A,B)$$

as $k \to \infty$. As T satisfies the proximal property, it follows that $\|x-y\| = d(A,B)$.

♣

Theorem 3.25

Let $\varphi : [0,\infty) \to [0,\infty)$ be a strictly increasing map. Also, let A and B be nonempty closed and convex subsets of a reflexive and strictly convex Banach space and $T : A\cup B \to A\cup B$ a cyclic φ-contraction map. If $(A-A)\cap(B-B) = \{0\}$, then there exists a unique $x \in A$ such that $T^2x = x$ and $\|x-y\| = d(A,B)$.

Proof: If $d(A,B) = 0$, the result follows from Theorem 1 in [9]. So, we assume that $d(A,B) > 0$. Since A is closed and convex, it is weakly closed. It follows from Theorem 3.23 that there exists $(x,y) \in A\times B$ such that $\|x-y\| = d(A,B)$. To show the uniqueness of (x,y), suppose that there exists another $(x',y') \in A\times B$ such that $\|x'-y'\| = d(A,B)$. Since $(A-A)\cap(B-B) = \{0\}$, we conclude that $x-y \neq x'-y'$. Since A and B are both convex, it follows from the strict convexity of X that

$$\|\frac{x+x'}{2} - \frac{y+y'}{2}\| = \|\frac{x-y}{2} + \frac{x'-y'}{2}\| < d(A,B),$$

a contradiction. As $\|Ty-Tx\| = \|Tx-Ty\| = \|x-y\| = d(A,B)$, we conclude, from uniqueness of (x,y), that $(Ty,Tx) = (x,y)$. Thus $Tx = y$, $Ty = x$ and $T^2x = x$.

♣

Theorem 3.26

Let $\varphi : [0,\infty) \to [0,\infty)$ be a strictly increasing map. Also, let A and B be nonempty subsets of a reflexive and strictly convex Banach space X such that A is closed and convex and $T : A\cup B \to A\cup B$ a cyclic φ-contraction map. Then, there exists a unique $x \in A$ such that $T^2x = x$ and $\|x-Ty\| = d(A,B)$ provided one of the following conditions is satisfied:

(a) T is weakly continuous on A.
(b) T satisfies the proximal property.

Proof: If $d(A,B) = 0$, the result follows from Theorem 1 in [9]. So, we assume that $d(A,B) > 0$. Since A is closed and convex, it is weakly closed. It follows from Theorem 3.24 that there exists $x \in A$ such that $\|x-Ty\| = d(A,B)$. In fact,

$T^2x = x$. To see this, assume that $T^2x - Tx \neq x - Tx$. It follows from the convexity of A and the strict convexity of X, that

$$0 < \|\frac{x+x'}{2} - \frac{y+y'}{2}\| = \|\frac{x-y}{2} + \frac{x'-y'}{2}\| < d(A,B),$$

To show the uniqueness of (x,y), suppose that there exists another $(x',y') \in A \times B$ such that $\|x' - y'\| = d(A,B)$. Since $(A-A) \cap (B-B) = \{0\}$, we conclude that $x - y \neq x' - y'$. Since A and B are both convex, it follows from the strict convexity of X that

$$\|\frac{T^2x+x}{2} - Tx\| = \|\frac{T^2x - Tx}{2} + \frac{x - Tx}{2}\| < d(A,B),$$

a contradiction. The uniqueness of x follows as in the proof of Theorem 3.25.

♣

Now, we are going to study some best proximity points results on ordered metric spaces.

Let X be a nonempty set and T a selfmap on X. We denote the set of all nonempty subsets of X by 2^X and the set of all invariant nonempty subsets of X by $I(T)$, that is

$$I(T) = \{Y \in 2^X : T(Y) \subseteq Y\}.$$

For each pair of sets X and Y and selfmaps $T : X \to X$ and $S : Y \to Y$, we define the selfmap $T \times S : X \times Y \to X \times Y$ by $T \times S(x,y) = (Tx, Sy)$. If $(X, \leq)$ is a partially ordered set, then we define

$$X_{\leq} = \{(x,y) \in X \times X : x \leq y \text{ or } y \leq x\}.$$

Let $(X, d, \leq)$ be an ordered metric space and $T : X \to X$ a selfmap on X. For each nonempty subset C of X and $x^* \in X$, we define

$$E_{T,C}(x^*) = \{x \in C : \lim_{n\to\infty} T^{2n}x = x^*\}.$$

We say that X has the property (C) whenever for each monotone sequence $\{x_n\}$ in X with $x_n \to x$ for some $x \in X$, there exists a subsequence $\{x_{n_k}\}$ of $\{x_n\}$ such that every element of $\{x_{n_k}\}$ is comparable with x. Also, X is called regular whenever every bounded monotone sequence in X is convergent. We say that a selfmap T : $X \to X$ is orbitally continuous whenever for each $x \in X$ and sequence $\{n(i)\}_{i\geq 1}$ with $T^{n(i)}x \to a$ for some $a \in X$, we have $T^{n(i)+1}x \to Ta$. Here, $T^{m+1} = T(T^m)$.

Theorem 3.27

Let $(X, d, \leq)$ be an ordered metric space, $A, B \in 2^X$ and T a decreasing selfmap on

$A \cup B$ such that $T(A) \subseteq B$ and $T(B) \subseteq A$. Suppose that there exists $x_0 \in A$ such that $x_0 \leq T^2x_0 \leq Tx_0$ and

$$d(Tx,Ty) \leq d(x,y) - \varphi(d(x,y)) + \varphi(d(A,B))$$

for all $x \in A$ and $y \in B$ with $x \leq y$, where $\varphi : [0,\infty) \to [0,\infty)$ is a strictly increasing map. If $x_{n+1} = Tx_n$ and $d_n = d(x_{n+1},x_n)$ for all $n \geq 0$, then $d_n \to d(A,B)$.

Proof: First note that we have

$$x_0 \leq x_2 \leq \cdots \leq x_{2n} \leq x_{2n+1} \leq \cdots \leq x_3 \leq x_1$$

for all $n \geq 1$. Thus, we obtain

$$0 \leq d_{n+1} \leq d_n - \varphi(d_n) + \varphi(d(A,B))$$

for all $n \geq 1$. Hence, the sequence $\{d_n\}$ is decreasing and bounded from below. If $d_{n_0} = 0$ for some n_0, then $d_n \to d(A,B) = 0$. Suppose that $d_n > 0$ for all $n \geq 1$ and $d_n \to t_0$ for some $t_0 \geq d(A,B)$. Since

$$\varphi(d(A,B)) \leq \varphi(d_n) \leq d_n - d_{n+1} + \varphi(d(A,B)),$$

we have $\varphi(d_n) \to \varphi(d(A,B))$. This implies that $\varphi(t_0) = \varphi(d(A,B))$. So, $t_0 = d(A,B)$ because φ is strictly increasing.

♣

Theorem 3.28

Let $(X,d,\leq)$ be a regular ordered metric space, $B \in 2^X$, A a closed nonempty subset of X and T a decreasing selfmap on $A \cup B$ such that $T(A) \subseteq B$ and $T(B) \subseteq A$. Suppose that there exists $x_0 \in A$ such that $x_0 \leq T^2x_0 \leq Tx_0$ and

$$d(Tx,Ty) \leq d(x,y) - \varphi(d(x,y)) + \varphi(d(A,B))$$

for all $x \in A$ and $y \in B$ with $x \leq y$, where $\varphi : [0,\infty) \to [0,\infty)$ is a strictly increasing map. If T is orbitally continuous or X has the property (C), then there exists $x \in A$ such that $d(x,Tx) = d(A,B)$.

Proof: Again, note that

$$x_0 \leq x_2 \leq \cdots \leq x_{2n} \leq x_1$$

for all $n \geq 1$. Since X is regular and A is closed, there exists $x \in A$ such that $x_{2n} \to x$. Also, note that

$$d(A,B) \leq d(x_{2n},Tx) = d(Tx_{2n-1},Tx) \leq d(Tx_{2n-1},Tx_{2n}) + d(Tx_{2n},Tx)$$

for all $n \geq 1$. If T is orbitally continuous, then $d(Tx_{2n}, Tx) \to 0$. Hence,

$$d(x,Tx) = d(A,B)$$

because $d(Tx_{2n-1}, Tx_{2n}) \to d(A,B)$ by Theorem 3.27. Now, suppose that X has the property (C). Since $\{x_{2n}\}$ is a bounded and increasing sequence, there exists a subsequence $\{x_{2n_k}\}$ of $\{x_{2n}\}$ such that $x_{2n_1} \leq x_{2n_2} \leq \cdots \leq x_{2n_k} \leq \cdots \leq x$. Therefore,

$$\begin{aligned} d(A,B) \leq d(x_{2n_k}, Tx) = d(Tx_{2n_k-1}, Tx) &\leq d(Tx_{2n_k-1}, Tx_{2n_k}) + d(Tx_{2n_k}, Tx) \\ &\leq d(Tx_{2n_k-1}, Tx_{2n_k}) + d(x_{2n_k}, x). \end{aligned}$$

Since $d(Tx_{2n_k-1}, Tx_{2n_k}) \to d(A,B)$, this implies that $d(x,Tx) = d(A,B)$.

♣

The following is another example for a cyclic φ-contraction. Note that, we should improve Example 3 in [9] because T is not cyclic φ-contraction in the example. For seeing this, it is sufficient that we put $x = \frac{-1}{2}$ and $y = \frac{1}{2}$. Then $\frac{2}{3} = d(Tx,Ty) > d(x,y) - \varphi(d(x,y)) + \varphi(d(A,B)) = \frac{1}{2}$. Now for improving, it is sufficient we change the function φ by $\varphi(t) = \frac{t^2}{2(1+t)}$.

Example 30 *Consider the Euclidian ordered metric space $X = \mathbb{R}$ with the usual norm. Suppose that $A = [-1,0]$, $B = [0,1]$ and $T : A \cup B \to A \cup B$ is defined by $Tx = \frac{-x}{3}$ for all $x \in A \cup B$. If $\varphi : [0,\infty) \to [0,\infty)$ is defined by $\varphi(t) = \frac{t}{2}$, then φ is strictly increasing and T is a cyclic φ-contraction map.*

The following example shows that Theorem 3.28 may be applied in situations where Theorem 8 in [9] does not work.

Example 31 *Consider the regular ordered metric space $X = L^1([0,1])$ with the norm $\|.\|_1$ and the order $f \leq g$ if and only if $f(t) \leq g(t)$ for almost all $t \in [0,1]$. Suppose that $A = \{f \in X : -1 \leq f \leq 0\}$, $B = \{g \in X : 0 \leq g \leq 1\}$ and $T : A \cup B \to A \cup B$ is defined by $Tf = \frac{-f}{3}$ for all $f \in A \cup B$. If $\varphi : [0,\infty) \to [0,\infty)$ is defined by $\varphi(t) = \frac{t}{2}$, then φ is strictly increasing and T is a decreasing cyclic φ-contraction map. Note that A is closed and convex, T is orbitally continuous and $T0 = 0$. But, X is not a reflexive Banach space.*

Theorem 3.29

Let $(X,d,\leq)$ be an ordered metric space, $A, B \in 2^X$ and T a selfmap on $A \cup B$ such that $T(A) \subseteq B$, $T(B) \subseteq A$ and $((A \times B) \cup (B \times A)) \cap X_{\leq} \in I(T \times T)$. Suppose that there exists $x_0 \in A$ such that $(x_0, Tx_0) \in X_{\leq}$ and

$$d(Tx,Ty) \leq d(x,y) - \varphi(d(x,y)) + \varphi(d(A,B))$$

for all $x \in A$ and $y \in B$ with $(x,y) \in X_{\leq}$, where $\varphi : [0,\infty) \to [0,\infty)$ is a strictly increasing map. If $x_{n+1} = Tx_n$ and $d_n = d(x_{n+1}, x_n)$ for all $n \geq 0$, then $d_n \to d(A,B)$.

Proof: First note that we have

$$d(T^{2n+1}x_0, T^{2n}x_0) \leq d(T^{2n}x_0, T^{2n-1}x_0) - \varphi(d(T^{2n}x_0, T^{2n-1}x_0)) + \varphi(d(A,B))$$

for all $n \geq 1$. Thus, we obtain

$$0 \leq d_{n+1} \leq d_n - \varphi(d_n) + \varphi(d(A,B))$$

for all $n \geq 1$. Hence, the sequence $\{d_n\}$ is decreasing and bounded from below. If $d_{n_0} = 0$ for some n_0, then $d_n \to d(A,B) = 0$. Suppose that $d_n > 0$ for all $n \geq 1$ and $d_n \to t_0$ for some $t_0 \geq d(A,B)$. Since

$$\varphi(d(A,B)) \leq \varphi(d_n) \leq d_n - d_{n+1} + \varphi(d(A,B)),$$

we have $\varphi(d_n) \to \varphi(d(A,B))$. This implies that $\varphi(t_0) = \varphi(d(A,B))$. So, $t_0 = d(A,B)$ because φ is strictly increasing.

♣

Theorem 3.30

Let $(X,d,\leq)$ be an ordered metric space, $A,B \in 2^X$ and T a selfmap on $A \cup B$ such that $T(A) = B$, $T(B) \subseteq A$ and $((A \times B) \cup (B \times A)) \cap X_{\leq} \in I(T \times T)$. Suppose that for each $x,y \in A$ there exists $z \in A$ such that $(x,z), (y,z) \in X_{\leq}$. Also, suppose that there exist $x_0, x^* \in A$ such that $x_0 \in E_{T,A}(x^*)$, $(x_0, Tx_0) \in X_{\leq}$ and

$$d(Tx,Ty) \leq d(x,y) - \varphi(d(x,y)) + \varphi(d(A,B))$$

for all $x \in A$ and $y \in B$ with $(x,y) \in X_{\leq}$, where $\varphi : [0,\infty) \to [0,\infty)$ is a strictly increasing map. Also, suppose that $y \in A$, $(x,y) \in X_{\leq}$ and $x \in E_{T,A}(x^*)$ imply that $y \in E_{T,A}(x^*)$. Then, $E_{T,A}(x^*) = A$ and the following statement holds:

$$E_{T,B}(Tx^*) = B \text{ and } d(x^*, Tx^*) = d(A,B) \Leftrightarrow T \text{ is orbitally continuous.}$$

Proof: Let $x \in A$. If $(x_0, x) \in X_{\leq}$, then $x \in E_{T,A}(x^*)$. If $(x_0, x) \notin X_{\leq}$, then there exists $z \in A$ such that $(x_0, z) \in X_{\leq}$ and $(x,z) \in X_{\leq}$. Hence, $x \in E_{T,A}(x^*)$. Thus, $E_{T,A}(x^*) = A$.

Now, suppose that T is orbitally continuous and $y \in B$. Choose $x' \in A$ such that $Tx' = y$. Since $E_{T,A}(x^*) = A$, $T^{2n}x' \to x^*$ and so $T^{2n+1}x' \to Tx^*$. Hence, we have $T^{2n}y \to Tx^*$. Thus, $E_{T,B}(Tx^*) = B$. If $d(x^*, Tx^*) \neq d(A,B)$, then $\{d(T^{2n+1}x_0,$

$T^{2n}x_0)\}$ is a decreasing sequence because $(x_0, Tx_0) \in X_{\leq}$. By Theorem 3.29, $d(T^{2n+1}x_0, T^{2n}x_0) \downarrow d(A,B)$. Choose a natural number n such that

$$d(A,B) \leq d(T^{2n+1}x_0, T^{2n}x_0) < d(x^*, Tx^*).$$

Put $x = T^{2n}x_0$ and $y = T^{2n+1}x_0$. Since $(x,y) \in X_{\leq}$, $(Tx, Ty) \in X_{\leq}$ and so $\{d(T^{2n}x, T^{2n}y)\}$ is a decreasing sequence and $d(T^{2n}x, T^{2n}y) \downarrow d(x^*, Tx^*)$. Hence,

$$d(x^*, Tx^*) \leq d(T^{2n+1}x_0, T^{2n}x_0) < d(x^*, Tx^*)$$

which is a contradiction. Therefore, $d(x^*, Tx^*) = d(A,B)$. Assume that $d(x^*, Tx^*) = d(A,B)$, $E_{T,B}(Tx^*) = B$, $x \in A \cup B$ and $T^{n(i)}x \to a$ for some $a \in A \cup B$. We shall show that $T^{n(i)+1}x \to Ta$. Put $A' = A \cap \{T^{n(i)}x\}$ and $B' = B \cap \{T^{n(i)}x\}$.

Case I. Let $d(A,B) = 0$. First suppose that $A' = \{T^{n_1(i)}x\}$ and $B' = \{T^{n_2(i)}x\}$ are subsequences of $\{T^{n(i)}x\}$. Since $\{T^{n_1(i)}x\}$ is a subsequence of $\{T^{2n}x\}$, $T^{n_1(i)}x \to x^*$. Also, we have $T^{n_1(i)+1}x \to Tx^*$ because $Tx \in B$ and $E_{T,B}(Tx^*) = B$. Since $\{T^{n_1(i)}x\}$ is a subsequence of $\{T^{n(i)}x\}$ and $T^{n(i)}x \to a$, $T^{n_1(i)}x \to a$. Thus, $a = x^*$ and so $a = x^* = Ta = Tx^*$. Since $\{T^{n_2(i)}x\}$ is a subsequence of $\{T^{2n+1}x\} = \{T^{2n}(Tx)\}$, $Tx \in B$ and $E_{T,B}(Tx^*) = B$, $T^{n_2(i)}x \to Tx^*$. Also, we have $T^{n_2(i)+1}x \to x^*$ because $T^2x \in A$, $E_{T,A}(x^*) = A$ and $\{T^{n_2(i)}x\}$ is a subsequence of $\{T^{2n+2}x\} = \{T^{2n}(T^2x)\}$. Hence, $T^{n(i)+1}x \to Ta$. Now, suppose that $B' = \{t_1, \cdots, t_k\}$ is finite. By using a similar argument, we have $T^{n_1(i)}x \to x^*$, $T^{n_1(i)+1}x \to Tx^*$ and $a = x^* = Ta = Tx^*$. Since $\{T^{n(i)+1}x\} = \{T^{n_1(i)+1}x\} \cup \{Tt_1, \cdots, Tt_k\}$, $T^{n(i)+1}x \to Ta$. If $A' = \{s_1, \cdots, s_m\}$ is finite, then $B' = \{T^{n_2(i)}x\}$ is a subsequence of $\{T^{n(i)}x\}$ and so $T^{n_2(i)}x \to a$. By using a similar argument, we have $T^{n_2(i)}x \to Tx^*$ and $T^{n_2(i)+1}x \to x^*$. Thus, $a = x^* = Ta = Tx^*$. Since

$$\{T^{n(i)+1}x\} = \{T^{n_2(i)+1}x\} \cup \{Ts_1, \cdots, Ts_m\},$$

we have $T^{n(i)+1}x \to Ta$.

Case II. Let $d(A,B) > 0$. We claim that A' or B' is finite. In fact, if A' and B' are infinite, then similar to the above case we have $T^{n_1(i)}x \to x^*$ and $T^{n_2(i)}x \to Tx^*$. Since $\{T^{n_1(i)}x\}$ and $\{T^{n_2(i)}x\}$ are subsequences of $\{T^{n(i)}x\}$ and $T^{n(i)}x \to a$, we obtain $a = x^* = Tx^*$. So, $d(A,B) = d(x^*, Tx^*) = 0$ which is a contradiction. Now, suppose that $B' = \{t_1, \cdots, t_k\}$ is finite. By using a similar argument in case I, we have $T^{n_1(i)}x \to x^*$, $T^{n_1(i)+1}x \to Tx^*$ and $a = x^*$. Since $\{T^{n(i)+1}x\} = \{T^{n_1(i)+1}x\} \cup \{Tt_1, \cdots, Tt_k\}$, $T^{n(i)+1}x \to Ta$. If $A' = \{s_1, \cdots, s_m\}$ is finite, then $B' = \{T^{n_2(i)}x\}$ is a subsequence of $\{T^{n(i)}x\}$ and so $T^{n_2(i)}x \to a$. By using a similar argument as in case I, we have $T^{n_2(i)}x \to Tx^*$. Thus, $a = Tx^*$. Also, we have $T^{n_2(i)+1}x \to x^*$ because $T^2x \in A$, $E_{T,A}(x^*) = A$ and $\{T^{n_2(i)}x\}$ is a subsequence of $\{T^{2n+2}x\} = \{T^{2n}(T^2x)\}$. Now, we show that $Ta = x^*$. In fact, $(x^*, x^*) \in X_{\leq}$ and

$$d(x^*, T^2x^*) \leq d(T^{2n}x^*, x^*) + d(T^{2n}x^*, T^2x^*).$$

Hence, by using the assumptions we have

$$d(T^{2n}x^*, T^2x^*) \leq d(T^{2n-2}x^*, x^*).$$

Thus, $d(x^*, T^2x^*) \leq d(T^{2n}x^*, x^*) + d(T^{2n-2}x^*, x^*)$. Since $E_{T,A}(x^*) = A$ and $x^* \in A$, we find that $T^{2n}x^* \to x^*$ and $T^{2n-2}x^* \to x^*$. Hence, $x^* = T^2x^*$. Since $a = Tx^*$, $Ta = x^*$. Thus, $T^{n_2(i)+1}x \to Ta$. Since $\{T^{n(i)+1}x\} = \{T^{n_2(i)+1}x\} \cup \{Ts_1, \cdots, Ts_m\}$, we have $T^{n(i)+1}x \to Ta$. This completes the proof.

♣

The following example shows that the assumption

$$d(Tx, Ty) \leq d(x, y) - \varphi(d(x, y)) + \varphi(d(A, B))$$

for all $x \in A$ and $y \in B$ with $(x, y) \in X_{\leq}$, does not imply the following assumption:

$$y \in A, (x, y) \in X_{\leq}, x \in E_{T,A}(x^*) \Rightarrow y \in E_{T,A}(x^*).$$

Example 32 *Consider the subsets*

$$A = \{x_1 = (6,3), x_2 = (1,3)\} \text{ and } B = \{y_1 = (2,0), y_2 = (0,4)\}$$

of $\mathbb{R}^2$ *via the following order:*

$$(a, b) \leq (c, d) \Leftrightarrow a \leq c \text{ and } b \leq d.$$

Define $T : A \cup B \to A \cup B$ *by* $Tx_1 = y_2$, $Tx_2 = y_1$, $Ty_1 = x_2$, $Ty_2 = x_1$. *Note that,* $x_2 \leq x_1$ *and* $y_1 \leq x_1$ *and other elements are not comparable. Also, we have* $d(Tx_1, Tx_2) = d(x_2, y_2) = d(A, B) = \sqrt{2}$ *and* $d(x_1, y_1) = \sqrt{25}$. *Consider the map* $\varphi : [0, \infty) \to [0, \infty)$ *by* $\varphi(x) = \frac{x}{2}$. *Then, we have*

$$d(Tx_1, Ty_1) \leq d(x_1, y_1) - \varphi(d(x_1, y_1)) + \varphi(d(A, B)),$$

while $T^{2n}x_1 \to x_1$ *and* $T^{2n}x_2 \to x_2$.

The following example shows that the assumptions of Theorem 3.30 do not imply orbital continuity of T.

Example 33 *Define* $T : \mathbb{R} \to \mathbb{R}$ *by* $x = \frac{-x}{3}$ *for all* $x \in \mathbb{R}$. *Put* $a_0 = -1$ *and define the sequences* $\{a_n\}_{n \geq 0}$ *and* $\{b_n\}_{n \geq 1}$ *by* $b_n = Sa_{n-1}$ *and* $a_n = Sb_n$ *for all* $n \geq 1$. *Now, define the sequences* $\{c_n\}_{n \geq 0}$ *and* $\{d_n\}_{n \geq 1}$ *as following:*

$$c_n = a_{2n+1} \text{ and } d_n = a_{2n} \text{ for all } n \geq 0.$$

Now, consider the subsets

$$A = \{(c_n, 0)\}_{n \geq 0} \cup \{(d_n, 0)\}_{n \geq 0} \cup \{(0, 0)\},$$

and

$$B = \{(b_{2n}, -1)\}_{n\geq 0} \cup \{(b_{2n+1}, -2)\}_{n\geq 1} \cup \{(0,-1)\}$$

of $\mathbb{R}^2$ *via the following order:*

$$(a,b) \leq (c,d) \Leftrightarrow a \leq c \text{ and } b \leq d.$$

Define $T : A\cup B \to A\cup B$ *by*

$$\begin{cases} T(c_n,0) = (b_{2n},-1) \\ T(c_n,0) = (b_{2n+1},-2) \\ T(b_{2n},-1) = (d_{n+1},0) \\ T(b_{2n+1},-2) = (c_{n+1},0) \\ T(0,0) = (0,-1) \\ T(0,-1) = (0,0). \end{cases}$$

If we define $\varphi : [0,\infty) \to [0,\infty)$ *by* $\varphi(t) = \frac{t}{2}$, *then it is easy to check that*

$$T(A) = B, T(B) \subseteq A, ((A\times B)\cup(B\times A))\cap X_{\leq} = \{(0,0),(-1,0)\} \in I(T\times T),$$

$$d(Tx,Ty) \leq d(x,y) - \varphi(d(x,y)) + \varphi(d(A,B))$$

for all $x \in A$ *and* $y \in B$ *with* $(x,y) \in X_{\leq}$ *and for each* $x,y \in A$ *there exists* $z \in A$ *such that* $(x,z),(y,z) \in X_{\leq}$. *If we put* $x_0 = x^* = (0,0)$, *then*

$$(x_0, Tx_0) = ((0,0),(0,-1)) \in X_{\leq},$$

and $y \in A$, $(x,y) \in X_{\leq}$ *and* $x \in E_{T,A}(x^*)$ *imply that* $y \in E_{T,A}(x^*)$. *Finally, note that* $T^{2n}x \to (0,0)$ *for all* $x \in A$, $T^{2n}x_0 \to x^*$, $E_{T,B}(Tx^*) = B$ *and* $d(x^*,Tx^*) = d(A,B)$ *while* $\lim_{n\to\infty} T^{2n+1}(c_m,0) = (0,-2) \neq Tx^* = (0,-1)$ *for all* $m \geq 1$. *This implies that* T *is not orbitally continuous.*

Theorem 3.31

Let $(X,d,\leq)$ *be an ordered metric space,* $A,B \in 2^X$ *and* T *a selfmap on* $A\cup B$ *such that* $T(A) = B$, $T(B) \subseteq A$ *and* $((A\times B)\cup(B\times A))\cap X_{\leq} \in I(T\times T)$. *Suppose that for each* $x,y \in A$ *there exists* $z \in A$ *such that* $(x,z),(y,z) \in X_{\leq}$. *Also, suppose that there exist* $x_0, x^* \in A$ *such that* $x_0 \in E_{T,A}(x^*)$ *and*

$$d(Tx,Ty) \leq d(x,y)$$

for all $x \in A$ *and* $y \in B$. *Also, suppose that* $y \in A$, $(x,y) \in X_{\leq}$ *and* $x \in E_{T,A}(x^*)$ *imply that* $y \in E_{T,A}(x^*)$. *Then,* $E_{T,A}(x^*) = A$ *and the following statement holds:*

$$E_{T,B}(Tx^*) = B \text{ and } d(x^*,Tx^*) = d(A,B) \Leftrightarrow T \text{ is orbitally continuous.}$$

Proof: Similar as in the proof of Theorem 3.30 we can show that $E_{T,A}(x^*) = A$ and $E_{T,B}(Tx^*) = B$ whenever T is orbitally continuous. If $d(x^*, Tx^*) \neq d(A,B)$, then there exists $x \in A$ and $y \in B$ such that

$$d(A,B) \leq d(x,y) < d(x^*, Tx^*).$$

Note that $\{d(T^{2n}x, T^{2n}y)\}$ is a decreasing sequence and $d(T^{2n}x, T^{2n}y) \downarrow d(x^*, Tx^*)$. Hence, $d(x^*, Tx^*) \leq d(x,y) < d(x^*, Tx^*)$ which is a contradiction. Thus,

$$d(x^*, Tx^*) = d(A,B).$$

Similar to the proof of Theorem 3.30 we can show that T is orbitally continuous whenever $E_{T,B}(Tx^*) = B$ and $d(x^*, Tx^*) = d(A,B)$.

♣

The following example shows that the assumption

$$d(Tx, Ty) \leq d(x,y)$$

for all $x \in A$ and $y \in B$, doesn't imply the following assumption in Theorem 3.31:

$$y \in A, (x,y) \in X_{\leq}, x \in E_{T,A}(x^*) \Rightarrow y \in E_{T,A}(x^*).$$

Example 34 *Consider the subsets*

$$A = \{x_1 = (0,0), x_2 = (0,1)\} \text{ and } B = \{y_1 = (1,0), y_2 = (1,1)\}$$

of $\mathbb{R}^2$ via the following order:

$$(a,b) \leq (c,d) \Leftrightarrow a \leq c \text{ and } b \leq d.$$

Define $T : A \cup B \to A \cup B$ by $Tx_1 = y_1$, $Tx_2 = y_2$, $Ty_1 = x_1$, $Ty_2 = x_2$. Note that,

$$d(Tx, Ty) \leq d(x,y)$$

for all $x \in A$ and $y \in B$, $T^{2n}x_1 \to x_1$ and $T^{2n}x_2 \to x_2$. Thus, the following assumption does not hold:

$$y \in A, (x,y) \in X_{\leq}, x \in E_{T,A}(x^*) \Rightarrow y \in E_{T,A}(x^*).$$

The following example shows that the following assumption is necessary in the Theorem 3.31:

$$d(Tx, Ty) \leq d(x,y)$$

for all $x \in A$ and $y \in B$.

Example 35 *Let $X = \mathbb{R}$, $A = [0,1]$ and $B = [2,3]$. Define $T : A \cup B \to A \cup B$ by $Tx = x+2$ for all $x \in A$ and $Tx = \frac{x-2}{2}$ for all $x \in B$. Note that, T is orbitally continuous and we have $T^{2n}x_0 = \frac{x_0}{2^n}$ and $T^{2n+1}x_0 = \frac{x_0}{2^n} + 2$ for all $x_0 \in A$ and $n \geq 0$. Thus, $T^{2n}x_0 \to 0$ and $T^{2n+1}x_0 \to 2$ for all $x_0 \in A$. But, note that the assumption does not hold because $d(T1,T2) \nleq d(1,2)$.*

The following example shows that the assumption

$$d(Tx,Ty) \leq d(x,y) \text{ for all } x \in A \text{ and } y \in B$$

could not substitute by the following assumption in Theorem 3.31:

$$d(Tx,Ty) \leq d(x,y) \text{ for all } x \in A \text{ and } y \in B \text{ with } (x,y) \in X_{\leq}.$$

Example 36 *Consider the subsets*

$$A = \{x_1 = (1,2), x_2 = (2,2)\} \text{ and } B = \{y_1 = (3,1), y_2 = (4,1)\}$$

of $\mathbb{R}^2$ via the following order:

$$(a,b) \leq (c,d) \Leftrightarrow a \leq c \text{ and } b \leq d.$$

Define $T : A \cup B \to A \cup B$ by $Tx_1 = y_1$, $Tx_2 = y_2$, $Ty_1 = Ty_2 = x_2$. Note that, $x_1 \leq x_2$ and $y_1 \leq y_2$ and other elements are not comparable. It is easy to check that $T(A) = B$, $T(B) \subseteq A$, $((A \times B) \cup (B \times A)) \cap X_{\leq} \in I(T \times T)$ and for each $x, y \in A$ there exists $z \in A$ such that $(x,z), (y,z) \in X_{\leq}$. Also, there exist $x_0, x^* \in A$ such that $x_0 \in E_{T,A}(x^*)$. Finally, $y \in A$, $(x,y) \in X_{\leq}$ and $x \in E_{T,A}(x^*)$ imply that $y \in E_{T,A}(x^*)$. Note that $T^{2n}x_i \to x_2$, $T^{2n+1}x_i \to y_2$, $T^{2n}y_i \to y_2$ and $T^{2n+1}y_i \to x_2$ for $i = 1,2$. Thus, T is orbitally continuous while $d(x_2, Tx_2) \neq d(A,B)$.

3.8 Approximate fixed points

At the end of section 3 in this chapter, we raised a question about completeness of a metric space. After you study this section, please think always about removing completeness in fixed point results for obtaining approximate fixed point results.

Let (X,d) be a metric space, T a selfmap on X and $\alpha : X \times X \to [0,\infty)$ a mapping. Recall we say that T is α-admissible whenever $\alpha(x,y) \geq 1$ implies $\alpha(Tx,Ty) \geq 1$. Also, we say that X has the property (H) whenever for each $x, y \in X$ there exists $z \in X$ such that $\alpha(x,z) \geq 1$ and $\alpha(y,z) \geq 1$ (see [94]). The selfmap T on X is called a generalized convex contraction whenever there exist a mapping $\alpha : X \times X \to [0,\infty)$ and $a, b \in (0,1)$ with $a+b < 1$ such that $\alpha(x,y)d(T^2x, T^2y) \leq ad(Tx,Ty) + bd(x,y)$ for all $x, y \in X$. We say that α is the based mapping. Also, we say that the selfmap T on X is a generalized convex contraction of order 2 whenever there exist a mapping $\alpha : X \times X \to [0,\infty)$ and $a_1, a_2, b_1, b_2 \in (0,1)$ with $a_1 + a_2 + b_1 + b_2 < 1$ such that

$$\alpha(x,y)d(T^2x,T^2y) \leq a_1 d(x,Tx) + a_2 d(Tx,T^2x) + b_1 d(y,Ty) + b_2 d(Ty,T^2y)$$

for all $x,y \in X$. Let $\varepsilon > 0$ be given. We say that $x_0 \in X$ is an ε-fixed point of the selfmap T on X whenever $d(x_0,Tx_0) < \varepsilon$ ([19]). Denote the set of all ε-fixed points of T by $F_\varepsilon(T)$. We say that T has an approximate fixed point (or T has the approximate fixed point property) whenever T has an ε-fixed point for all $\varepsilon > 0$ ([19]). It is known that there are many selfmaps which have approximate fixed points while have no fixed points. We need the following result in our main results.

Lemma 3.3

[19] Let (X,d) be a metric space and T an asymptotic regular selfmap on X, that is, $d(T^n(x),T^{n+1}(x)) \to 0$ for all $x \in X$. Then T has the approximate fixed point property.

Theorem 3.32

Let (X,d) be a metric space and T a generalized convex contraction on X with the based mapping α. Suppose that T is α-admissible and there exists $x_0 \in X$ such that $\alpha(x_0,Tx_0) \geq 1$. Then T has an approximate fixed point. Moreover, T has a fixed point whenever T is continuous and (X,d) is a complete metric space and also T has a unique fixed point whenever X has the property (H).

Proof: Let $x_0 \in X$ be such that $\alpha(x_0,Tx_0) \geq 1$. Define the sequence $\{x_n\}$ by $x_{n+1} = T^{n+1}x_0$ for all $n \geq 0$. If $x_n = x_{n+1}$ for some n, then we have nothing to prove. Assume that $x_n \neq x_{n+1}$ for all $n \geq 0$. Since T is α-admissible, it is easy to check that $\alpha(x_n,x_{n+1}) \geq 1$ for all n. Let $v = d(Tx_0,T^2x_0) + d(x_0,Tx_0)$ and $\lambda = a+b$. Then, we have $d(Tx_0,T^2x_0) \leq v$. Now, put $x = Tx_0$ and $y = x_0$. Then

$$d(T^3x_0,T^2x_0) \leq \alpha(Tx_0,x_0)d(T^3x_0,T^2x_0) \leq a\, d(T^2x_0,Tx_0) + b\, d(x_0,Tx_0) \leq \lambda v.$$

By continuing this process and using a similar technique in proof of Theorem 3 in [11], it is easy to see that $d(T^{m+1}x_0,T^mx_0) \leq 2\lambda^{l-1}v$, where $m = 2l$ or $m = 2l-1$ for all $l \geq 2$. This implies that $d(T^{m+1}x_0,T^mx_0) \to 0$. By using Lemma 3.3, T has an approximate fixed point. Also by using a similar technique in proof of Theorem 3 in [11], it is easy to see that $d(T^mx_0,T^nx_0) \leq \frac{4\lambda^l}{1-\lambda}v$ for all $n > m$. This shows that $\{x_n\}$ is a Cauchy sequence. If T is continuous and (X,d) is a complete metric, then there exists $x^* \in X$ such that $x_n \to x^*$. Thus, $Tx_n \to Tx^*$ and so $Tx^* = x^*$. Now, suppose that X has also the property (H). We show that T

has a unique fixed point. Let x^* and y^* be fixed points of T. Choose $z \in X$ such that $\alpha(x^*,z) \geq 1$ and $\alpha(y^*,z) \geq 1$. Since T is α-admissible, $\alpha(x^*,T^m z) \geq 1$ and $\alpha(y^*,T^m z) \geq 1$ for all $m \geq 1$. Put $\lambda = a+b$ and $v = d(x^*,T^2 z)+d(x^*,Tz)$. Then, we have

$$\begin{aligned} d(x^*,T^3 z) = d(T^2 x^*,T^2(Tz)) &\leq \alpha(x^*,Tz)d(T^2 x^*,T^2(Tz)) \\ &\leq ad(x^*,T^2 z)+bd(x^*,Tz) \leq \lambda v \end{aligned}$$

and

$$\begin{aligned} d(x^*,T^4 z) = d(T^2 x^*,T^2(T^2 z)) &\leq \alpha(x^*,T^2 z)d(T^2 x^*,T^2(T^2 z)) \\ \leq ad(x^*,T^3 z)+bd(x^*,T^2 z) &= a^2 d(x^*,T^2 z)+abd(x^*,Tz)+bd(x^*,T^2 z) \\ \leq ad(x^*,T^2 z)+bd(x^*,T^2 z)&+ad(x^*,Tz)+bd(x^*,Tz) = \lambda v. \end{aligned}$$

Also, we have

$$\begin{aligned} d(x^*,T^5 z) &= d(T^2 x^*,T^2(T^3 z)) \leq \alpha(x^*,T^3 z)d(T^2 x^*,T^2(T^3 z)) \\ &\leq ad(x^*,T^4 z)+bd(x^*,T^3 z) \leq a^3 d(x^*,T^2 z) \\ &+a^2 bd(x^*,Tz)+2abd(x^*,T^2 z)+b^2 d(x^*,Tz) \\ &= (a^3+2ab)d(x^*,T^2 z)+(a^2 b+b^2)d(x^*,Tz) \leq \lambda^2 v \end{aligned}$$

and one can easily get that $d(x^*,T^6 z) \leq 2\lambda^2 v$. By continuing this process, we obtain $d(x^*,T^m z) \leq 2\lambda^{l-1} v$, where $m = 2l$ or $m = 2l-1$ for all $l \geq 2$. Hence, $T^m z \to x^*$. Similarly, we can show that $T^m z \to y^*$. Thus, we get $x^* = y^*$ and so T has a unique fixed point.

♣

In 2011, Haghi, Rezapour and Shahzad proved that some fixed point generalizations are not real generalization ([45]). But the following examples show that the notion of generalized convex contractions is a real generalization for the notions of convex contractions and ordered convex contractions which have been provided respectively in [11] and [42].

Example 37 *Let $X = \{1,3,5\}$, $d(x,y) = |x-y|$ and T be a selfmap on X defined by $T1 = 3$, $T3 = 1$ and $T5 = 5$. Then, by putting $a = \frac{1}{4}$, $b = \frac{1}{4}$, $x = 1$ and $y = 3$ we have $2 = d(T^2 1,T^2 3) > ad(T1,T3)+bd(1,3) = 1$. Thus, T is not a convex contraction while by putting $\alpha(x,y) = \frac{1}{4}$ whenever $x \leq y$ and $\alpha(x,y) = 0$ otherwise, $a = \frac{1}{4}$ and $b = \frac{1}{4}$, it is easy to see that T is a generalized convex contractions.*

Example 38 *Let $X = \{1,3,5\}$, $d(x,y) = |x-y|$. Define the order $\leq$ on X by*

$$\leq = \{(1,1),(3,3),(5,5),(1,3),(1,5),(3,5)\}$$

and define the selfmap T on X by $T1 = 3$, $T3 = 1$ *and* $T5 = 5$. *Then, by putting* $a = \frac{1}{4}$ *and* $b = \frac{1}{4}$, $x = 1$ *and* $y = 3$ *we have*

$$2 = d(T^2 1, T^2 3) > ad(T1, T3) + bd(1,3) = 1.$$

If $x = 3$ *and* $y = 5$ *or* $x = 1$ *and* $y = 5$, *then we have*

$$2 = d(T^2 3, T^2 5) > ad(T3, T5) + bd(3,5) = 1.5$$

and

$$4 = d(T^2 1, T^2 5) > ad(T1, T5) + bd(1,5) = 1.5.$$

Thus, T does not satisfy the condition of Theorem 2.4 in [42]. If we put $a = \frac{1}{4}$ *and* $b = \frac{1}{4}$ *and define* $\alpha(x,y) = \frac{1}{4}$ *whenever* $x \leq y$ *and* $\alpha(x,y) = 0$ *otherwise, then it is easy to see that T is a generalized convex contractions.*

Theorem 3.33

Let (X,d) *be a metric space and T a generalized convex contraction of order 2 on X with the based mapping* α. *Suppose that T is* α*-admissible and there exists* $x_0 \in X$ *such that* $\alpha(x_0, Tx_0) \geq 1$. *Then T has an approximate fixed point. Moreover, T has a fixed point whenever T is continuous and* (X,d) *is a complete metric space and also T has a unique fixed point whenever X has the property (H).*

Proof: Let $x_0 \in X$ be such that $\alpha(x_0, Tx_0) \geq 1$. Define the sequence $\{x_n\}$ by $x_{n+1} = T^{n+1}x_0$ for all $n \geq 0$. If $x_n = x_{n+1}$ for some n, then we have nothing to prove. Assume that $x_n \neq x_{n+1}$ for all $n \geq 0$. Since T is α-admissible, it is easy to check that $\alpha(x_n, x_{n+1}) \geq 1$ for all n. Let $v = d(Tx_0, T^2x_0) + d(x_0, Tx_0)$, $\beta = 1 - b_2$ and $\lambda = a_1 + a_2 + b_1$. Then, we have

$$d(T^3x_0, T^2x_0) \leq \alpha(Tx_0, x_0) d(T^3x_0, T^2x_0)$$
$$\leq a_1 d(x_0, Tx_0) + a_2 d(Tx_0, T^2x_0) + b_1 d(Tx_0, T^2x_0) + b_2 d(T^3x_0, T^2x_0)$$
$$\leq a_1 v + (a_2 + b_1)v + b_2 d(T^3x_0, T^2x_0).$$

Hence, $d(T^3x_0, T^2x_0) \leq (\frac{\lambda}{\beta})v$. Now, put $x = Tx_0$ and $y = T^2x_0$. Then,

$$d(T^3x_0, T^4x_0) \leq \alpha(T^2x_0, Tx_0) d(T^3x_0, T^4x_0)$$
$$\leq a_1 d(Tx_0, T^2x_0) + a_2 d(T^2x_0, T^3x_0) + b_1 d(T^2x_0, T^3x_0) + b_2 d(T^3x_0, T^4x_0)$$
$$\leq a_1 v + (a_2 + b_1)\frac{a_1 + a_2 + b_1}{1 - b_2} v + b_2 d(T^3x_0, T^4x_0).$$

Hence, $d(T^3x_0, T^4x_0) \leq (\frac{\lambda}{\beta})v$. Similarly, we obtain

$$d(T^5x_0, T^4x_0) \leq (\frac{\lambda}{\beta})^2 v$$

and $d(T^5x_0, T^6x_0) \leq (\frac{\lambda}{\beta})^2 v$. By continuing this process and using a similar technique in proof of Theorem 4 in [11], it is easy to see that $d(T^{m+1}x_0, T^m x_0) \leq (\frac{\lambda}{\beta})^l v$, where $m = 2l$ or $m = 2l + 1$ for $l \geq 1$ or $d(T^{m+1}x_0, T^m x_0) \leq (\frac{\lambda}{\beta})^{l-1} v$, where $m = 2l$ or $m = 2l - 1$ for $l \geq 2$. Thus, $d(T^{m+1}x_0, T^m x_0) \to 0$. By using Lemma 3.3, T has an approximate fixed point. Now, suppose that T is continuous and (X, d) is a complete metric space. Then, By using a similar technique in proof of Theorem 4 in [11], it is easy to see that $\{x_n\}$ is a Cauchy sequence. Choose $x^* \in X$ such that $x_n \to x^*$. Since T is continuous, $Tx_n \to Tx^*$ and so $Tx^* = x^*$. If X has the property (H), then by using similar technique in proof of Theorem 3.32, we can prove uniqueness of the fixed point of T.

♣

Again, the following examples show that the notion of generalized convex contractions of order 2 is a real generalization for the notions of convex contractions of order 2 and ordered convex contractions of order 2 which have been provided respectively in [11] and [42].

Example 39 *Let $X = \{1, 3, 5\}$, $d(x, y) = |x - y|$ and T be a selfmap on X defined by $T1 = 3$, $T3 = 1$ and $T5 = 5$. Then, by putting $a_1 = a_2 = b_1 = b_2 = \frac{1}{8}$, $x = 1$ and $y = 3$, we have*

$$2 = d(T^2 1, T^2 3) > a_1 d(1, T1) + a_2 d(T1, T^2 1) + b_1 d(3, T3) + b_2 d(T3, T^2 3) = 0.8.$$

Thus, T is not a convex contraction of order 2 while by putting $\alpha(x, y) = \frac{1}{4}$ whenever $x \leq y$ and $\alpha(x, y) = 0$ otherwise and $a_1 = a_2 = b_1 = b_2 = \frac{1}{8}$, It is easy to see that T is a generalized convex contraction of order 2.

Example 40 *Let $X = \{1, 3, 5\}$, $\leq = \{(1, 1), (3, 3), (5, 5), (1, 3)\}$, $d(x, y) = |x - y|$ and T be a selfmap on X defined by $T1 = 3$, $T3 = 1$ and $T5 = 5$. Then, by putting $a_1 = a_2 = b_1 = b_2 = \frac{1}{8}$, $x = 1$ and $y = 3$, we have*

$$2 = d(T^2 1, T^2 3) > a_1 d(1, T1) + a_2 d(T1, T^2 1) + b_1 d(3, T3) + b_2 d(T3, T^2 3) = 0.8.$$

Thus, T is not an ordered convex contraction of order 2 which has been used in Theorem 2.5 of [42], while by putting $\alpha(x, y) = \frac{1}{4}$ whenever $x \leq y$ and $\alpha(x, y) = 0$ otherwise and $a_1 = a_2 = b_1 = b_2 = \frac{1}{8}$, It is easy to check that the selfmap T is a generalized convex contraction of order 2.

In 1972, Zamfirescu defined the following notion. Now, we provide the notion of weakly Zamfirescu mappings here (see [16]).

Definition 3.4 Let (X, d) be a metric space and T a self map on X. Then, T is called weakly Zamfirescu whenever there exists $\gamma : X \times X \to [0, 1]$ with

$$\theta(a, b) := sup\{\gamma(x, y) : a \leq d(x, y) \leq b\} < 1$$

for all $0 < a \le b$, such that $d(T(x),T(y)) \le \gamma(x,y)M_T(x,y)$ for all $x,y \in X$, where

$$M_T(x,y) = max\{d(x,y), \frac{1}{2}[d(x,T(y)) + d(y,T(x))], \frac{1}{2}[d(x,T(x)) + d(y,T(y))]\}.$$

Now, by using the main idea of this chapter, we define α-weakly Zamfirescu self-maps as following. Let (X,d) be a metric space, $\alpha : X \times X \to [0,\infty)$ a function and T a self map on X. Then, T is called α-weakly Zamfirescu whenever there exists a map $\gamma : X \times X \to [0,1]$ with $\theta(a,b) := sup\{\gamma(x,y) : a \le d(x,y) \le b\} < 1$ for all $0 < a \le b$, such that $\alpha(x,y)d(T(x),T(y)) \le \gamma(x,y)M_T(x,y)$ for all $x,y \in X$.

Theorem 3.34

Let (X,d) be a metric space, $\alpha : X \times X \to [0,\infty)$ a function and T an α-weakly Zamfirescu selfmap on X. Suppose that T is α-admissible and there exists $x_0 \in X$ such that $\alpha(x_0,Tx_0) \ge 1$. Then T has an approximate fixed point. Moreover, T has a fixed point whenever T is continuous and (X,d) is a complete metric space.

Proof: Let $x_0 \in X$ be such that $\alpha(x_0,Tx_0) \ge 1$. Define the sequence $\{x_n\}$ by $x_{n+1} = T^{n+1}x_0$ for all $n \ge 0$. We show that $d(x_n,x_{n+1}) \le \gamma(x_{n-1},x_n)d(x_{n-1},x_n)$ for all $n \ge 0$. Since T is α-admissible, it is easy to check that $\alpha(x_n,x_{n+1}) \ge 1$ for all n. But, for each n we have

$$\begin{aligned} M_T(x_{n-1},x_n) &= max\{d(x_{n-1},x_n), \frac{1}{2}[d(x_{n-1},T(x_{n-1})) + d(x_n,T(x_n))], \\ &\frac{1}{2}[d(x_{n-1},T(x_n)) + d(x_n,T(x_{n-1}))]\} \\ &= \max\{d(x_{n-1},x_n), \frac{1}{2}[d(x_{n-1},x_n) + d(x_n,x_{n+1})], \frac{1}{2}[d(x_{n-1},x_{n+1}) + d(x_n,x_n)]\} \\ &= \max\{d(x_{n-1},x_n), \frac{1}{2}[d(x_{n-1},x_n) + d(x_n,x_{n+1})], \frac{1}{2}d(x_{n-1},x_{n+1})\}. \end{aligned}$$

If $M_T(x_{n-1},x_n) = d(x_{n-1},x_n)$, then

$$d(x_n,x_{n+1}) \le \alpha(x_{n-1},x_n)d(T(x_{n-1}),T(x_n) \le \gamma(x_{n-1},x_n)d(x_{n-1},x_n).$$

If $M_T(x_{n-1},x_n) = \frac{1}{2}[d(x_{n-1},x_n) + d(x_n,x_{n+1})]$, then

$$\alpha(x_{n-1},x_n)d(x_n,x_{n+1}) \le \gamma(x_{n-1},x_n)\frac{[d(x_{n-1},x_n) + d(x_n,x_{n+1})]}{2}$$

$$\begin{aligned} d(x_n,x_{n+1}) &\le \frac{\gamma(x_{n-1},x_n)}{2\alpha(x_{n-1},x_n) - \gamma(x_{n-1},x_n)}d(x_{n-1},x_n) \\ &\le \gamma(x_{n-1},x_n)d(x_{n-1},x_n). \end{aligned}$$

If $M_T(x_{n-1},x_n) = \frac{1}{2}d(x_{n-1},x_{n+1})$, then

$$\alpha(x_{n-1},x_n)d(x_n,x_{n+1}) \leq \frac{\gamma(x_{n-1},x_n)}{2}\frac{1}{2}d(x_{n-1},x_{n+1})$$

$$\leq \frac{\gamma(x_{n-1},x_n)}{2}[d(x_{n-1},x_n)+d(x_n,x_{n+1})],$$

and so

$$d(x_n,x_{n+1}) \leq \frac{\gamma(x_{n-1},x_n)}{2\alpha(x_{n-1},x_n)-\gamma(x_{n-1},x_n)}d(x_{n-1},x_n) \leq \gamma(x_{n-1},x_n)d(x_{n-1},x_n).$$

Thus, the claim is proved. This implies that the sequence $\{d(x_n,x_{n+1})\}$ is non-increasing and so converges to the real number $d = \inf_{n\geq 1} d(x_{n-1},x_n)$. We have to show that $d = 0$. Let $d > 0$. Since $0 < d \leq d(x_n,x_{n+1}) \leq d(x_0,x_1)$ for all n, $\gamma(x_{n-1},x_n) \leq \theta$ for all n, where $\theta = \theta(d,d(x_0,x_1))$. Hence,

$$d \leq d(x_n,x_{n+1}) \leq \theta^n d(x_0,x_1)$$

for all n. But, this is impossible because $d > 0$ and $0 \leq \theta < 1$. Thus, T has an approximate fixed point. Now, suppose that (X,d) is a complete metric space and T is continuous. By using a similar proof in Theorem 28 of [16], we can show that $\{x_n\}$ is a Cauchy sequence. This implies easily that T has a fixed point.

♣

Now, we give the following example which shows that there exist α-weakly Zamfirescu maps which are not weakly Zamfirescu.

Example 41 *Let $X = [0,1]$, $d(x,y) = |x-y|$ and the selfmap f on X be defined by $f(x) = \frac{4}{3}x$ whenever $0 \leq x < 1$ and $f(x) = 0$ whenever $x = 1$. Now, by putting $\alpha(x,y) = \frac{1}{9}$ whenever $x \leq y$ and $\alpha(x,y) = 0$ otherwise. It is easy to check that f is α-weakly Zamfirescu, but it is not weakly Zamfirescu.*

Now, denote by $\mathcal{R}$ the set of all continuous mappings $g\colon [0,\infty)^5 \longrightarrow [0,\infty)$ satisfying the following conditions, for more details see [11]:

a) $g(1,1,1,2,0) = g(1,1,1,0,2) = h \in (0,1)$,

b) $g(\alpha x_1,\alpha x_2,\alpha x_3,\alpha x_4,\alpha x_5) \leq \alpha g(x_1,x_2,x_3,x_4,x_5)$, for all (x_1,x_2,x_3,x_4,x_5) in $[0,\infty)^5$ and $\alpha \geq 0$,

c) if $x_i,y_i \in [0,\infty)$ and $x_i < y_i$ for $i = 1,\ldots,4$, then

$$g(x_1,x_2,x_3,x_4,0) < g(y_1,y_2,y_3,y_4,0), \qquad g(x_1,x_2,x_3,0,x_4) < g(y_1,y_2,y_3,0,y_4).$$

This class of mappings has the property in the following proposition which is necessary for developing our next results.

Proposition 3.1

(see [10]) If $g \in \mathcal{R}$ and $u, v \in [0, \infty)$ are such that

$$u \leq \max\{g(v,v,u,v+u,0), g(v,v,u,0,v+u), g(v,u,v,v+u,0), g(v,u,v,0,v+u)\},$$

then $u \leq hv$.

Here,we introduce the notion of generalized α-contractive mapping. We say that the selfmap T on the metric space X is a *generalized α-contractive mapping* whenever there exists $g \in \mathcal{R}$ such that

$$\alpha(x,y)d(Tx,Ty) \leq g(d(x,y), d(x,Tx), d(y,Ty), d(x,Ty), d(y,Tx))$$

for all $x, y \in X$.

Theorem 3.35

Let (X,d) be a metric space and T a generalized α-contractive and an α-admissible selfmap on X such that $\alpha(x_0, Tx_0) \geq 1$, for some $x_0 \in X$. Then T has an approximate fixed point.

Proof: Fix $1 > r > h$ and $x_0 \in X$ such that $\alpha(x_0, Tx_0) \geq 1$. Define the sequence $\{x_n\}$ by $x_{n+1} = T^{n+1}x_0$ for all $n \geq 0$.

If $x_n = x_{n+1}$, for some n, then we have nothing to prove.

Assume that $x_n \neq x_{n+1}$ for all $n \geq 0$. Since T is α-admissible, it is easy to check that $\alpha(x_n, x_{n+1}) \geq 1$ for all n.

Since

$$\begin{aligned} d(x_1,x_2) = d(Tx_0,Tx_1) &\leq \alpha(x_0,x_1)d(Tx_0,Tx_1) \\ \leq g(d(x_0,x_1), d(x_1,Tx_1), &d(x_0,Tx_0), d(x_0,Tx_1), d(x_1,Tx_0)) \\ \leq g(d(x_0,x_1), d(x_1,x_2), &d(x_0,x_1), d(x_0,x_1)+d(x_1,x_2), 0), \end{aligned}$$

by using Proposition 3.1, we obtain $d(x_1,x_2) \leq hd(x_0,x_1) < rd(x_0,x_1)$.

Since

$$\begin{aligned} d(x_2,x_3) = d(Tx_1,Tx_2) &\leq \alpha(x_1,x_2)d(Tx_1,Tx_2) \\ \leq g(d(x_1,x_2), d(x_2,Tx_2), &d(x_1,Tx_1), d(x_1,Tx_2), d(x_2,Tx_1)) \\ \leq g(d(x_1,x_2), d(x_2,x_3), &d(x_1,x_2), d(x_1,x_2)+d(x_2,x_3), 0), \end{aligned}$$

again by using Proposition 3.1, we obtain

$$d(x_2,x_3) \leq hd(x_1,x_2) < rd(x_1,x_2) < r^2d(x_0,x_1).$$

By continuing this process, we obtain $d(x_n, x_{n+1}) < r^n d(x_0, x_1)$ for all n. Hence, $d(T^{n+1}x_0, T^n x_0) \to 0$, and from Lemma 3.3, we conclude that T has the approximate fixed point property.

♣

The following example shows that there are generalized α-contractive selfmaps on non-complete metric spaces satisfying Theorem 3.32 which have no fixed point while have approximate fixed points.

Example 42 Let $X = (0,1)$, $\alpha(x,y) = 1$ whenever $x^2 = y$ and $\alpha(x,y) = \frac{1}{20}$ otherwise and $d(x,y) = |x-y|$. Define the selfmap T on X by $Tx = x^2$ for all $x \in X$. Also, define $g \in \mathcal{R}$ by $g(x_1,x_2,x_3,x_4,x_5) = \frac{9}{20}(x_2+x_3)$. Then, it is easy to check that T is generalized α-contractive and α-admissible selfmap on X such that $\alpha(x_0, Tx_0) \geq 1$ for $x_0 = \frac{1}{2}$.

The following corollaries show us that there are different types of generalized α-contractive mappings satisfying Theorem 3.32. One can provide many examples for each type of such mappings. We shall provide some examples for each type of the maps.

Corollary 3.2

Let (X,d) be a complete metric space and T a continuous, generalized α-contractive and α-admissible selfmap on X such that $\alpha(x_0, Tx_0) \geq 1$, for some $x_0 \in X$. Then T has a fixed point.

Proof: By using a similar argument in proof of Theorem 3.32, we obtain

$$d(x_n, x_{n+1}) \leq hd(x_{n-1}, x_n) < rd(x_{n-1}, x_n) < r^n d(x_0, x_1),$$

for all n. If suppose $m < n$, then it is easy to see that

$$d(x_m, x_n) \leq (r^m + r^{m+1} + \cdots + r^{n-1}) d(x_0, x_1) < \frac{r^m}{1-r} d(x_0, x_1).$$

Thus, $\{x_n\}$ is a Cauchy sequence. Since (X,d) is complete, there exists $x^* \in X$ such that $x_n \to x^*$. Moreover, T is continuous, $Tx_n \to Tx^*$ and so $Tx^* = x^*$.

♣

As we mentioned before, Banach proved his contraction principle result in 1922. By following above direction, we say that the selfmap T is α-contractive whenever exists $\lambda \in (0,1)$ such that $\alpha(x,y)d(Tx,Ty) \leq \lambda d(x,y)$ for all $x, y \in X$. Note that each α-contractive selfmap is a contraction when $\alpha(x,y) = 1$.

Corollary 3.3

Let (X,d) be a metric space and T an α-contractive and α-admissible selfmap on X such that $\alpha(x_0,Tx_0) \geq 1$ for some $x_0 \in X$. Then T has the approximate fixed point property.

Proof: Consider $g \in \mathcal{R}$ given by $g(x_1,x_2,x_3,x_4,x_5) = \lambda x_1$. Then, T is a generalized α-contractive and α-admissible selfmap on X such that $\alpha(x_0,Tx_0) \geq 1$, for some $x_0 \in X$. Hence by Theorem 3.32, T has the approximate fixed point property.

♣

Next example shows that last result is the real one.

Example 43 *Let $X = [0,\infty)$ and $d(x,y) = |x-y|$. Define the selfmap T on X by $Tx = \frac{4}{3}x$ for all $x \in X$ and put $\lambda = \frac{1}{2}$. Then, the selfmap T is not contractive. Let $\lambda \in [0,1)$. Define $\alpha(x,y) = \mu$ for all $x,y \in X$, where $\mu \leq \frac{3\lambda}{4}$. It is easy to see that T is α-contractive.*

In 1968, Kannan contraction was introduced. Now, we say that the selfmap T on a metric space is α-Kannan mapping whenever there exists $\beta \in (0,\frac{1}{2})$ such that $\alpha(x,y)d(Tx,Ty) \leq \beta(d(x,Tx)+d(y,Ty))$ for all $x,y \in X$.

Corollary 3.4

Let (X,d) be a metric space and T an α-Kannan and α-admissible selfmap on X such that $\alpha(x_0,Tx_0) \geq 1$, for some $x_0 \in X$. Then T has the approximate fixed point property.

Proof: Consider $g \in \mathcal{R}$ by the formula $g(x_1,x_2,x_3,x_4,x_5) = \beta(x_2+x_3)$. Then, T is a generalized α-contractive and α-admissible selfmap on X such that $\alpha(x_0,Tx_0) \geq 1$ for some $x_0 \in X$. By using Theorem 3.32, T has the approximate fixed point property.

♣

Corollary 3.5

Let (X,d) be a complete metric space and T a continuous, α-Kannan and α-admissible selfmap on X such that $\alpha(x_0,Tx_0) \geq 1$, for some $x_0 \in X$. Then T has a fixed point.

Proof: Define $g \in \mathcal{R}$ by $g(x_1,x_2,x_3,x_4,x_5) = \beta(x_2+x_3)$. Then, by using Corollary 3.2, it follows that T has a fixed point.

♣

The following example shows that there exist α-Kannan mappings which are not Kannan map.

Example 44 *Let $X = [0,\infty)$ and $d(x,y) = |x-y|$. Define the selfmap T on X by $Tx = \frac{5}{4}x$ for all $x \in X$. Put $\beta = \frac{1}{4}$ and $\alpha(x,y) = \mu$ for all $x,y \in X$, where $\mu \leq \frac{1}{20}$. Then, it is easy to check that the selfmap T is not Kannan map whereas T is an α-Kannan map.*

In 1972, the notion of Chatterjea-contraction was introduced. Now, we say that the selfmap T is α-Chatterjea mapping whenever exists $\beta \in (0,\frac{1}{2})$ such that

$$\alpha(x,y)d(Tx,Ty) \leq \beta(d(x,Ty)+d(y,Tx))$$

for all $x,y \in X$.

Corollary 3.6

Let (X,d) be a metric space, T an α-Chatterjea and α-admissible selfmap on X such that $\alpha(x_0,Tx_0) \geq 1$, for some $x_0 \in X$. Then T has the approximate fixed point property.

Proof: Let $g \in \mathcal{R}$ given by the formula $g(x_1,x_2,x_3,x_4,x_5) = \beta(x_4+x_5)$. Then, T is a generalized α-contractive and α-admissible selfmap on X such that $\alpha(x_0,Tx_0) \geq 1$ for some $x_0 \in X$. By using Theorem 3.32, T has the approximate fixed point property.

♣

The next corollary generalizes the results in the literature about fixed point of Chatterjea selfmaps.

Corollary 3.7

Let (X,d) be a complete metric space and T a continuous, α-Chatterjea and α-admissible selfmap on X such that $\alpha(x_0,Tx_0) \geq 1$, for some $x_0 \in X$. Then T has a fixed point.

Proof: Define $g \in \mathcal{R}$ by $g(x_1,x_2,x_3,x_4,x_5) = \beta(x_4+x_5)$. Then by using Corollary 3.2, T has a fixed point.

♣

The following example shows that there exist α-Chatterjea mappings which are not Chatterjea selfmap.

Example 45 *Let $X = [0,\infty)$ and $d(x,y) = |x-y|$. Define the selfmap T on X by $Tx = \frac{4}{3}x$ for all $x \in X$. For each $\beta \in (0,\frac{1}{2})$, put $x = \beta$ and $y = \frac{3}{4}\beta$. Then, it is easy to see that T is not a Chatterjea mapping. If we put $\beta = \frac{1}{4}$ and define $\alpha(x,y) = \mu$ for all $x,y \in X$, where $\mu \leq \frac{3}{80}$, then one can easily check that T is an α-Chatterjea selfmap.*

In 1972, the notion of Zamfirescu-contractions introduced. Now, we say that the selfmap T is α-Zamfirescu mapping whenever exists $\beta \in (0,1)$ such that

$$\alpha(x,y)d(Tx,Ty) \leq \beta M_T(x,y)$$

for all $x,y \in X$, where

$$M_T(x,y) = \max\{d(x,y), \frac{1}{2}[d(x,Ty)+d(y,Tx)], \frac{1}{2}[d(x,Tx)+d(y,Ty)]\}.$$

Corollary 3.8

Let (X,d) be a metric space and T an α-Zamfirescu and α-admissible selfmap on X such that $\alpha(x_0,Tx_0) \geq 1$, for some $x_0 \in X$. Then T has the approximate fixed point property.

Proof: Define $g \in \mathcal{R}$ by $g(x_1,x_2,x_3,x_4,x_5) = \beta \max\{x_1, \frac{1}{2}[x_4+x_5], \frac{1}{2}[x_2+x_3]\}$. Then, T is a generalized α-contractive and α-admissible selfmap on X such that $\alpha(x_0,Tx_0) \geq 1$ for some $x_0 \in X$. Hence by using Theorem 3.32, T has the approximate fixed point property.

♣

Next corollary generalizes the results in the literature about fixed point of Zamfirescu selfmaps.

Corollary 3.9

Consider (X,d) be a complete metric space and T a continuous, α-Zamfirescu and α-admissible selfmap on X such that $\alpha(x_0,Tx_0) \geq 1$ for some $x_0 \in X$. Then T has a fixed point.

Proof: Define $g \in \mathcal{R}$ by $g(x_1,x_2,x_3,x_4,x_5) = \beta \max\{x_1, \frac{1}{2}[x_4+x_5], \frac{1}{2}[x_2+x_3]\}$. Then by using Corollary 3.2, T has a fixed point.

♣

The following example shows that there exist α-Zamfirescu mappings which are not Zamfirescu selfmap.

Example 46 *Let $X = [0,\infty)$ and $d(x,y) = |x-y|$. Define the selfmap T on X by $Tx = \frac{4}{3}x$ for all $x \in X$. For each $\beta \in (0,1]$, put $x = \beta$ and $y = \frac{3}{4}\beta$. Then, it is easy to see that T is not a Zamfirescu mapping. If we put $\beta = \frac{1}{4}$ and define $\alpha(x,y) = \mu$ for all $x,y \in X$, where $\mu \leq \frac{3}{16}$, then one can easily check that T is an α-Zamfirescu selfmap.*

In 1971, the notion of Reich-contraction was introduced. Following this idea, we say that the selfmap T is α-Reich mapping whenever there exists nonnegative real numbers α, β, γ with $\alpha + \beta + \gamma < 1$ such that

$$\alpha(x,y)d(Tx,Ty) \leq \alpha d(x,y) + \beta d(x,Tx) + \gamma d(y,Ty)$$

for all $x, y \in X$.

Corollary 3.10

Let (X,d) be a metric space and T an α-Reich and α-admissible selfmap on X such that $\alpha(x_0,Tx_0) \geq 1$, for some $x_0 \in X$. Then T has an approximate fixed point.

Proof: Define $g \in \mathcal{R}$ by $g(x_1,x_2,x_3,x_4,x_5) = \alpha x_1 + \beta x_2 + \gamma x_3$. Then, T is a generalized α-contractive and α-admissible selfmap on X such that $\alpha(x_0,Tx_0) \geq 1$ for some $x_0 \in X$. Hence by using Theorem 3.32, T has the approximate fixed point property.

♣

Corollary 3.11

Let (X,d) be a complete metric space and T a continuous, α-Reich and α-admissible selfmap on X such that $\alpha(x_0,Tx_0) \geq 1$, for some $x_0 \in X$. Then T has a fixed point.

Proof: Define $g \in \mathcal{R}$ by $g(x_1,x_2,x_3,x_4,x_5) = \alpha x_1 + \beta x_2 + \gamma x_3$. Then by using Corollary 3.2, we get that T has a fixed point.

♣

The following example shows that there exist α-Reich mappings which are not Reich selfmap.

Example 47 *Let $X = [0,\infty)$ and $d(x,y) = |x-y|$. Define the selfmap T on X by $Tx = 2x$ for all $x \in X$. For each $\alpha, \beta, \gamma \in (0,1)$ with $\alpha + \beta + \gamma < 1$, put $x = 0$ and*

$y = 1$. Then, it is easy to see that T is not a Reich mapping. If we put $\alpha = \beta = \gamma = \frac{1}{4}$ and define $\alpha(x,y) = \mu$ for all $x,y \in X$, where $\mu \leq \frac{1}{8}$, then one can easily check that T is an α-Reich selfmap.

In 1972, the notion of Ciric-contraction introduced. Now, we say that the selfmap T is α-Ciric mapping whenever there exists $\lambda \in (0,1)$ such that

$$\alpha(x,y)d(Tx,Ty) \leq \lambda M_T(x,y)$$

for all $x, y \in X$, where

$$M_T(x,y) = \max\{d(x,y), d(x,Tx), d(y,Ty), \frac{1}{2}[d(x,Ty) + d(y,Tx)]\}.$$

Corollary 3.12

Let (X,d) be a metric space and T an α-Ciric and α-admissible selfmap on X such that $\alpha(x_0, Tx_0) \geq 1$, for some $x_0 \in X$. Then T has the approximate fixed point property.

Proof: Let $g \in \mathcal{R}$ given by $g(x_1,x_2,x_3,x_4,x_5) = \lambda \max\{x_1,x_2,x_3,\frac{1}{2}[x_4 + x_5]\}$. Then, T is a generalized α-contractive and α-admissible selfmap on X with $\alpha(x_0,Tx_0) \geq 1$, for some $x_0 \in X$. Hence by using Theorem 3.32, T has the approximate fixed point property.

♣

Corollary 3.13

Let (X,d) be a complete metric space and T a continuous, α-Ciric and α-admissible selfmap on X such that $\alpha(x_0,Tx_0) \geq 1$, for some $x_0 \in X$. Then T has a fixed point.

Proof: Define $g \in \mathcal{R}$ by $g(x_1,x_2,x_3,x_4,x_5) = \beta \max\{x_1,x_2,x_3,\frac{1}{2}[x_4 + x_5]\}$. Then by using Corollary 3.2, T has a fixed point.

♣

The following example shows that there exist α-Ciric mappings which are not Ciric selfmap.

Example 48 *Let $X = [0,\infty)$ and $d(x,y) = |x-y|$. Define the selfmap T on X by $Tx = \frac{3}{2}x$ for all $x \in X$. For each $\lambda \in (0,1)$, put $x = \lambda$ and $y = \frac{2}{3}\lambda$. Then, it is easy to see that T is not a Ciric mapping. If $\lambda \in (0,1)$ and we define $\alpha(x,y) = \mu$ for all $x,y \in X$, where $\mu \leq \frac{2}{3}\lambda$, then one can easily check that T is an α-Ciric selfmap.*

In 1972, the notion of Bianchini-contractions was introduced. Following this idea, we say that the selfmap T is α-Bianchini mapping whenever there exists $h \in (0,1)$ such that $\alpha(x,y)d(Tx,Ty) \leq h\max\{d(x,Tx),d(y,Ty)\}$ for all $x,y \in X$.

Corollary 3.14

Let (X,d) be a metric space and T an α-Bianchini and α-admissible selfmap on X such that $\alpha(x_0,Tx_0) \geq 1$, for some $x_0 \in X$. Then T has the approximate fixed point property.

Proof: Define $g \in \mathcal{R}$ by $g(x_1,x_2,x_3,x_4,x_5) = h\max\{x_2,x_3\}$. Then, T is a generalized α-contractive and α-admissible selfmap on X such that $\alpha(x_0,Tx_0) \geq 1$ for some $x_0 \in X$. Hence by using Theorem 3.32, T has the approximate fixed point property.

♣

The next corollary generalizes the results in the literature about the fixed point of Bianchini selfmaps.

Corollary 3.15

Let (X,d) be a complete metric space, T a continuous, α-Bianchini and α-admissible selfmap on X such that $\alpha(x_0,Tx_0) \geq 1$ for some $x_0 \in X$. Then T has a fixed point.

Proof: Define $g \in \mathcal{R}$ by $g(x_1,x_2,x_3,x_4,x_5) = h\max\{x_2,x_3\}$. Then by using Corollary 3.2, T has a fixed point.

♣

The following example shows that there exist α-Bianchini mappings which are not Bianchini selfmap.

Example 49 *Let $X = [0,\infty)$ and $d(x,y) = |x-y|$. Define the selfmap T on X by $Tx = \frac{8}{7}x$ for all $x \in X$. For each $h \in (0,1)$, put $x = h$ and $y = \frac{7}{8}h$. Then, it is easy to see that T is not a Bianchini mapping. If $h = \frac{1}{4}$ and we define $\alpha(x,y) = \mu$ for all $x,y \in X$, where $\mu \leq \frac{1}{64}$, then one can easily check that T is an α-Bianchini selfmap.*

In 1973, the notion of Hardy-Rogers-contractions was introduced. Now, we say that the selfmap T is α-Hardy-Rogers mapping whenever there exist nonnegative real numbers a,b,c,e,f with $a+b+c+2\max\{e,f\}<1$ such that

$$\alpha(x,y)d(Tx,Ty) \leq ad(x,y)+bd(x,Tx)+cd(y,Ty)+ed(y,Tx)+fd(x,Ty)$$

for all $x,y \in X$.

Corollary 3.16

Let (X,d) be a metric space and T an α-Hardy-Rogers and α-admissible selfmap on X such that $\alpha(x_0,Tx_0) \geq 1$ for some $x_0 \in X$. Then T has the approximate fixed point property.

Proof: Define $g \in \mathcal{R}$ by $g(x_1,x_2,x_3,x_4,x_5) = ax_1+bx_2+cx_3+ex_4+fx_5$. Then, T is a generalized α-contractive and α-admissible selfmap on X such that $\alpha(x_0,Tx_0) \geq 1$ for some $x_0 \in X$. Hence by using Theorem 3.32, T has the approximate fixed point property.

♣

The following corollary generalizes the results in the literature about the fixed point of Hardy-Rogers selfmaps.

Corollary 3.17

Let (X,d) be a complete metric space and T a continuous, α-Hardy-Rogers and α-admissible selfmap on X such that $\alpha(x_0,Tx_0) \geq 1$ for some $x_0 \in X$. Then T has a fixed point.

Proof: Define $g \in \mathcal{R}$ by $g(x_1,x_2,x_3,x_4,x_5) = ax_1+bx_2+cx_3+ex_4+fx_5$. Then by using Corollary 3.2, T has a fixed point.

♣

The following example shows that there exist α-Hardy-Rogers mappings which are not Hardy-Rogers selfmap.

Example 50 *Let $X=[0,\infty)$ and $d(x,y)=|x-y|$. Define the selfmap T on X by $Tx=3x$ for all $x \in X$. For each nonnegative real numbers a,b,c,e,f with $a+b+c+2\max\{e,f\}<1$, put $x=0$ and $y=1$. Then, it is easy to see that T is not a Hardy-Rogers mapping. If we put $a=b=c=e=f=\frac{1}{8}$ and define $\alpha(x,y)=\mu$ for all $x,y \in X$, where $\mu \leq \frac{1}{24}$, then one can easily check that T is an α-Hardy-Rogers selfmap.*

Chapter 4

Fixed point results for set-valued mappings

4.1 Introduction

By presenting a completely slang example, we intend to show the advantage of the concept of set-value maps over functions. Consider two persons, Mike and James. James only pays you one dollar a day, but Mike pays you one hundred dollars a day. Think a little. Do you prefer Mike or James?

Those who prefer Mike should read this chapter more carefully.

In fact, James plays role of a function $f : X \to Y$ which each element of X is assigned to only one element of Y while Mikes plays role of a set-valued mapping $T : X \multimap Y$ which each element of X is assigned to a subset of Y (for example subset via 100 elements). In this book, we may be using the notations $T : X \to Y$, $T : X \multimap Y$ or even $T : X \twoheadrightarrow Y$ for set-valued mappings (or multifunctions). You need to pay attention to the nature of the mapping you are reading, not the symbol used. Also for values of a set-valued map $T : X \multimap Y$, we will use both notations Tx and $T(x)$.

As you will see in this and the next chapter, the advantage of set-valued mappings is that we can mix different ideas from distinct areas for obtaining new results on set-valued maps.

4.2 Banach principal contraction for set-valued mappings

In 1969, Nadler proved the Banach Principal contraction for set-valued mappings ([74]). We first provide some preliminaries.

Let A be a nonempty subset of a metric space X. For $x \in X$, define

$$d(x,A) = \inf\{d(x,y) : y \in A\}.$$

Let $CB(X)$ denote the set of nonempty closed bounded subsets of X and $K(X)$ (or $\mathcal{H}(X)$) denote the set of nonempty compact subsets of X. It is clear that $K(X)$ is included in $CB(X)$. For $A, B \in CB(X)$, define

$$H(A,B) = \max\{\sup_{a\in A} d(a,B), \sup_{b\in B} d(b,A)\}.$$

The function H is called the Hausdorff metric which is a metric on $\mathcal{H}(X)$ and is a generalized metric on $CB(X)$ (that is, H is a metric on $CB(X)$ with the difference that it can get values ∞). In this book, we will also use the notation d_h for the Hausdorff metric.
use

Let $A, B \in CB(X)$ and $a \in A$. One can prove that for$\varepsilon > 0$, there must exist a $b \in B$ such that $d(a,b) \leq H(A,B) + \varepsilon$. Let (X,d) be a metric space and $T : X \to CB(X)$ be a set-valued mapping. We say that T is Lipschitzian whenever there exists a constant $k > 0$ such that $H(Tx,Ty) \leq kd(x,y)$ for all $x,y \in X$. The constant k is called Lipschitz constant. A set-valued Lipschitzian mapping T is said to be contraction whenever its Lipschitz constant is $k < 1$ (we say that k is the contractive constant). Also, a set-valued Lipschitzian mapping T is called nonexpansive whenever its Lipschitz constant is $k = 1$. Let $F(T)$ denote the set of fixed points of T, that is, $F(T) = \{x \in X : x \in Tx\}$.

Theorem 4.1

(Nadler's fixed point theorem) Let (X,d) be a complete metric space and $T : X \to CB(X)$ be a contractive set-valued mapping. Then T has a fixed point in X.

Proof: Let $0 < k < 1$ be the Lipschitz constant of T, $x_0 \in X$ and $x_1 \in Tx_0$. Choose an element $x_2 \in Tx_1$ such that $d(x_1,x_2) \leq H(Tx_0,Tx_1) + k$. Similarly, there exists $x_3 \in Tx_2$ such that $d(x_2,x_3) \leq H(Tx_1,Tx_2) + k_2$. Thus, there exists a sequence $\{x_n\}_{n\geq 1}$ in X such that $x_{n+1} \in Tx_n$ and $d(x_n,x_{n+1}) \leq H(Tx_{n-1},Tx_n) + k^n$ for all $n \geq 1$. Note that for each n, $x_{n+1} \in Tx_n$ and so $d(x_n,x_{n+1}) \leq H(Tx_{n-1},Tx_n) + k^n \leq kd(x_{n-1},x_n) + k^n$ and so by induction we get $d(x_n,x_{n+1}) \leq k^n d(x_0,x_1)$ for all n.

Since $\sum_{n=1}^{\infty} k^n < \infty$ and $\sum_{n=1}^{\infty} nk^n < \infty$, we conclude that $\sum_{n=1}^{\infty} d(x_n, x_{n+1}) \leq d(x_0, x_1) \sum_{n=1}^{\infty} k^n + \sum_{n=1}^{\infty} nk^n < \infty$. Hence, $\{x_n\}_{n \geq 1}$ is a Cauchy sequence. By completeness of X, there exists $v \in X$ such that $x_n \to v$. Thus, $\lim_{n \to \infty} H(Tx_n, Tv) = 0$ because T is a contractive set valued map. Since $x_{n+1} \in Tx_n$, $d(x_{n+1}, Tv) \to 0$ which implies that $d(v, Tv) = 0$. Since Tv is closed, we conclude that $v \in Tv$.

♣

Remember that each contraction on a metric space has a unique fixed point. Next example shows that contractive set-valued mappings may have more than one fixed point.

Example 51 *Let $X = [0,1]$ with the usual Euclidean metric and $f : [0,1] \to [0,1]$ be the mapping defined by $f(x) = \frac{x}{2} + \frac{1}{2}$ whenever $0 \leq x \leq \frac{1}{2}$ and $f(x) = \frac{-x}{2} + 1$ whenever $\frac{1}{2} \leq x \leq 1$. Define $T : X \to CB(X)$ by $Tx = \{f(x)\} \cup \{0\}$ for all $x \in X$. Then T is a contractive set-valued mapping with $F(T) = \{0, \frac{2}{3}\}$.*

4.3 Banach type contraction for set-valued mappings

In this section, we are going to provide some recent fixed point results for set-valued maps. We want to show young researchers that by using considerable changes on the Banach principal contraction for set-valued maps, one can obtain new results.

Let (X,d) be a metric space. Suppose that $\mathcal{H}(X)$ is the set of all compact subsets of (X,d) and d_h stands for the Hausdorff metric respect to d.

Theorem 4.2

Assume that K is a non-empty closed subset of a complete metric space (X,d) and $T : K \to \mathcal{H}(X)$ satisfies the inequality $d_h(Tx, Ty) \leq cd(x,y)$ for all $x, y \in X$, where $0 \leq c < 1$ is the contractive constant. Suppose that K_0 is a bounded subset of K and $\{x_n\}_{n \geq 1}$ is a sequence in K_0 such that $\bigcup_{i=1}^{n} T^i x_n \subseteq K$ for all $n \geq 1$. Then T has a fixed point in K.

Proof: First we claim that

$$d_h(T^{i+1}x, T^{i+1}y) \leq cd_h(T^i x, T^i y), \tag{4.1}$$

for all $i \geq 1$ and all $x, y \in K$ with $T^i x, T^i y \subseteq K$. This claim is obtained from a well-known remark of Nadler [74] which states

$$d_h(T(A), T(B)) \leq d_h(A, B)$$

for all compact subsets A and B of a metric space. Since K_0 is bounded, there exist $\theta \in K$ and $c_0 > 0$ such that

$$d(\theta, z) \leq c_0, \tag{4.2}$$

for all $z \in K_0$. Now, we continue the proof in several steps.

Step I. For each $\varepsilon > 0$ there exists $n_0 \geq 1$ such that

$$d_h(T^i x_n, T^{i+1} x_n) \leq \varepsilon, \tag{4.3}$$

for all $n > n_0$ and $n_0 \leq i < n$.

Proof of Step I. If (4.3) does not hold, then for each $m \geq 1$ there exist n_m and i_m such that $m \leq i_m < n_m$ and

$$d_h(T^{i_m} x_{n_m}, T^{i_m+1} x_{n_m}) > \varepsilon, \tag{4.4}$$

for some $\varepsilon > 0$. Choose a natural number m such that $m > \frac{2c_0 + d_h(\{\theta\}, T\theta)}{(1-c)\varepsilon}$. Since $c < 1$, by using (5.1) and (4.4) we have $d_h(T^i x_{n_m}, T^{i+1} x_{n_m}) > \varepsilon$ for all $i = 1, 2, \cdots, i_m$. Since $d_h(T^{i+2} x_{n_m}, T^{i+1} x_{n_m}) \leq c d_h(T^{i+1} x_{n_m}, T^i x_{n_m})$ for all $i = 1, 2, \cdots, i_m - 1$, by (4.4) we get

$$\begin{aligned} &d_h(T^{i+2} x_{n_m}, T^{i+1} x_{n_m}) - d_h(T^{i+1} x_{n_m}, T^i x_{n_m}) \\ &\leq (c-1) d_h(T^{i+1} x_{n_m}, T^i x_{n_m}) < -(1-c)\varepsilon. \end{aligned}$$

On the other hand, $\varepsilon < d_h(T^2 x_{n_m}, T x_{n_m}) \leq c d_h(\{x_{n_m}\}, T x_{n_m}) \leq d_h(\{x_{n_m}\}, T x_{n_m})$. Hence,

$$\begin{aligned} &d_h(T^2 x_{n_m}, T x_{n_m}) - d_h(\{x_{n_m}\}, T x_{n_m}) \\ &\leq (c-1) d_h(\{x_{n_m}\}, T x_{n_m}) < -(1-c)\varepsilon. \end{aligned}$$

Thus,

$$\begin{aligned} -d_h(\{x_{n_m}\}, T x_{n_m}) &\leq d_h(T^{i_m+1} x_{n_m}, T^{i_m} x_{n_m}) - d_h(\{x_{n_m}\}, T x_{n_m}) \\ &\leq \sum_{i=1}^{i_m - 1} [d_h(T^{i+2} x_{n_m}, T^{i+1} x_{n_m}) - d_h(T^{i+1} x_{n_m}, T^i x_{n_m})] \\ &\quad + [d_h(T^2 x_{n_m}, T x_{n_m}) - d_h(\{x_{n_m}\}, T x_{n_m})] \\ &< -(c-1)\varepsilon i_m \leq -m(1-c)\varepsilon. \end{aligned} \tag{4.5}$$

Since d_h is a metric on $\mathcal{H}(X)$, from (4.2) and (4.5) we have

$$m(1-c)\varepsilon \leq d_h(\{x_{n_m}\}, T x_{n_m})$$

$$\leq d_h(\{x_{n_m}\},\{\theta\})+d_h(\{\theta\},T\theta)+d_h(T\theta,Tx_{n_m})$$
$$\leq d(x_{n_m},\theta)+d_h(\{\theta\},T\theta)+cd(\theta,x_{n_m})\leq 2c_0+d_h(\{\theta\},T\theta)$$

which is a contradiction. Therefore (4.3) holds.

Step II. For each $\delta>0$ there exists $n_0\geq 1$ such that

$$d_h(T^ix_n,T^jx_n)\leq\delta, \tag{4.6}$$

for all $n>n_0$ and $n_0\leq i,j<n$.

Proof of Step II. Let $\varepsilon<\frac{1}{4}\delta(1-c)$ and choose $n_0\geq 1$ such that (4.3) holds for all $n>n_0$ and $n_0\leq i<n$. Let i, j and n be natural numbers so that $n_0\leq i,j<n$. We claim that (4.6) holds. If $d_h(T^ix_n,T^jx_n)>\delta$, then

$$d_h(T^ix_n,T^jx_n)\leq d_h(T^ix_n,T^{i+1}x_n)+d_h(T^{i+1}x_n,T^{j+1}x_n)+d_h(T^{j+1}x_n,T^jx_n)$$
$$\leq 2\varepsilon+d_h(T^{i+1}x_n,T^{j+1}x_n)\leq 2\varepsilon+cd_h(T^ix_n,T^jx_n).$$

Hence, $\delta<d_h(T^ix_n,T^jx_n)\leq\frac{2\varepsilon}{1-c}$ which is a contradiction.

Step III. For each $\varepsilon>0$ there exists $n_0\geq 1$ such that

$$d_h(T^{n_0}x_{n_1},T^{n_0}x_{n_2})\leq\varepsilon, \tag{4.7}$$

for all $n_1,n_2>n_0$.

Proof of Step III. Choose a natural number m such that $m>\frac{4c_0}{\varepsilon(1-c)}$. Let $n_1,n_2>m$. We claim that

$$d_h(T^mx_{n_1},T^mx_{n_2})\leq\varepsilon. \tag{4.8}$$

If (4.8) does not hold, then $d_h(T^ix_{n_1},T^ix_{n_2})>\varepsilon$ for all $i=1,2,\cdots,m$ because $c<1$. Note that, $\varepsilon<d_h(Tx_{n_1},Tx_{n_2})\leq cd(x_{n_1},x_{n_2})\leq d(x_{n_1},x_{n_2})$ and so $-d(x_{n_1},x_{n_2})<-\varepsilon$. Since $d_h(T^{i+1}x_{n_1},T^{i+1}x_{n_2})\leq cd_h(T^ix_{n_1},T^ix_{n_2})$ for all $i=1,2,\cdots,m-1$,

$$d_h(T^{i+1}x_{n_1},T^{i+1}x_{n_2})-d_h(T^ix_{n_1},T^ix_{n_2})$$
$$\leq(c-1)d_h(T^ix_{n_1},T^ix_{n_2})<-(1-c)\varepsilon$$

for all $i=1,2,\cdots,m-1$. This implies that

$$-d(x_{n_1},x_{n_2})\leq d_h(T^mx_{n_1},T^mx_{n_2})-d(x_{n_1},x_{n_2})$$
$$\leq\sum_{i=1}^{m-1}[d_h(T^{i+1}x_{n_1},T^{i+1}x_{n_2})-d_h(T^ix_{n_1},T^ix_{n_2})]$$

$$+[d_h(Tx_{n_1},Tx_{n_2})-d(x_{n_1},x_{n_2})]\leq -m(1-c)\varepsilon.$$

Hence, $m(1-c)\varepsilon \leq d(x_{n_1},x_{n_2}) \leq d(x_{n_1},\theta)+d(\theta,x_{n_2}) \leq 2c_0$ and so $m \leq \frac{2c_0}{\varepsilon(1-c)}$. This contradiction shows that (4.8) holds.

Step IV. For each $\varepsilon > 0$ there exists $m(\varepsilon) \geq 1$ such that

$$d_h(T^ix_{n_1},T^jx_{n_2})\leq \varepsilon$$

for all $n_1,n_2 > m(\varepsilon)$ and all natural numbers $i \in [m(\varepsilon),n_1)$ and $j \in [m(\varepsilon),n_2)$.

Proof of Step IV. Let $\varepsilon > 0$ be given. By using (4.7), choose $m_1 \geq 1$ such that

$$d_h(T^{m_1}x_{n_1},T^{m_1}x_{n_2})\leq \frac{\varepsilon}{4}$$

for all $n_1,n_2 > m_1$. Also by using (4.6), choose $m_2 \geq 1$ such that $d_h(T^ix_n,T^jx_n) \leq \frac{\varepsilon}{4}$ for all $n > m_2$ and $m_2 \leq i,j < n$. Now, let $n_1,n_2 > m(\varepsilon) := m_1+m_2$, $i \in [m(\varepsilon),n_1)$ and $j \in [m(\varepsilon),n_2)$. Then, we have $d_h(T^{m(\varepsilon)}x_{n_1},T^{m(\varepsilon)}x_{n_2}) \leq d_h(T^{m_1}x_{n_1}, T^{m_1}x_{n_2}) \leq \frac{\varepsilon}{4}$. Also, $d_h(T^{m(\varepsilon)}x_{n_1},T^ix_{n_1}) \leq \frac{\varepsilon}{4}$, $d_h(T^{m(\varepsilon)}x_{n_2},T^jx_{n_2}) \leq \frac{\varepsilon}{4}$. Thus,

$$d_h(T^ix_{n_1},T^jx_{n_2})\leq d_h(T^{m(\varepsilon)}x_{n_1},T^ix_{n_1})$$

$$+d_h(T^{m(\varepsilon)}x_{n_1},T^{m(\varepsilon)}x_{n_2})+d_h(T^{m(\varepsilon)}x_{n_2},T^jx_{n_2})<\varepsilon.$$

This completes the proof of the step.

Now, we complete the proof of the theorem. Consider the sequences $\{T^{n-2}x_n\}_{n\geq 3}$ and $\{T^{n-1}x_n\}_{n\geq 2}$. For each $\varepsilon > 0$ take $N = m(\varepsilon)+2$. Let $m,n \geq N$, $i = m-2$, $j = n-2$, $n_1 = m$ and $n_2 = n$. Then $i \in [m(\varepsilon),n_1)$ and $j \in [m(\varepsilon),n_2)$. Hence by step (iv), $d_h(T^{m-2}x_m,T^{n-2}x_n) < \varepsilon$. Thus, $\{T^{n-2}x_n\}_{n\geq 3}$ is a Cauchy sequence. A similar argument shows that $\{T^{n-1}x_n\}_{n\geq 2}$ is a Cauchy sequence and $\lim_{n\to\infty} d_h(T^{n-2}x_n,T^{n-1}x_n) = 0$. Note that, the sequences $\{T^{n-2}x_n\}_{n\geq 3}$ and $\{T^{n-1}x_n\}_{n\geq 2}$ lie in K and $(\mathcal{H}(K),d_h)$ is a complete metric space. Hence, there exists $A \in \mathcal{H}(K)$ such that

$$\lim_{n\to\infty} d_h(T^{n-2}x_n,A)=\lim_{n\to\infty} d_h(A,T^{n-1}x_n)=0.$$

Since

$$d_h(T^{n-1}x_n,T(A))$$

$$=\max\{\sup_{a\in T^{n-2}x_n,\,a_1\in Ta}\ \inf_{b\in A,\,b_1\in Tb} d(a_1,b_1),\ \sup_{d_1\in Td,\,d\in A}\ \inf_{c_1\in Tc,\,c\in T^{n-2}x_n} d(c_1,d_1)\}$$

$$=\max\{\sup_{a\in T^{n-2}x_n}\inf_{b\in A}\sup_{a_1\in Ta}\inf_{b_1\in Tb} d(a_1,b_1),\sup_{d\in A}\inf_{c\in T^{n-2}x_n}\sup_{d_1\in Td}\inf_{c_1\in Tc} d(c_1,d_1)\}$$

$$\leq \max\{\sup_{a\in T^{n-2}x_n}\inf_{b\in A} d_h(Ta,Tb), \sup_{d\in A}\inf_{c\in T^{n-2}x_n} d_h(Tc,Td)\}$$

$$\leq c\max\{\sup_{a\in T^{n-2}x_n}\inf_{b\in A} d(a,b), \sup_{d\in A}\inf_{c\in T^{n-2}x_n} d(c,d)\}$$

$$= cd_h(T^{n-2}x_n, A) \to 0,$$

we get $A = T(A)$. Thus, $T|_A : A \to \mathcal{H}(A)$ is a contraction multifunction and $(A, d|_{A\times A})$ is a complete metric space. Therefore by using Theorem 6 in [70], $T|_A$ has a fixed point in A, that is, there exists $x_0 \in A \subseteq K$ such that $x_0 \in Tx_0$.

♣

Now for clarity of the subject, we provide two examples to illustrate Theorem 4.2.

Example 52 *Let $X = [0,\infty)$ with the Euclidean norm, $m \geq 4$ a fixed natural number, $K = \{\frac{1}{2}, \frac{1}{3}, \frac{1}{4}, \cdots\} \cup \{0,1\}$ and $K_0 = \{\frac{1}{2}, \frac{1}{3}, \frac{1}{4}, \cdots\}$. Define $T : K \to \mathcal{H}(X)$ by*

$$Tx = \{\frac{x}{2}, \frac{x}{3}, \cdots, \frac{x}{m}\}$$

for all $x \in K$. Note that, the values of T are compact and $d_h(Tx,Ty) \leq \frac{1}{2}|x-y|$ for all $x,y \in K$. Since $Tx \subseteq K$ for all $x \in K_0$, it is easy to see that if $\{x_n\}_{n\geq 1}$ is a sequence in K_0, then $\bigcup_{i=1}^{n} T^i x_n \subseteq K$ for all $n \geq 1$. Now by using Theorem 4.2, T has a fixed point. Note that, $x_0 = 0$ is the unique fixed point of T.

We first provided an easy example. Now, we give next considerable one.

Example 53 *Let $X = C[0,1]$ with the supremum norm, $K = \{f \in X : f \geq 0\}$ and $K_0 = \{f \in K : 0 < \|f\|_\infty < 1\}$. Define $T : K \multimap \mathcal{H}(X)$ by*

$$T(f)(t) = \bigcup_{n=1}^{\infty}\{1 - \frac{1}{2}\int_{\frac{t}{n}}^{1} f(x)dx\} \bigcup \{1 - \frac{1}{2}\int_{0}^{1} f(x)dx\}$$

for all $t \in [0,1]$. Note that the values of T are compact and $d_h(T(f),T(g)) \leq \frac{1}{2}\|f-g\|_\infty$ for all $f,g \in K$. If $f \in K_0$, then $0 \leq f(x) < 1$ and so $0 < 1 - \frac{1}{2}\int_{\frac{t}{n}}^{1} f(x)dx < \frac{1}{2}$ for all $t \in [0,1]$ and $n \geq 1$. Hence, $Tf \subseteq K$ for all $f \in K_0$. Thus, $\bigcup_{i=1}^{n} T^i f_n \subseteq K$ for all $n \geq 1$ whenever $\{f_n\}_{n\geq 1}$ is a sequence in K_0. Now by using Theorem 4.2, T has a fixed point. Consider the function $g_0(t) = e^{\frac{1}{2}(t-1)}$ in X. Since $1 - \frac{1}{2}\int_{\frac{t}{1}}^{1} g_0(x)dx = g_0(t)$ for $t \in [0,1]$, $g_0 \in Tg_0$, that is, g_0 is a fixed point of T.

4.4 An iterative scheme for closed-valued multifunctions

Study of iterative schemes for various classes of nonexpansive mappings is a central topic in nonlinear functional analysis. It began with the classical Banach theorem on the existence of a unique fixed point for a strict contraction. This celebrated result also yields convergence of iterative to the unique fixed point. Since Banach's seminal result, many developments have taken place in this area. There are many works in existence and approximation results regarding fixed points of those nonexpansive mappings which are not necessarily strictly contractive. Also, some authors have provided some results on existence and approximation of fixed points of some multifunctions. In 2008, Suzuki published a series of papers and provided some new methods for mappings and multifunctions. In 2010, Reich and Zaslavski introduced a new inexact iterative scheme for fixed points of nonexpansive set-valued maps ([83]).

We say that a multifunction $T : X \longrightarrow 2^Y$ is onto whenever for each $A \in 2^Y$ there exists $x \in X$ such that $Tx = A$.

Theorem 4.3

Let (X,d) be a complete metric space, $T : X \to C(X)$ a set-valued map and $\{\varepsilon_i\}_{i=0}^{\infty}$ and $\{\delta_i\}_{i=0}^{\infty}$ two sequences in $(0,\infty)$ with $\sum_{i=0}^{\infty} \varepsilon_i < \infty$ and $\sum_{i=0}^{\infty} \delta_i < \infty$. Suppose that there exist $\alpha, \beta \in (0,1)$ such that $\alpha(3-2\alpha+\beta) \leq 1$ and

$$\alpha d(x,Tx) \leq d(x,y) \text{ implies } H(Tx,Ty) \leq \beta d(x,y)$$

for all $x,y \in X$. Assume that for each integer $i \geq 0$, the multifunction $T_i : X \to 2^X$ satisfies $H(Tx,T_ix) \leq \varepsilon_i$ for all $x \in X$. Let $x_0 \in X$. If $\frac{\varepsilon_i}{1-\alpha} \leq d(x_i,x_{i+1}) \leq d(x_i,T_ix_i) + \delta_i$ and $x_{i+1} \in T_ix_i$ for all $i \geq 0$, then $\{x_i\}_{i=0}^{\infty}$ converges to a fixed point of T.

Proof: We first show that $\{x_i\}_{i=0}^{\infty}$ is a Cauchy sequence. Let $i \geq 0$ be an integer. Then,

$$\begin{aligned} d(x_{i+1},x_{i+2}) &\leq d(x_{i+1},T_{i+1}x_{i+1}) + \delta_{i+1} \\ &\leq d(x_{i+1},Tx_{i+1}) + H(Tx_{i+1},T_{i+1}x_{i+1}) + \delta_{i+1} \\ &\leq H(T_ix_i,Tx_{i+1}) + \varepsilon_{i+1} + \delta_{i+1} \\ &\leq H(T_ix_i,Tx_i) + H(Tx_i,Tx_{i+1}) + \varepsilon_{i+1} + \delta_{i+1}. \end{aligned}$$

Hence,

$$d(x_{i+1},x_{i+2}) \leq H(Tx_i,Tx_{i+1}) + \varepsilon_i + \varepsilon_{i+1} + \delta_{i+1} \tag{4.9}$$

for all $i \geq 0$. Since $\alpha(2-\alpha) < 1$ and

$$d(Tx_i,x_i) \leq d(x_i,T_ix_i) + H(T_ix_i,Tx_i) \leq d(x_i,x_{i+1}) + \varepsilon_i$$

$$\leq d(x_i,x_{i+1})+(1-\alpha)d(x_i,x_{i+1})=(2-\alpha)d(x_i,x_{i+1}),$$

we have $\alpha d(Tx_i,x_i)<d(x_i,x_{i+1})$ and so

$$H(Tx_i,Tx_{i+1})\leq \beta d(x_i,x_{i+1}). \tag{4.10}$$

Now, by using (4.9) and (4.10) we obtain

$$d(x_{i+1},x_{i+2})\leq \beta d(x_i,x_{i+1})+\varepsilon_i+\varepsilon_{i+1}+\delta_{i+1} \tag{4.11}$$

for all $i\geq 0$. Thus,

$$d(x_1,x_2)\leq \beta d(x_0,x_1)+\varepsilon_0+\varepsilon_1+\delta_1 \tag{4.12}$$

and

$$d(x_2,x_3)\leq \beta^2 d(x_0,x_1)+\beta(\varepsilon_0+\varepsilon_1+\delta 1)+\varepsilon_1+\varepsilon_2+\delta_2. \tag{4.13}$$

Now, we show by induction that for each $n\geq 1$, we have

$$d(x_n,x_{n+1})\leq \beta^n d(x_0,x_1)+\sum_{i=0}^{n-1}\beta^i(\varepsilon_{n-i}+\delta_{n-i}+\varepsilon_{n-i-1}). \tag{4.14}$$

In view of (4.12) and (4.13), inequality (4.14) holds for $n=1,2$. Assume that $k\geq 1$ is an integer and that (4.14) holds for $n=k$. When combined with (4.11), this implies that

$$\begin{aligned}
d(x_{k+1},x_{k+2}) &\leq \beta d(x_k,x_{k+1})+\varepsilon_k+\varepsilon_{k+1}+\delta_{k+1}\\
&\leq \beta^{k+1}d(x_0,x_1)+\beta\sum_{i=0}^{k-1}\beta^i(\varepsilon_{k-i}+\delta_{k-i}+\varepsilon_{k-i-1})+\varepsilon_k+\varepsilon_{k+1}+\delta_{k+1}\\
&= \beta^{k+1}d(x_0,x_1)+\sum_{i=1}^{k}\beta^i(\varepsilon_{k-i+1}+\delta_{k-i+1}+\varepsilon_{k-i})+\varepsilon_k+\varepsilon_{k+1}+\delta_{k+1}\\
&= \beta^{k+1}d(x_0,x_1)+\sum_{i=0}^{k}\beta^i(\varepsilon_{k-i+1}+\delta_{k-i+1}+\varepsilon_{k-i}).
\end{aligned}$$

This implies that (4.14) holds for all $n\geq 1$. Now, by using (4.14) we obtain

$$\begin{aligned}
\sum_{n=1}^{\infty}d(x_n,x_{n+1}) &\leq \sum_{n=1}^{\infty}\beta^n d(x_0,x_1)+\sum_{n=1}^{\infty}\sum_{i=0}^{n-1}\beta^i(\varepsilon_{n-i}+\delta_{n-i}+\varepsilon_{n-i-1})\\
&= \sum_{n=1}^{\infty}\beta^n d(x_0,x_1)+\sum_{n=1}^{\infty}\sum_{i=1}^{n}\beta^{n-i}(\varepsilon_i+\delta_i+\varepsilon_{i-1})
\end{aligned}$$

$$\leq \sum_{n=1}^{\infty} \beta^n d(x_0,x_1) + \beta^0(\varepsilon_1+\delta_1+\varepsilon_0) + \beta^0(\varepsilon_2+\delta_2+\varepsilon_1) + \beta^1(\varepsilon_1+\delta_1+\varepsilon_0)$$

$$+\beta^0(\varepsilon_3+\delta_3+\varepsilon_2) + \beta^1(\varepsilon_2+\delta_2+\varepsilon_1) + \beta^2(\varepsilon_1+\delta_1+\varepsilon_0) + \ldots$$

$$= \sum_{n=1}^{\infty} \beta^n d(x_0,x_1) + (\beta^0+\beta^1+\beta^2+\ldots)(\varepsilon_1+\delta_1+\varepsilon_0)$$

$$+(\beta^0+\beta^1+\beta^2+\ldots)(\varepsilon_2+\delta_2+\varepsilon_1) + (\beta^0+\beta^1+\beta^2+\ldots)(\varepsilon_3+\delta_3+\varepsilon_2) + \ldots$$

$$= (\sum_{n=1}^{\infty} \beta^n)(d(x_0,x_1) + \sum_{i=1}^{\infty} \varepsilon_i + \delta_i + \varepsilon_{i-1}) < \infty.$$

Thus, $\{x_i\}_{i=0}^{\infty}$ is a Cauchy sequence and so exists $x \in X$ such that $x = \lim_{n\to\infty} x_n$. Now, we claim that for each $n \geq 1$ either

$$\alpha d(x_n, Tx_n) \leq d(x_n, x)$$

or

$$\alpha d(x_{n+1}, Tx_{n+1}) \leq d(x_{n+1}, x)$$

hold. If $\alpha d(x_n,Tx_n) > d(x_n,x)$ and $\alpha d(x_{n+1},Tx_{n+1}) > d(x_{n+1},x)$ for some $n \geq 1$, then we obtain

$$\begin{aligned}
d(x_{n+1},x_n) &\leq d(x_{n+1},x) + d(x,x_n) \\
&< \alpha d(x_{n+1},Tx_{n+1}) + \alpha d(x_n,Tx_n) \\
&\leq \alpha(H(T_nx_n,Tx_{n+1}) + (2-\alpha)d(x_n,x_{n+1})) \\
&\leq \alpha(H(T_nx_n,Tx_n) + H(Tx_n,Tx_{n+1}) + (2-\alpha)d(x_n,x_{n+1})) \\
&\leq \alpha(\varepsilon_n + \beta d(x_n,x_{n+1}) + (2-\alpha)d(x_n,x_{n+1})) \\
&\leq \alpha((1-\alpha)d(x_n,x_{n+1}) + \beta d(x_n,x_{n+1}) + (2-\alpha)d(x_n,x_{n+1})) \\
&= \alpha(3-2\alpha+\beta)d(x_n,x_{n+1}).
\end{aligned}$$

It implies that $\alpha(3-2\alpha+\beta) > 1$ which is a contradiction. Therefore, our claim is proved. Thus, by using the assumption, for each $n \geq 1$ either $H(Tx_n,Tx) \leq \beta d(x_n,x)$ or $H(Tx_{n+1},Tx) \leq \beta d(x_{n+1},x)$ hold. Therefore, one of the following cases holds:

(i) There exists an infinite subset $I \subseteq \mathbb{N}$ such that $H(Tx_n,Tx) \leq \beta d(x_n,x)$ for all $n \in I$,

(ii) There exists an infinite subset $J \subseteq \mathbb{N}$ such that $H(Tx_{n+1},Tx) \leq \beta d(x_{n+1},x)$ for all $n \in J$.

In the case (i) we obtain

$$\begin{aligned}
d(x,Tx) \leq d(x,x_n) + d(x_n,Tx) &\leq d(x,x_n) + d(x_n,Tx_n) + H(Tx_n,Tx) \\
&\leq d(x,x_n) + \frac{1}{\alpha} d(x_n,x_{n+1}) + \beta d(x,x_n),
\end{aligned}$$

for all $n \in I$, and so $d(x,Tx) = 0$. Hence $x \in Tx$. Similar to (i) we obtain $x \in Tx$, from case (ii). This completes the proof.

♣

Next example shows that Theorem 4.3 is a generalization of Theorem 2.1 in [83].

Example 54 *Put* $X = [5,6] \cup [-6,-5] \cup \{0\}$ *and define a* $T : X \to C(X)$ *by*

$$T(x) = \begin{cases} [5, \frac{7(-x)-9}{-x}] & -6 \le x < -5.4 \\ \{0\} & x \in [5,5.4] \bigcup \{0\} \bigcup [-5.4,-5] \\ \{-\frac{7x-9}{x}\} & 5.4 < x \le 6 \end{cases}$$

We show that T *satisfies the assumption of Theorem 4.3 for* $\alpha = \frac{1}{4}$ *and* $\beta = \frac{100}{101}$ *while* T *is not a contractive multifunction. If* $5.4 < x \le 6$*, then* $5 < 5 + \frac{2x-9}{x} = \frac{7x-9}{x} \le \frac{11}{2}$ *and* $\frac{100}{101}x - \frac{7x-9}{x} > 0$*. If* $-6 \le x < -5.4$*, then* $d(0,Tx) = 5 < -\frac{100}{101}x = \frac{100}{101}|x|$*,* $Tx \subset [5,5.5]$ *and* $H(\{0\},Tx) = \frac{7(-x)-9}{-x} < -\frac{100}{101}x = \frac{100}{101}|x|$*. If* $5.4 < x \le 6$*, then* $Tx \subset [-5.5,-5)$ *and* $H(\{0\},Tx) = d(0,Tx) = \frac{7x-9}{x} < \frac{100}{101}x = \frac{100}{101}|x|$*. Thus,* $H(Tx,Ty) < \frac{100}{101}d(x,y)$ *holds whenever* $x = 0$ *and* $y \neq 0$ *or* $y = 0$ *and* $x \neq 0$*. If* $x \in [5,6]$ *and* $y \in [-6,-5]$*, then* $\frac{1}{4}d(x,Tx) < \frac{11.5}{4} < 10 \le d(x,y)$ *and so*

$$H(Tx,Ty) \le d(Tx,0) + H(Ty,\{0\}) \le \frac{100}{101}(|x|+|y|) = \frac{100}{101}d(x,y).$$

If $x,y \in [5,5.4]$ *or* $x,y \in [-5.4,-5]$*, then* $\frac{1}{4}d(x,Tx) \ge \frac{5}{4} > d(x,y)$*. If* $x \in [5,5.4]$ *and* $y \in [5.4,6]$*, or* $x \in [-5.4,-5]$ *and* $y \in [-6,-5.4]$*, or* $y \in [5,5.4]$ *and* $x \in [5.4,6]$*, or* $y \in [-5.4,-5]$ *and* $x \in [-6,-5.4]$*, then we have* $\frac{1}{4}d(x,Tx) \ge \frac{5}{4} > 1 \ge d(x,y)$*. If* $x = 5$ *and* $y = 6$*, then* $H(Tx,Ty) = \frac{11}{2} > \frac{100}{101}d(x,y)$*.*

Theorem 4.4

Let (X,d) *be a complete metric space,* F *a nonempty closed subset of* X*,* $T : X \longrightarrow C(X)$ *a multifunction and* $\{\varepsilon_i\}_{i=0}^{\infty}$ *a sequence in* $(0,\infty)$ *such that* $\sum_{i=0}^{\infty}\varepsilon_i < \infty$*. Suppose that* $T\,T(F) \subset F$*,* $d(x,Ty) \le d(x,y)$ *for every* $x \in F^c, y \in F$ *and there exists* $\alpha \in (0,\frac{1}{2})$ *such that* $\alpha d(x,Tx) \le d(x,y)$ *implies* $H(Tx,Ty) < d(x,y)$ *for all* $x,y \in X$*. Let* $T_i : X \to 2^X$ *satisfy, for each integer* $i \ge 0$*,* $H(Tx,T_ix) \le \varepsilon_i$ *for all* $x \in X$*. Assume that for each* $x \in X$ *there exists a sequence* $\{x_i\}_{i=0}^{\infty}$ *in* X *such that*

$$x_0 = x,\ x_{i+1} \in Tx_i \text{ for all } i \ge 0 \text{ and } \lim_{i\to\infty} d(x_i,F) = 0.$$

Then for each $x \in X$ *there exists a sequence* $\{x_i\}_{i=0}^{\infty}$ *in* X *such that*

$$x_0 = x,\ x_{i+1} \in T_ix_i \text{ for all } i \ge 0 \text{ and } \lim_{i\to\infty} d(x_i,F) = 0.$$

We prove it by using next two Lemmas.

Lemma 4.1

Let $x \in X$, p a natural number, $\delta > 0$, $\{x_0, x_1, \cdots, x_p\} \subset X$ such that $x_0 = x$ and $x_{i+1} \in T_i x_i$ $(i = 0, 1, ..., p-1)$. Then there is a natural number $q > p$ and

$$\{x_p, x_{p+1}, \cdots, x_q\} \subset X$$

such that $x_{i+1} \in T_i x_i$ $(i = p, ..., q-1)$ and $d(x_q, F) \leq \delta$.

*Proof:*Choose a natural number $p_1 > p$ such that $\sum_{i=p_1}^{\infty} \varepsilon_i < \frac{\delta}{8}$ and a sequence

$$\{x_p, x_{p+1}, \cdots, x_{p_1}\} \subset X$$

such that $x_{i+1} \in T_i x_i$ $(i = p, ..., p_1 - 1)$. By using the assumptions, there is a sequence $\{y_i\}_{i=p_1}^{\infty} \subset X$, such that $y_{p_1} = x_{p_1}$, $y_{i+1} \in Ty_i$ for all$i \geq p_1$ and $\lim_{i \to \infty} d(y_i, F) = 0$. We are now going to define by induction a sequence $\{x_i\}_{i=p_1}^{\infty} \subset X$. To this end, assume that $k \geq p_1$ is an integer and that we have already defined $x_i \in X$, $i = p_1, ..., k$, such that $x_{i+1} \in T_i x_i$ $(i = p_1, ..., k-1)$ and $\min\{d(x_k, y_k), d(x_k, y_{k+1})\} \leq \sum_{i=p_1}^{k-1} \varepsilon_i$. Clearly this assumption holds for $k = p_1$. Since $y_{k+1} \in Ty_k$, by using a similar proof in Theorem 4.3, it is easy to prove that for each k, one of the following cases hold:
(i) $\alpha d(y_k, Ty_k) \leq d(y_k, x_k)$,
(ii) $\alpha d(y_{k+1}, Ty_{k+1}) \leq d(y_{k+1}, x_k)$.
From the case (i), we obtain $H(Ty_k, Tx_k) \leq d(x_k, y_k)$ and so $d(y_{k+1}, Tx_k) \leq d(x_k, y_k)$. Hence, there exists $\widetilde{y}_{k+1} \in Tx_k$ such that $d(y_{k+1}, \widetilde{y}_{k+1}) \leq d(x_k, y_k) + \varepsilon_k$. This implies that $d(\widetilde{y}_{k+1}, T_k x_k) \leq H(Tx_k, T_k x_k) \leq \varepsilon_k$ and so there exists $x_{k+1} \in T_k x_k$ such that

$$d(\widetilde{y}_{k+1}, x_{k+1}) \leq 2\varepsilon_k.$$

Thus, we have

$$\begin{aligned} d(x_{k+1}, y_{k+1}) &\leq d(x_{k+1}, \widetilde{y}_{k+1}) + d(y_{k+1}, \widetilde{y}_{k+1}) \\ &\leq 2\varepsilon_k + d(x_k, y_k) + \varepsilon = 3\varepsilon_k + d(x_k, y_k). \end{aligned}$$

From (ii), we obtain $H(Ty_{k+1}, Tx_k) \leq d(x_k, y_{k+1})$ and so $d(y_{k+2}, Tx_k) \leq d(x_k, y_{k+1})$. Hence, there exists $\widetilde{y}_{k+1} \in Tx_k$ such that $d(y_{k+2}, \widetilde{y}_{k+1}) \leq d(x_k, y_{k+1}) + \varepsilon_k$. This implies that $d(\widetilde{y}_{k+1}, T_k x_k) \leq H(Tx_k, T_k x_k) \leq \varepsilon_k$ and so there exists $x_{k+1} \in T_k x_k$ such that

$$d(\widetilde{y}_{k+1}, x_{k+1}) \leq 2\varepsilon_k.$$

Thus, we have

$$\begin{aligned} d(x_{k+1}, y_{k+2}) &\leq d(x_{k+1}, \widetilde{y}_{k+1}) + d(y_{k+1}, \widetilde{y}_{k+2}) \\ &\leq 2\varepsilon_k + d(x_k, y_{k+1}) + \varepsilon = 3\varepsilon_k + d(x_k, y_{k+1}). \end{aligned}$$

Thus, by considering the cases, we have

$$\min\{d(x_{k+1},y_{k+2}),d(x_{k+1},y_{k+1})\} \le \min\{d(x_k,y_k),d(x_k,y_{k+1})\}+3\varepsilon$$

$$\le \sum_{i=p_1}^{k-1} \varepsilon_i + 3\varepsilon_k = 3\sum_{i=p_1}^{k} \varepsilon_i.$$

Therefor, we have indeed defined by induction a sequence $\{x_i\}_{i=p_1}^{\infty} \subset X$ such that $x_{i+1} \in Tx_i$ $(i = p_1, \cdots)$ and $\min\{d(x_k,y_k),d(x_k,y_{k+1})\} \le \sum_{i=p_1}^{k-1} \varepsilon_i$. Hence, there exists an integer $q > p_1+2$ such that $d(y_q,F) < \frac{\delta}{4}$ and $d(y_{q+1},F) < \frac{\delta}{4}$. Thus, we obtain

$$d(x_q,F) \le d(x_q,y_q)+d(y_q,F) \le d(x_q,y_q)+\frac{\delta}{4} \tag{4.15}$$

and

$$d(x_q,F) \le d(x_q,y_{q+1})+d(y_{q+1},F) \le d(x_q,y_{q+1})+\frac{\delta}{4}. \tag{4.16}$$

When combined with (4.15) and (4.16), this implies that

$$d(x_q,F) \le \min\{d(x_q,y_q),d(x_q,y_{q+1})\}+\frac{\delta}{4} \le \sum_{i=p_1}^{\infty} \varepsilon_i + \frac{\delta}{4} < \frac{\delta}{8}+\frac{\delta}{4}.$$

This completes the proof.

♣

Lemma 4.2

Let $\{x_i\}_{i=0}^{\infty}$ be a sequence in X, $x_{i+1} \in T_i x_i$ for all $i \ge 0$, $\delta > 0$, p a natural number, $d(x_p,F) \le \delta$ and $\sum_{i=p}^{\infty} \varepsilon_i < \delta$. Then $d(x_i,F) \le 3\delta$ for all $i \ge p$.

Proof: We intend to show by induction that

$$d(x_n,F) \le \delta + \sum_{i=p}^{n-1} 2\varepsilon_i \tag{4.17}$$

for all $n \ge p$. It is clear that (4.17) holds for $n = p$. Now, assume that (4.17) holds for $n \ge p$. Then there exists $y_n \in F$ such that $d(x_n,y_n) \le \delta + \sum_{i=p}^{n-1} 2\varepsilon_i + \frac{\varepsilon_n}{4}$. By assumption, $d(Tx_n,x_{n+1}) \le \varepsilon_n$ and so there exists $\widetilde{x}_{n+1} \in Tx_n$ such that $d(\widetilde{x}_{n+1},x_{n+1}) \le \frac{3}{2}\varepsilon_n$. If $x_n \in F$, then $d(x_{n+1},F) \le d(x_{n+1},Tx_n) \le \varepsilon_n \le \delta + \sum_{i=p}^{n} 2\varepsilon_i$. On the other hand

$$\alpha d(y_n,Ty_n) \le \alpha(d(x_n,y_n)+d(x_n,Ty_n)) \le 2\alpha d(x_n,y_n) < d(x_n,y_n),$$

and so $H(Tx_n, Ty_n) \leq d(x_n, y_n)$. Hence, there exists $\widetilde{y}_{n+1} \in Ty_n$ such that

$$d(\widetilde{y}_{n+1}, \widetilde{x}_{n+1}) \leq d(x_n, y_n) + \frac{\varepsilon_n}{4}$$

and so

$$\begin{aligned} d(x_{n+1}, F) &\leq d(x_{n+1}, \widetilde{y}_{n+1}) \leq d(x_{n+1}, \widetilde{x}_{n+1}) + d(\widetilde{y}_{n+1}, \widetilde{x}_{n+1}) \\ &\leq \frac{7}{4}\varepsilon_n + d(x_n, y_n) \leq \delta + \sum_{i=p}^{n} 2\varepsilon_i. \end{aligned}$$

This implies that $d(x_n, F) \leq \delta + 2\sum_{i=p}^{n-1} \varepsilon_i \leq \delta + 2\sum_{i=p}^{\infty} \varepsilon_i \leq \delta + 2\delta = 3\delta$ for all $n \geq p$.

♣

Now, we Complete the proof of Theorem 4.4. *Proof:* Let $x \in X$, By using Lemma 4.1, there exists a sequence $\{x_i\}_{i=0}^{\infty}$ and a strictly increasing sequence of natural numbers $\{n_k\}_{k=1}^{\infty}$ such that $x_{i+1} \in T_i x_i$ for all $i \geq 0$ and for each $k \geq 1$ $d(x_{n_k}, F) \leq 2^{-k}$ and $\sum_{i=n_k}^{\infty} \varepsilon_i < 2^{-k}$. Hence, by using Lemma 4.2 we have $\lim_{n\to\infty} d(x_n, F) = 0$. This completes the proof.

♣

The following example of Suzuki in [98] shows that T satisfies the assumption of Theorem 4.4 for $\alpha = \frac{1}{2}$ while T is not a nonexpansive multifunction.

Example 55 *Let $X = \{(0,0), (4,0), (0,4), (4,5), (5,4)\}$ and its metric d defined by $d((x_1, x_2), (y_1, y_2)) = |x_1 - y_1| + |x_2 - y_2|$. Define a mapping T on X by*

$$T(x) = \begin{cases} \{(x_1, 0)\} & x_1 \leq x_2 \\ \{(0, x_2)\} & x_2 < x_1 \end{cases}$$

Put $F = \{(0,0)\}$. Then T satisfies the assumptions of Theorem 4.4 while T is not a nonexpansive multifunction. First note that, $H(Tx, Ty) \leq d(x, y)$ if $(x, y) \neq ((4,5), (5,4))$ and $(x, y) \neq ((5,4), (4,5))$. Also, for each $x \in X$ there exists a sequence $\{x_i\}_{i=0}^{\infty}$ in X such that $x_0 = x$, $x_{i+1} \in Tx_i$ for all $i \geq 0$, $\lim_{i\to\infty} d(x_i, F) = 0$ and $d(x, Ty) \leq d(x, y)$ for all $x \in F^c$ and $y \in F$. Thus,

$$\frac{1}{2}d((4,5), T(4,5)) > \frac{5}{2} > 2 = d((4,5), (5,4)),$$

$$\frac{1}{2}d((5,4), T(5,4)) > \frac{5}{2} > 2 = d((5,4), (4,5)).$$

Hence, T satisfies the assumptions of Theorem 4.4 while T is not a nonexpansive multifunction because $d(T(4,5), T(5,4)) = 8 > 5 = d((4,5), (5,4))$.

4.5 Generalizations of the Suzuki's method for closed-valued multifunctions

In this section, our aim is to show young researchers how they can mix some ideas on set-valued mappings to obtain a generalization result.

As we said before, Suzuki introduced a new type of mappings and a generalization of the Banach contraction principle in which the completeness can be also characterized by the existence of fixed points of these mappings ([98]). Consider the decreasing function θ form $[0,1)$ onto $(\frac{1}{2},1]$ defined by

$$\theta(r)=\begin{cases} 1 & 0\leq r\leq \frac{\sqrt{5}-1}{2} \\ (1-r)r^{-2} & \frac{\sqrt{5}-1}{2}\leq r\leq 2^{\frac{-1}{2}} \\ (1+r)^{-1} & 2^{\frac{-1}{2}}\leq r<1. \end{cases}$$

Suzuki proved the following result in 2008 ([98]).

Theorem 4.5

For a metric space (X,d), X is complete if and only if every mapping T on X such that there exists $r\in[0,1)$, $\theta(r)d(x,Tx)\leq d(x,y)$ implies $d(Tx,Ty)\leq rd(x,y)$ for all $x,y\in X$, has a fixed point.

Then, Kikkawa and Suzuki extended Theorem 4.5 to multivalued mappings ([67]).

Theorem 4.6

Let (X,d) be a complete metric space and $F:X\to CB(X)$ a multifunction. Assume that there exists $r\in[0,1)$ such that $\frac{1}{1+r}d(x,Fx)\leq d(x,y)$ implies $H(Fx,Fy)\leq rd(x,y)$ for all $x,y\in X$. Then F has a fixed point.

Later, Mot and Petrusel extended Theorem 4.6 as following (see [73]).

Theorem 4.7

Let (X,d) be a complete metric space and $F:X\to CB(X)$ a multifunction. Assume *that there exist* $a,b,c \in [0,1)$ *such that* $a+b+c<1$ *and*

$$\frac{1-b-c}{1+a} d(x,Fx) \le d(x,y)$$

implies $H(Fx,Fy) \le ad(x,y)+bd(x,Fx)+cd(y,Fy)$ *for all* $x,y \in X$. *Then* F *has a fixed point.*

Finally, Dhompongsa and Yingtaweesittikul generalized Theorem 4.7 as following (see [34]).

Theorem 4.8

For a metric space (X,d), X *is complete if and only if for each* $r \in [0,1)$, *every multifunction* $F : X \longrightarrow 2^X$ *with closed values such that*

$$\theta(r)d(x,Fx) \le d(x,y) \text{ implies } H(Fx,Fy) \le r\max\{d(x,y),d(x,Fx),d(y,Fy)\}$$

for all $x,y \in X$ *and the function* $x \mapsto d(x,Fx)$ *is lower semicontinuous, has a fixed point.*

We provide a new condition on T which guarantees the existence of its fixed point. Our new result will generalize above results. In this way, consider a continuous function $g : [0,\infty)^5 \longrightarrow [0,\infty)$ satisfying the following conditions:
a) $g(1,1,1,2,0) = g(1,1,1,0,2) = h \in (0,1)$,
b) g is sub-homogeneous, that is, $g(\alpha x_1, \alpha x_2, \alpha x_3, \alpha x_4, \alpha x_5) \le \alpha g(x_1,x_2,x_3,x_4,x_5)$ for all $\alpha \ge 0$ and all $(x_1,x_2,x_3,x_4,x_5) \in [0,\infty)^5$,
c) If $x_i, y_i \in [0,\infty)$ and $x_i < y_i$ for $i = 1,...,4$, then $g(x_1,x_2,x_3,x_4,0) < g(y_1,y_2,y_3,y_4,0)$ and $g(x_1,x_2,x_3,0,x_4) < g(y_1,y_2,y_3,0,y_4)$.

In this case, we write $g \in \mathcal{R}$. We appeal the following results in the sequel (see Lemma 1.3 in [32]).

Proposition 4.1

If $g \in \mathcal{R}$ *and* $u,v \in [0,\infty)$ *are such that*

$$u \le \max\{g(v,v,u,v+u,0),\ g(v,v,u,0,v+u),\ g(v,u,v,v+u,0),\ g(v,u,v,0,v+u)\},$$

then $u \le hv$.

Now, we give the following result about common fixed points of two multifunction.

Lemma 4.3

Let (X,d) be a complete metric space and $F,G: X \longrightarrow C(X)$ two multifunctions. Suppose that there exits $\alpha \in (0,\infty)$ and $g \in \mathcal{R}$ such that $\alpha d(x,Fx) \leq d(x,y)$ or $\alpha d(y,Gy) \leq d(x,y)$ implies

$$H(Fx,Gy) \leq g(d(x,y),d(x,Fx),d(y,Gy),d(x,Gy),d(y,Fx))$$

for all $x,y \in X$. Then $\mathcal{F}(F) = \mathcal{F}(G)$, where $\mathcal{F}(F)$ is the set of fixed points of F.

Proof: If $x \in Tx$, then $\alpha d(x,Tx) = 0 = d(x,x)$. Hence,

$$d(x,Gx) \leq H(Tx,Gx) \leq g(d(x,x),d(x,Fx),d(x,Gx),d(x,Gx),d(x,Fx))$$
$$\leq g(0,0,d(x,Gx),d(x,Gx),0).$$

By using Proposition 4.1, we have $d(x,Gx) \leq h0 = 0$ and so $x \in \mathcal{F}(G)$. Thus, $F(T) \subseteq F(G)$. Similarly, we can obtain $F(G) \subseteq F(T)$.

♣

Theorem 4.9

Let (X,d) be a complete metric space and $F,G: X \to C(X)$ two set-valued mappings. Suppose that there exist $\alpha \in (0,1)$ and $g \in \mathcal{R}$ such that $\alpha(h+1) \leq 1$ and $\alpha d(x,Fx) \leq d(x,y)$ or $\alpha d(y,Gy) \leq d(x,y)$ implies

$$H(Fx,Gy) \leq g(d(x,y),d(x,Fx),d(y,Gy),d(x,Gy),d(y,Fx))$$

for all $x,y \in X$. Then $F(T) = F(G)$ and $F(T)$ is non-empty.

Proof: By Lemma 4.3, $\mathcal{F}(F) = \mathcal{F}(G)$. Fix $1 > r > h$ and $x_0 \in X$. Choose $x_1 \in Fx_0$ such that $\alpha d(x_0,Fx_0) < d(x_0,x_1)$. Then, we have

$$d(x_1,Gx_1) \leq H(Fx_0,Gx_1)$$
$$\leq g(d(x_0,x_1),d(x_0,Fx_0),d(x_1,Gx_1),d(x_0,Gx_1),d(x_1,Fx_0))$$
$$\leq g(d(x_0,x_1),d(x_0,x_1),d(x_1,Gx_1),d(x_0,x_1)+d(x_1,Gx_1),0).$$

By using Proposition 4.1, we obtain $d(x_1,Gx_1) \leq hd(x_0,x_1) < rd(x_0,x_1)$. Now, choose a number μ such that $\inf_{y \in Gx_1} d(x_1,y) = d(x_1,Gx_1) < \mu < rd(x_0,x_1)$. Thus, there exists $x_2 \in Gx_1$ such that $d(x_1,x_2) < \mu < rd(x_0,x_1)$. Since $\alpha d(x_1,Gx_1) < d(x_1,x_2)$, by using the assumption we have

$$d(x_2,Fx_2) \leq H(Fx_2,Gx_1)$$
$$\leq g(d(x_1,x_2),d(x_2,Fx_2),d(x_1,Gx_1),d(x_2,Gx_1),d(x_1,Fx_2))$$

$$\leq g(d(x_1,x_2),d(x_2,Fx_2),d(x_1,x_2),0,d(x_1,x_2)+d(x_2,Fx_2)).$$

By using Proposition 4.1, we obtain $d(x_2,Fx_2) \leq hd(x_1,x_2) < rd(x_1,x_2)$. Now by using a similar method, we can choose $x_3 \in Fx_2$ such that

$$d(x_2,x_3) < rd(x_1,x_2) < r^2d(x_0,x_1).$$

By continuing this process, we obtain a sequence $\{x_n\}_{n\geq 1}$ in X such that

$$x_{2n-1} \in Fx_{2n-2}, x_{2n} \in Gx_{2n-1}, d(x_{2n-1},x_{2n}) < r^n d(x_0,x_1),$$

$$d(x_{2n},Fx_{2n}) \leq hd(x_{2n-1},x_{2n}) \text{ and } d(x_{2n-1},Gx_{2n-1}) \leq hd(x_{2n-2},x_{2n-1}).$$

If $x_m = x_{m+1}$ for some $m \geq 1$, then F and G have a fixed point. Suppose that $x_n \neq x_{n+1}$ for all $n \geq 1$. Since $\{x_n\}_{n\geq 1}$ is a Cauchy sequence, there exists $x \in X$ such that $x_n \to x$. Now, we claim that for each $n \geq 1$ one of the relations $\alpha d(x_{2n},Fx_{2n}) \leq d(x_{2n},x)$ and $\alpha d(x_{2n+1},Gx_{2n+1}) \leq d(x_{2n+1},x)$ hold. If $\alpha d(x_{2n},Fx_{2n}) > d(x_{2n},x)$ and $\alpha d(x_{2n+1},Gx_{2n+1}) > d(x_{2n+1},x)$ for some $n \geq 1$, then we obtain

$$\begin{aligned} d(x_{2n},x_{2n+1}) &\leq d(x_{2n},x)+d(x,x_{2n+1}) \\ &< \alpha d(x_{2n},Fx_{2n})+\alpha d(x_{2n+1},Gx_{2n+1}) \\ &\leq \alpha d(x_{2n},x_{2n+1})+\alpha hd(x_{2n},x_{2n+1}). \end{aligned}$$

Thus, we get $\alpha(h+1) > 1$ which is a contradiction. Therefore, our claim is proved. Now by using the assumption, for each $n \geq 1$ either

$$H(Fx_{2n},Gx) \leq g(d(x_{2n},x),d(x_{2n},Fx_{2n}),d(x,Gx),d(x_{2n},Gx),d(x,Fx_{2n}))$$

or

$$H(Fx_{2n+1},Gx) \leq g(d(x_{2n+1},x),d(x,Fx),d(x_{2n+1},Gx_{2n+1}),d(x,Gx_{2n+1}),d(x_{2n+1},Fx))$$

hold. Therefore, one of the following cases holds:
(i) There exists an infinite subset $I \subseteq \mathbb{N}$ such that

$$\begin{aligned} &d(x_{2n+1},Gx) \\ &\leq H(Fx_{2n},Gx) \leq g(d(x_{2n},x),d(x_{2n},Fx_{2n}),d(x,Gx),d(x_{2n},Gx),d(x,Fx_{2n})) \end{aligned}$$

for all $n \in I$,
(ii) There exists an infinite subset $J \subseteq \mathbb{N}$ such that

$$\begin{aligned} &d(Fx,x_{2n+2}) \leq H(Fx,Gx_{2n+1}) \\ &\leq g(d(x_{2n+1},x),d(x,Fx),d(x_{2n+1},Gx_{2n+1}),d(x,Gx_{2n+1}),d(x_{2n+1},Fx)) \end{aligned}$$

for all $n \in J$.
In the case (i), we obtain

$$d(x,Gx) \leq d(x,x_{2n+1})+d(x_{2n+1},Gx)$$

$$\leq d(x,x_{2n}) + g(d(x_{2n},x), d(x_{2n},Fx_{2n}), d(x,Gx), d(x_{2n},Gx), d(x,Fx_{2n}))$$
$$\leq d(x,x_{2n}) + g(d(x_{2n},x), d(x_{2n},x_{2n+1}), d(x,Gx), d(x_{2n},x) + d(Gx,x), d(x,x_{2n+1}))$$

for all $n \in I$. Since g is continuous we obtain

$$d(x,Gx) \leq g(0,0,d(x,Gx),0+d(x,Gx),0).$$

Now by using Proposition 4.1, we have $d(x,Fx) = 0$ and so $x \in Gx$.
In the case (ii), we obtain

$$d(x,Fx) \leq d(x,x_{2n+2}) + d(Fx,x_{2n+2}) \leq d(x,x_{2n+2})$$
$$+g(d(x_{2n+1},x), d(x,Fx), d(x_{2n+1},Gx_{2n+1}), d(x,Gx_{2n+1}), d(x_{2n+1},Fx))$$
$$\leq d(x,x_{2n+2}) + g(d(x_{2n+1},x), d(x,Fx), d(x_{2n+1},x_{2n+2}), d(x,x_{2n+2}), d(x,Fx)$$
$$+d(x,x_{2n+1}))$$

for all $n \in J$. Since g is continuous, we obtain

$$d(x,Fx) \leq g(0,d(x,Fx),0,0,d(x,Fx)+0).$$

Now by using Proposition 4.1, we have $d(x,Fx) = 0$ and so $x \in Fx$. This completes the proof.

♣

The following result is a consequence of Theorem 4.9.

Theorem 4.10

Let (X,d) be a complete metric space and $T : X \to C(X)$ a set-valued map. Assume that there exist $\alpha \in (0,1)$, $g \in \mathcal{R}$ with $h = g(1,1,1,2,0)$ and $\alpha(h+1) \leq 1$ and $\alpha d(x,Tx) \leq d(x,y)$ implies

$$H(Tx,Ty) \leq g(d(x,y), d(y,Ty), d(x,Tx), d(x,Ty), d(y,Tx))$$

for all $x,y \in X$. Then T has a fixed point.

It is easy to see that the Nadler's fixed point theorem and Theorem 2 in [34] follow directly from Theorem 4.10.

Corollary 4.1

Theorem 4.7 is a special case of Theorem 4.10.

Proof: Define $g \in \mathcal{R}$ by $g(x_1,x_2,x_3,x_4,x_5) = ax_1 + cx_2 + bx_3$. Put $\alpha = \frac{1-b-c}{1+a}$. Since $h = a+b+c$ and $\alpha(1+h) \leq 1$, by using Theorem 4.10, T has a fixed point.

♣

Corollary 4.2

Let (X,d) be a complete metric space and $T : X \to CB(X)$ a set-valued map. Assume that there exists $r \in [2^{\frac{-1}{2}}, 1)$ such that $\theta(r)d(x,Tx) \leq d(x,y)$ implies $H(Tx,Ty) \leq r\max\{d(x,y), d(x,Tx), d(y,Ty)\}$ for all $x,y \in Y$. Then T has a fixed point.

Proof: Define $g \in \mathcal{R}$ by $g(x_1,x_2,x_3,x_4,x_5) = r\max\{x_1,x_2,x_3\}$. Put $\alpha = \theta(r)$. Since $h = r$ and $\alpha(1+h) \leq 1$, by using Theorem 4.10, T has a fixed point.

♣

Corollary 4.3

Let (X,d) be a complete metric space and $T : X \to CB(X)$ a set-valued map. Assume that there exist $\beta, \gamma \in [0,1)$ such that

$$\frac{1}{2\beta+\gamma+1}d(x,Tx) \leq d(x,y) \text{ implies } H(Tx,Ty) \leq \gamma d(x,y) + \beta d(x,Tx) + \beta d(y,Ty)\}$$

for all $x,y \in Y$. Then T has a fixed point.

Proof: Define $g \in \mathcal{R}$ by $g(x_1,x_2,x_3,x_4,x_5) = \gamma x_1 + \beta x_2 + \beta x_3\}$. Put $\alpha = \frac{1}{2\beta+\gamma+1}$. Since $h = 2\beta+\gamma$ and $\alpha(1+h) \leq 1$, by using Theorem 4.10, T has a fixed point.

♣

Here, we present multifunction version results of the stated results in section 5 of chapter 3. Similar to mapping case, we say that $T : X \to CB(X)$ on a metric spaces X is a generalized set-valued α-ψ-contraction map whenever $\alpha(x,y)H(Tx,Ty) \leq \psi(M(x,y))$ for all $x,y \in X$, where H is the Hausdorff metric, $\psi \in \Psi$ is sublinear and

$$M(x,y) = \max\{d(x,y), d(x,Tx), d(y,Ty), \frac{1}{2}\max\{d(x,Ty), d(y,Tx)\}\}.$$

Recall that X X has the property (B) respect to α whenever $\{x_n\}$ is a sequence in X such that $\alpha(x_n,x_{n+1}) \geq 1$ for all $n \geq 1$ and $x_n \to x$, then $\alpha(x_n,x) \geq 1$ for all $n \geq 1$.

Theorem 4.11

Let (X,d) be a complete metric space, $\alpha : X \times X \to [0,+\infty)$ a mapping, $\psi \in \Psi$ and $T : X \to CB(X)$ a generalized α-ψ-contractive multifunction such that $\alpha(x,y) \geq 1$ implies $\alpha(u,v) \geq 1$ for all $u \in Tx$ and $v \in Ty$. Suppose that there exists $x_0 \in X$ and

$x_1 \in Tx_0$ such that $\alpha(x_0,x_1) \geq 1$. If X has the property (B) respect to α, then T has a fixed point.

Proof: Take $x_0 \in X$ and $x_1 \in Tx_0$ such that $\alpha(x_0,x_1) \geq 1$. If $x_0 = x_1$, then x_0 is a fixed point for T. Let $x_0 \neq x_1$. Then,

$$H(Tx_0,Tx_1) \leq \alpha(x_0,x_1)H(Tx_0,Tx_1) \leq \psi(M(x_0,x_1)),$$

where $M(x_0,x_1) = \max\{d(x_0,x_1),d(x_1,Tx_1),\frac{1}{2}d(x_0,Tx_1)\}$ because $x_1 \in Tx_0$. Let $x_1 \notin Tx_1$. Then $M(x_0,x_1) \neq d(x_1,Tx_1)$. Thus, $M(x_0,x_1) = \max\{d(x_0,x_1),\frac{1}{2}d(x_0, Tx_1)\}$. If $M(x_0,x_1) = d(x_0,x_1)$, then $d(x_1,Tx_1) \leq \psi(d(x_0,x_1))$. If $M(x_0,x_1) = \frac{1}{2}d(x_0,Tx_1)$, then $d(x_1,Tx_1) \leq \psi(\frac{1}{2}d(x_0,Tx_1)) \leq \frac{1}{2}\psi(d(x_0,x_1))+\frac{1}{2}\psi(d(x_1,Tx_1))$ and so

$$d(x_1,Tx_1) < \frac{1}{2}\psi(d(x_0,x_1))+\frac{1}{2}d(x_1,Tx_1) \Rightarrow d(x_1,Tx_1) \leq \psi(d(x_0,x_1))$$

because ψ is sublinear. Thus there exists $x_2 \in Tx_1$ such that $d(x_1,x_2) \leq \psi(d(x_0, x_1))$. Let $x_1 \neq x_2$. Then,

$$H(Tx_x,Tx_2) \leq \alpha(x_1,x_2)H(Tx_1,Tx_2) \leq \psi(M(x_1,x_2))$$

and $\alpha(x_1,x_2) \geq 1$, where $M(x_1,x_2) = \max\{d(x_1,x_2),d(x_2,Tx_2),\frac{1}{2}d(x_1,Tx_2)\}$ because $x_2 \in Tx_1$. Let $x_2 \notin Tx_2$. Thus,

$$M(x_1,x_2) = \max\{d(x_1,x_2),\frac{1}{2}d(x_1,Tx_2)\}.$$

If $M(x_1,x_2) = d(x_1,x_2)$, then $d(x_2,Tx_2) \leq \psi(d(x_1,x_2))$. If $M(x_1,x_2) = \frac{1}{2}d(x_1,Tx_2)$, then $d(x_2,Tx_2) < \frac{1}{2}\psi(d(x_1,x_2))+\frac{1}{2}d(x_2,Tx_2)$ because ψ is sublinear. Thus, we get $d(x_2,Tx_2) \leq \psi(d(x_1,x_2))$. This implies that there exists $x_3 \in Tx_2$ such that $d(x_2,x_3) \leq \psi(d(x_1,x_2))$ and so $d(x_2,x_3) \leq \psi^2(d(x_0,x_1))$. By continuing this steps, we obtain a sequence $\{x_n\}_{n\geq 0}$ in X such that $x_{n+1} \in Tx_n$, $\alpha(x_n,x_{n+1}) \geq 1$ and $d(x_n,x_{n+1}) \leq \psi^n d(x_0,x_1)$ for all $n \geq 0$. Fix $\varepsilon > 0$ and choose $n(\varepsilon) \geq 1$ such that $\sum_{n\geq n(\varepsilon)} \psi^n(d(x_0,x_1)) < \varepsilon$. Let $m > n > n(\varepsilon)$. By using the triangular inequality, we obtain

$$d(x_n,x_m) \leq \sum_{k=n}^{m-1} d(x_k,x_{k+1}) \leq \sum_{k=n}^{m-1} \psi^k(d(x_0,x_1)) \leq \sum_{n\geq n(\varepsilon)} \psi^n(d(x_0,x_1)) < \varepsilon.$$

Thus, $\{x_n\}_{n\geq 0}$ is a Cauchy sequence. Hence, there exists $x^* \in X$ such that $x_n \to x^*$. Note that, $\alpha(x_n,x^*) \geq 1$ for all n. Moreover, we have

$$H(Tx_n,Tx^*) \leq \alpha(x_n,x^*)H(Tx_n,Tx^*) \leq \psi(M(x_n,x^*))$$

for all n, where

$$M(x_n,x^*) = \max\{d(x_n,x^*), d(x_n,Tx_n), d(x^*,Tx^*), \frac{1}{2}\max\{d(x_n,Tx^*), d(x^*,Tx_n)\}\}.$$

If $M(x_n,x^*) = d(x_n,x^*)$, then $d(x_{n+1},Tx^*) \leq H(Tx_n,Tx^*) \leq \psi(d(x_n,x^*))$. In the case $M(x_n,x^*) = d(x_n,Tx_n)$, we have

$$d(x_{n+1},Tx^*) \leq H(Tx_n,Tx^*) \leq \psi(d(x_n,Tx_n)) \leq \psi(d(x_n,x_{n+1})).$$

If $M(x_n,x^*) = d(x^*,Tx^*)$, then $x^* \in Tx^*$. In fact, if $x^* \notin Tx^*$, then $d(x^*,Tx^*) > 0$ and so $d(x_{n+1},Tx^*) \leq H(Tx_n,Tx^*) \leq \psi(d(x^*,Tx^*))$. Hence,

$$d(x^*,Tx^*) = \lim_{n\to+\infty} d(x_{n+1},Tx^*) \leq \psi(d(x^*,Tx^*)) < d(x^*,Tx^*)$$

which is a contradiction. If $M(x_n,x^*) = \frac{1}{2}d(x_n,Tx^*)$, then

$$d(x_{n+1},Tx^*) \leq H(Tx_n,Tx^*) \leq \psi(\frac{1}{2}d(x_n,Tx^*)) \leq \frac{1}{2}d(x_n,Tx^*).$$

If $M(x_n,x^*) = \frac{1}{2}d(x^*,Tx_n)$, then we have

$$d(x_{n+1},Tx^*) \leq H(Tx_n,Tx^*) \leq \psi(\frac{1}{2}d(x^*,Tx_n)) \leq \frac{1}{2}\psi(d(x^*,x_{n+1})).$$

Since ψ is continuous at $t = 0$, we get $d(Tx^*,x^*) = 0$ and so $x^* \in Tx^*$.

♣

Next example illustrates last result.

Example 56 *Let $X = [-2,-1] \cup \{0\} \cup [1,\frac{5}{2}]$ and $d(x,y) = |x-y|$ for all $x,y \in X$. Now, consider the set-valued map $T : X \to CB(X)$ defined by*

$$Tx = \begin{cases} [-\frac{x}{4}+2, \frac{5}{2}] & x \in [-2,-\frac{3}{2}) \\ \{0\} & x \in \{-1,0,1\} \cup (2,\frac{5}{2}] \cup [-\frac{3}{2},-1) \\ [-\frac{3}{2}, -\frac{x}{4}-1] & x \in (1,2], \end{cases}$$

Assume that $\psi(t) = \frac{4t}{5}$ for all $t \geq 0$ and $\alpha : X \times X \to [0,+\infty)$ is defined by

$$\alpha(x,y) = \begin{cases} 2 & x,y \in [-2,-\frac{3}{2}) \text{ or } x,y \in (1,2] \\ 1 & x \in [-2,-\frac{3}{2}) \cup (1,2] \text{ and } y \in \{-1,0,1\} \cup (2,\frac{5}{2}] \cup [-\frac{3}{2},-1) \\ \frac{3}{2} & x,y \in \{-1,0,1\} \cup (2,\frac{5}{2}] \cup [-\frac{3}{2},-1) \\ \frac{1}{2} & x \in [-2,-\frac{3}{2}) \text{ and } y \in (1,2] \end{cases}$$

with $\alpha(x,y) = \alpha(y,x)$ for all $x,y \in X$. One can check that (X,d) is a complete metric space, X has the property (B) respect to α and T is a closed and bounded valued generalized α-ψ-contractive multifunction on X. Note that, for $x_0 = -2$ we have $Tx_0 = \{\frac{5}{2}\}$ and $\alpha(-2, \frac{5}{2}) = 1$. Therefore, T satisfy the conditions Theorem 4.11.

Now by mixing our idea with the Suzuki's idea, we provide the next result. We say that the multifunction $T : X \to CB(X)$ on a metric spaces X is a Suzuki-generalized α-ψ-contraction if $\theta(r)d(x,Tx) \leq d(x,y)$ implies $\alpha(x,y)H(Tx,Ty) \leq \psi(M(x,y))$ for all $x,y \in X$, where H is the Hausdorff metric and

$$\theta(r) = \begin{cases} \frac{1}{2} & 0 \leq r \leq \frac{1}{2}(\sqrt{5}-1), \\ \frac{1-r}{r^2} & \frac{1}{2}(\sqrt{5}-1) \leq r \leq \frac{1}{\sqrt{2}}, \\ \frac{1}{1+2r} & \frac{1}{\sqrt{2}} \leq r < 1. \end{cases}$$

Finally, we say that X has the property (C) respect to α whenever for each sequence $\{x_n\}_{n\geq 1}$ in X and $x \in X$ such that $\alpha(x_n,x) \geq 1$ for all n and $x_n \to x^* \in X$ we have $\alpha(x^*,x) \geq 1$.

Theorem 4.12

Let (X,d) be a complete metric space, $\alpha : X \times X \to [0,+\infty)$ a mapping, $\psi \in \Psi$ and $T : X \to CB(X)$ a Suzuki-generalized α-ψ-contractive multifunction such that $\alpha(x,y) \geq 1$ implies $\alpha(u,v) \geq 1$ for all $u \in Tx$ and $v \in Ty$. Suppose that there exists $x_0 \in X$ and $x_1 \in Tx_0$ such that $\alpha(x_0,x_1) \geq 1$. If X has the property (C) respect to α, then, T has a fixed point.

Proof: Take $x_0 \in X$ and $x_1 \in Tx_0$ such that $\alpha(x_0,x_1) \geq 1$. If $x_0 = x_1$, then x_0 is a fixed point of T. Let $x_1 \neq x_0$. Since $\theta(r) \leq 1$, $\theta(r)d(x_0,Tx_0) \leq d(x_0,Tx_0) \leq d(x_0,x_1)$ and so $d(x_1,Tx_1) \leq H(Tx_0,Tx_1) \leq \alpha(x_0,x_1)H(Tx_0,Tx_1) \leq \psi(M(x_0,x_1))$, where $M(x_0,x_1) = \max\{d(x_0,x_1), \frac{1}{2}d(x_1,Tx_1), \frac{1}{2}d(x_0,Tx_1)\}$ because $x_1 \in Tx_0$. Now, let $x_1 \notin Tx_1$. Then $M(x_0,x_1) \neq \frac{1}{2}d(x_1,Tx_1)$. If $M(x_0,x_1) = d(x_0,x_1)$, then $d(x_1,Tx_1) \leq H(Tx_0,Tx_1) \leq \psi(d(x_0,x_1))$. If $M(x_0,x_1) = \frac{1}{2}d(x_0,Tx_1)$, then

$$d(x_1,Tx_1) \leq H(Tx_0,Tx_1) \leq \psi(\frac{1}{2}d(x_0,Tx_1)) \leq \frac{1}{2}\psi(d(x_0,x_1)) + \frac{1}{2}\psi(d(x_1,Tx_1))$$

and so $d(x_1,Tx_1) < \frac{1}{2}\psi(d(x_0,x_1)) + \frac{1}{2}d(x_1,Tx_1)$. Hence,

$$d(x_1,Tx_1) \leq \psi(d(x_0,x_1)).$$

Thus, there exists $x_2 \in Tx_1$ such that $d(x_1,x_2) \leq \psi(d(x_0,x_1))$. By continuing this process, we obtain a sequence $\{x_n\}_{n\geq 0}$ in X such that $x_{n+1} \in Tx_n$, $\alpha(x_n,x_{n+1}) \geq 1$

and $d(x_n,x_{n+1}) \leq \psi^n d(x_0,x_1)$ for all $n \geq 0$. Fix $\varepsilon > 0$ and choose $n(\varepsilon) \geq 1$ such that

$$\sum_{n \geq n(\varepsilon)} \psi^n(d(x_0,x_1)) < \varepsilon.$$

Let $m > n > n(\varepsilon)$. By using the triangular inequality, we obtain

$$d(x_n,x_m) \leq \sum_{k=n}^{m-1} d(x_k,x_{k+1}) \leq \sum_{k=n}^{m-1} \psi^k(d(x_0,x_1)) \leq \sum_{n \geq n(\varepsilon)} \psi^n(d(x_0,x_1)) < \varepsilon.$$

Thus, $\{x_n\}_{n \geq 0}$ is a Cauchy sequence. Therefore, there exists $x^* \in X$ such that $x_n \to x^*$. By using the assumption, we get $\alpha(x_n,x^*) \geq 1$ for all n. Now, we show that $d(x^*,Tx) \leq \psi(d(x^*,x))$ for all $x \in X \setminus \{x^*\}$ with $\alpha(x_n,x) \geq 1$ for all n. Suppose that $x \in X \setminus \{x^*\}$ with $\alpha(x_n,x) \geq 1$ for all n. Since $x_n \to x^*$, there exists $n_0 \in N$ such that $d(x_n,x^*) \leq \frac{1}{3} d(x,x^*)$ for all $n \geq n_0$. Then,

$$\begin{aligned}\theta(r)d(x_n,Tx_n) &\leq d(x_n,Tx_n) \leq d(x_n,x_{n+1}) \leq d(x_n,x^*) + d(x^*,x_{n+1}) \\ &= d(x_n,x^*) + d(x^*,x_{n+1}) \leq \frac{2}{3} d(x,x^*) = d(x,x^*) - \frac{1}{3} d(x,x^*) \\ &\leq d(x,x^*) - d(x^*,x_n) \leq d(x,x_n) \qquad (*)\end{aligned}$$

and so $H(Tx_n,Tx) \leq \alpha(x_n,x)H(Tx_n,Tx) \leq \psi(M(x_n,x))$, where

$$M(x_n,x) = \max\{d(x_n,x), \frac{1}{2}\max\{d(x_n,Tx_n),d(x,Tx)\}, \frac{d(x,Tx_n)+d(x_n,Tx)}{2}\}.$$

Thus, $d(x_{n+1},Tx) \leq \psi(M(x_n,x))$ for all $n \geq n_0$. Therefore, if $M(x_n,x) = d(x_n,x)$ or $M(x_n,x) = \frac{1}{2} d(x_n,Tx_n)$, then by using (*) we have $d(x_{n+1},Tx) \leq \psi(d(x_n,x))$ and so

$$\lim_{n \to \infty} d(x_{n+1},Tx) \leq \lim_{n \to \infty} \psi(d(x_n,x)) \Rightarrow d(x^*,Tx) \leq \psi(d(x^*,x)).$$

If $M(x_n,x) = \frac{1}{2} d(x,Tx)$, then $d(x_{n+1},Tx) \leq \psi(\frac{1}{2} d(x,Tx)) \leq \frac{1}{2}\psi(d(x,x_n)) + \frac{1}{2} d(x_n, Tx)$ and so $\lim_{n \to \infty} d(x_{n+1},Tx) \leq \frac{1}{2} \lim_{n \to \infty} \psi(d(x,x_n)) + \frac{1}{2} \lim_{n \to \infty} d(x_n,Tx)$. Hence,

$$d(x^*,Tx) \leq \psi(d(x^*,x)).$$

If $M(x_n,x) = \frac{d(x,Tx_n)+d(x_n,Tx)}{2}$, then

$$d(x_{n+1},Tx) \leq \psi(\frac{d(x,Tx_n)+d(x_n,Tx)}{2}) \leq \frac{1}{2}\psi(d(x,x_{n+1}) + \frac{1}{2} d(x_n,Tx)$$

and so $d(x^*,Tx) \leq \psi(d(x^*,x))$. Therefore, we proved the claim. Again by using the assumption, we get $\alpha(x,x^*) \geq 1$ and so $H(Tx,Tx^*) \leq \alpha(x,x^*)H(Tx,Tx^*)$.

Now, we show that $H(Tx,Tx^*) \leq \psi(M(x,x^*))$. If $x \neq x^*$, then we get three cases. First, suppose that $0 \leq r \leq \frac{1}{2}(\sqrt{5}-1)$. Then, $\theta(r) = \frac{1}{2}$ and

$$d(x,Tx) \leq d(x,x^*) + d(x^*,Tx) \leq d(x,x^*) + \psi(d(x^*,x)) < 2d(x,x^*)$$

Hence, $\theta(r)d(x,Tx) \leq d(x,x^*)$ and so $H(Tx,Tx^*) \leq \psi(M(x,x^*))$. Now, suppose that $\frac{1}{2}(\sqrt{5}-1) \leq r \leq \frac{1}{\sqrt{2}}$. Then, $\theta(r) = \frac{1-r}{2r^2}$. In this case, for each $n \geq 1$ there exists $y_n \in Tx$ such that $d(x^*,y_n) \leq d(x^*,Tx) + \frac{1}{n}d(x^*,x)$. Thus,

$$d(x,Tx) \leq d(x,y_n) \leq d(x,x^*) + d(x^*,y_n) \leq d(x,x^*) + d(x^*,Tx) + \frac{1}{n}d(x^*,x)$$

$$\leq d(x,x^*) + \psi(d(x,x^*)) + \frac{1}{n}d(x^*,x) < (2 + \frac{1}{n})d(x^*,x)$$

for all $n \geq 1$ and so

$$\theta(r)d(x,Tx) = \frac{1}{2}d(x,Tx) \leq d(x^*,x) \Rightarrow H(Tx,Tx^*) \leq \psi(M(x,x^*)).$$

Finally, suppose that $\frac{1}{\sqrt{2}} \leq r < 1$. Then, $\theta(r) = \frac{1}{1+2r}$. For each $n \geq 1$, there exists an element $z_n \in Tx$ such that $d(x^*,z_n) \leq d(x^*,Tx) + (\frac{1}{4} + \frac{1}{n})d(x^*,x)$. Hence,

$$d(x,Tx) \leq d(x,z_n) \leq d(x,x^*) + d(x^*,z_n) \leq d(x,x^*) + d(x^*,Tx) + (\frac{1}{4} + \frac{1}{n})d(x^*,x)$$

$$\leq d(x,x^*) + \psi(d(x,x^*)) + (\frac{1}{4} + \frac{1}{n})d(x^*,x) < (2 + (\frac{1}{4} + \frac{1}{n}))d(x^*,x)$$

for all n. Thus, $d(x,Tx) \leq (2 + \frac{1}{4})d(x^*,x) = \frac{9}{4}d(x^*,x)$. This implies that

$$\theta(r)d(x,Tx) = \frac{1}{1+2r}d(x,Tx) \leq \frac{4}{9}d(x,Tx) \leq d(x^*,x)$$

and so $H(Tx,Tx^*) \leq \psi(M(x,x^*))$. Thus,

$$d(x^*,Tx^*) = \lim_{n\to\infty} d(x_{n+1},Tx^*) \leq \lim_{n\to\infty} H(Tx_n,Tx^*) \leq \lim_{n\to\infty} \psi(M(x_n,x^*)).$$

If $M(x_n,x^*) = d(x_n,x^*)$, then $d(x^*,Tx^*) \leq \lim_{n\to\infty} \psi(d(x_n,x^*)) = 0$ and so $d(x^*, Tx^*) = 0$. If $M(x_n,x^*) = \frac{1}{2}d(x_n,Tx_n)$, then

$$d(x^*,Tx^*) \leq \lim_{n\to\infty} \psi(\frac{1}{2}d(x_n,Tx_n)) \leq \lim_{n\to\infty} \psi(d(x_n,x_{n+1})) = 0.$$

If $M(x_n,x^*) = \frac{1}{2}d(x^*,Tx^*)$, then $d(x^*,Tx^*) \leq \lim_{n\to\infty} \psi(\frac{1}{2}d(x^*,Tx^*)) < \frac{1}{2}d(x^*,Tx^*)$ which is a contradiction. If $M(x_n,x^*) = \frac{d(x_n,Tx^*)+d(x^*,Tx_n)}{2}$, then

$$d(x^*,Tx^*) \leq \lim_{n\to\infty} \psi(\frac{d(x_n,Tx^*) + d(x^*,Tx_n)}{2}) \leq \frac{1}{2} \lim_{n\to\infty} \psi(d(x_n,x^*))$$

$$+\frac{1}{2}\lim_{n\to\infty}\psi(d(x^*,Tx^*))+\frac{1}{2}\lim_{n\to\infty}\psi(d(x^*,x_{n+1}))\leq\frac{1}{2}\psi(d(x^*,Tx^*)).$$

Thus, $d(x^*,Tx^*)=0$ and so $x^*\in Tx^*$.

♣

4.6 Semi-convex and hemi-convex set-valued mappings

Similar to the last section, our aim is to provide some methods to young researchers to learn how to mix some ideas on set-valued mappings for obtaining new results.

Throughout this section, we suppose that X and Y are Banach spaces and M is a nonempty convex subset of X. Let $T:X\to 2^Y$ be a multifunction. The graph of T is defined by $Gr(T)=\{(x,y):x\in X,y\in T(x)\}$. The multifunction T is called closed (resp., convex) whenever GrT is closed (resp., convex). Also, T is called upper semi-continuous (lower semi-continuous) whenever

$$\{x\in X:T(x)\subset A\}$$

($\{x\in X:T(x)\bigcap A\neq\emptyset\}$) is open for all open subsets A of Y. In 1980, Yanagi has defined the notion of semi-convex multifunctions ([105]). Also, he proved the following results (Theorem 1 and Corollary 2 in [105]).

Theorem 4.13

Let K be a weakly compact and convex subset of a Banach space X, $T:K\longrightarrow 2^K$ a nonexpansive multifunction with compact values and $Tx\subseteq\overline{I_K(x)}$ for all $x\in K$, where $I_K(x)=\{x+\lambda(y-x):y\in K,\lambda\geq 1\}$. If $I-T$ is demiclosed or semi-convex, then T has a fixed point.

Corollary 4.4

Let K be a compact and convex subset of X, $T:K\longrightarrow 2^K$ a nonexpansive multifunction with compact values and $Tx\subseteq\overline{I_K(x)}$ for all $x\in K$. Then T has a fixed point.

Then Bae and Park reviewed some fixed point theorems for multivalued mappings in Banach spaces by using the notion of semi-convex type multifunctions ([18]). They proved the following results (Theorems 1 and 2 in [18]).

Theorem 4.14

Let K be a closed and convex subset of a Banach space X, $T : K \to C(X)$ a contraction multifunction such that $P_T(x) \cap \overline{I_K(x)} \neq \emptyset$ for all $x \in K$, where $P_T(x) = \{y \in K : d(x,y) = d(x,Tx)\}$. Then T has a fixed point.

Theorem 4.15

Let K be a weakly compact and convex subset of a Banach space X, $T : K \to C(X)$ a nonexpansive multifunction such that $P_T(x) \cap \overline{I_K(x)} \neq \emptyset$ for all $x \in K$. If $I - T$ is semi-convex type, then T has a fixed point.

Our aim in this section is to give the notion of hemi-convex multifunctions which is weaker than the convexity of multifunctions. We shall show that this notion is independent of the notion of semi-convex multifunctions. Also, we shall present some results about fixed point of hemi-convex multifunctions by providing some conditions.

Definition 4.1 Let M be a convex subset of a Banach space X and $r > 0$. We say that the multifunction $T : M \longrightarrow 2^M$ is r-hemi-convex whenever

$$d(\lambda x + (1-\lambda)y, T(\lambda x + (1-\lambda)y)) \leq r$$

for all $\lambda \in [0,1]$ and $x, y \in M$ with $d(x, T(x)) < r$ and $d(y, T(y)) < r$. We say that T is hemi-convex whenever T is r-hemi-convex for all $r > 0$.

It is clear that each convex multifunction on a Banach space is a hemi-convex multifunction. Now, by providing the following example we show that the converse is not true.

Example 57 *Define the multifunction $T : \mathbb{R} \longrightarrow 2^{\mathbb{R}}$ by $T(x) = [2x, 3x]$ if $x \geq 0$ and $T(x) = [3x, 2x]$ if $x < 0$. Then T is not convex whereas T is hemi-convex. In fact, $(1,2), (-1,-3) \in Gr(T)$, but for $\lambda = \frac{1}{2}$ we have*

$$\lambda(1,2) + (1-\lambda)(-1,-3) \notin Gr(T).$$

Since $d(x, T(x)) = |x|$ for all $x \in \mathbb{R}$, T is a hemi-convex multifunction.

Let M be a convex subset of a Banach space X. We say that $T : M \to CB(X)$ is semi-convex whenever for each $x, y \in M$, $z = \lambda x + (1-\lambda)y$, where $\lambda \in [0,1]$, and any $x_1 \in Tx$, $y_1 \in Ty$, there exists $z_1 \in Tz$ such that $\|z_1\| \leq \max\{\|x_1\|, \|y_1\|\}$ ([105]). Now by providing next examples, we show that the notions semi-convexity and hemi-convexity are independent although both extend the notion of convexity of multifunctions.

Example 58 *Define the multifunction $T : \mathbb{R} \longrightarrow 2^{\mathbb{R}}$ by $T(x) = \{-x+1\}$ if $x \geq 0$ and $T(x) = [x+1, x+2]$ if $x < 0$. Then T is hemi-convex whereas T is not semi-convex. In fact, let $x = -1$, $y = 1$, $z = \frac{1}{2}x + \frac{1}{2}y = 0$, $x_1 = 0$, $y_1 = 0 \in Ty = \{0\}$ and $z_1 = 1 \in Tz = \{1\}$. Then, the relation $\|z_1\| \leq \max\{\|x_1\|, \|y_1\|\}$ does not hold. Hence, T is not semi-convex. On the other hand, $d(x, T(x)) = |2x-1|$ if $x \geq 0$ and $d(x, T(x)) = 1$ if $x < 0$. Without loss of generality, suppose that $x < y$ and $r > 0$. If $x, y \geq 0$, $d(x, T(x)) < r$ and $d(y, T(y)) < r$, then*

$$d(\lambda x + (1-\lambda)y, T(\lambda x + (1-\lambda)y)) \leq r.$$

If $x, y < 0$, then $d(x, T(x)) = 1$, $d(y, T(y)) = 1$ and

$$d(\lambda x + (1-\lambda)y, T(\lambda x + (1-\lambda)y)) = 1.$$

If $x < 0$, $y \geq 0$ and $\lambda x + (1-\lambda)y < 0$, then $d(x, T(x)) = 1$ and

$$d(\lambda x + (1-\lambda)y, T(\lambda x + (1-\lambda)y)) = 1.$$

If $d(y, T(y)) < r$ and $r \geq 1$, then

$$d(\lambda x + (1-\lambda)y, T(\lambda x + (1-\lambda)y)) \leq \min\{1, r\}.$$

If $x < 0$, $y \geq 0$ and $\lambda x + (1-\lambda)y \geq 0$, then $d(x, T(x)) = 1$, $d(y, T(y)) = |2y-1|$ and $-1 \leq 2(\lambda x + (1-\lambda)y) - 1 \leq 2y - 1$. Thus, the relation

$$d(\lambda x + (1-\lambda)y, T(\lambda x + (1-\lambda)y)) = |2(\lambda x + (1-\lambda)y) - 1| \leq \max\{1, |2y-1|\}$$

implies that T is hemi-convex.

Example 59 *Define the multifunction $T : \mathbb{R} \longrightarrow 2^{\mathbb{R}}$ by $T(x) = [x, x+1]$ if $x > 0$ and $T(x) = \{\sqrt[3]{x}\}$ if $x \leq 0$. Then T is semi-convex whereas T is not hemi-convex. In fact, let $x = 1$, $y = -1$ and $z = \frac{1}{4}x + \frac{3}{4}y = \frac{-1}{2}$. Then, $d(x, T(x)) = d(y, T(y)) = 0$ while*

$$d(z, T(z)) = d(\frac{-1}{2}, -\sqrt[3]{\frac{1}{2}}) > 0.$$

Hence, T is not hemi-convex. Now, without loss of generality suppose that $x < y$. If $x, y > 0$ or $x, y < 0$ and $z = (\lambda x + (1-\lambda)y)$, it is easy to see that for each $x_1 \in Tx$ and $y_1 \in Ty$, there exists $z_1 \in Tz$ such that $\|z_1\| \leq \max\{\|x_1\|, \|y_1\|\}$. If $x \leq 0$, $y > 0$ and $z = \lambda x + (1-\lambda)y \leq 0$, then for each $x_1 = \sqrt[3]{x} \in Tx$ and $y_1 \in Ty$ we have $\sqrt[3]{x} = x_1 \leq z_1 = \{\sqrt[3]{\lambda x + (1-\lambda)y}\} \leq 0 < y_1$. Hence, $\|z_1\| \leq \max\{\|x_1\|, \|y_1\|\}$. If $x \leq 0$, $y > 0$ and $z = \lambda x + (1-\lambda)y > 0$, then for each $x_1 = \sqrt[3]{x} \in Tx$ and $y_1 \in Ty$, there exists $z_1 \in Tz$ such that $\sqrt[3]{x} = x_1 \leq 0 < z_1 \leq y_1$. Hence, $\|z_1\| \leq \max\{\|x_1\|, \|y_1\|\}$. Therefore, T is semi-convex.

Theorem 4.16

Let $T, T_n : M \longrightarrow CB(M)$ be given. If T_n is a hemi-convex set-valued mapping for all $n \geq 1$ and $H(T_n(x), T(x)) \longrightarrow 0$ for all $x \in M$, then T is a hemi-convex multifunction.

Proof: Fix $\varepsilon > 0, r > 0, 0 \leq \lambda \leq 1$ and $x, y \in M$ with $d(x, T(x)) < r, d(y, T(y)) < r$. Choose a natural number N such that

$$H(T_n(x), T(x)) < \varepsilon, H(T_n(y), T(y)) < \varepsilon, H(T_n(\lambda x + (1-\lambda)y), T(\lambda x + (1-\lambda)y)) < \varepsilon$$

for all $n \geq N$. Then, for each $n \geq N$ we have

$$d(x, T_n(x)) \leq d(x, T(x)) + H(T_n(x), T(x)) < r + \varepsilon$$

and

$$d(y, T_n(y)) \leq d(y, T(y)) + H(T_n(y), T(y)) < r + \varepsilon.$$

Thus, $d(\lambda x + (1-\lambda)y, T_n(\lambda x + (1-\lambda)y)) \leq r + \varepsilon$. Hence, for each $n \geq N$ we have

$$d(\lambda x + (1-\lambda)y, T(\lambda x + (1-\lambda)y)) \leq$$

$$d(\lambda x + (1-\lambda)y, T_n(\lambda x + (1-\lambda)y)) + H(T_n(\lambda x + (1-\lambda)y), T(\lambda x + (1-\lambda)y)) < r + 2\varepsilon.$$

Since ε was arbitrary, we obtain $d(\lambda x + (1-\lambda)y, T(\lambda x + (1-\lambda)y)) \leq r$. Therefore, T is a hemi-convex multifunction.

♣

Theorem 4.17

Let $T : M \to CB(M)$ be an upper-semi continuous hemi-convex multifunction. Then the set of fixed points of T is convex and closed.

Proof: Set $F = \{x : x \in T(x)\}$. For each $x, y \in F$ we have $d(x, Tx) = 0$ and

$$d(y, Ty) = 0.$$

Thus, $d(T(\lambda x + (1-\lambda)y), \lambda x + (1-\lambda)y) = 0$ and so $\lambda x + (1-\lambda)y \in F$ for all $\lambda \in [0, 1]$, because T is a closed-valued multifunction. Since T is upper-semi continuous and closed-valued, $Gr(T)$ is closed. Let $\{x_n\}_{n \geq 1}$ be a sequence in F with $x_n \to x$. Since $x_n \in Tx_n$, $(x_n, x_n) \in Gr(T)$. Hence, $(x, x) \in Gr(T)$ and so $x \in F$.

♣

Definition 4.2 Let M be a convex subset of a Banach space X and $r > 0$. We say that the function $f : X \longrightarrow \mathbb{R}$ is r-hemi-convex on M whenever

$$f(\lambda x + (1-\lambda)y) < r$$

for all $\lambda \in [0, 1]$ and $x, y \in M$ with $f(x) < r$ and $f(y) < r$. We say that f is hemi-convex on M whenever f is r-hemi-convex on M for all $r > 0$.

Lemma 4.4

Let M be a convex subset of a Banach space X, $\delta > 0$, $m \geq 2$ and also $f : X \to \mathbb{R}$ a hemi-convex function on M. If $x_1, \cdots, x_m \in M$ with $f(x_i) < \delta$ for $i = 1, \cdots, m$ and $\lambda_1, \cdots, \lambda_m \in [0, \infty)$ with $\sum_{i=1}^{m} \lambda_i = 1$, then $f(\sum_{i=1}^{m} \lambda_i x_i) < \delta$.

Proof: We prove this by induction. For $m = 2$ we have nothing to prove. Suppose that this Lemma holds for each $1 \leq k \leq m-1$. We have to prove it for m. Note that, one can assume $\lambda_1 \neq 0$ and so

$$f(\sum_{i=1}^{m} \lambda_i x_i) = f(\lambda_1 x_1 + \sum_{i=2}^{m} \lambda_i x_i) = f(\lambda_1 x_1 + (1-\lambda_1) \sum_{i=2}^{m} \frac{\lambda_i}{(1-\lambda_1)} x_i).$$

Put $y = \sum_{i=2}^{m} \frac{\lambda_i}{(1-\lambda_1)} x_i$. Since $\sum_{i=2}^{m} \frac{\lambda_i}{(1-\lambda_1)} = 1$, by assumption of the induction, $f(y) < \delta$. Now, by the case of $m = 2$, we obtain

$$f(\sum_{i=1}^{m} \lambda_i x_i) = f(\lambda_1 x_1 + (1-\lambda_1) \sum_{i=2}^{m} \frac{\lambda_i}{(1-\lambda_1)} x_i) = f(\lambda_1 x_1 + (1-\lambda_1) y) < \delta.$$

This completes the proof.

♣

Theorem 4.18

Let M be a weakly compact subset of X, $T : M \to CB(X)$ a set-valued mapping and $\inf_{x \in M} d(x, Tx) = 0$. If the function $f : M \to [0, \infty)$, defined by $f(x) = d(x, Tx)$, is lower semi-continuous and hemi-convex on M, then T has a fixed point in M.

Proof: Choose a sequence $\{x_n\}_{n \geq 1}$ in M such that $d(x_n, Tx_n) \longrightarrow 0$. Since M is weakly compact, there exists a subsequence $\{z_n\}_{n \geq 1}$ of $\{x_n\}_{n \geq 1}$ such that $z_n \xrightarrow{w} x_0$ for some $x_0 \in M$. Since f is a lower semi-continuous function, for each $\varepsilon > 0$ choose $\delta > 0$ such that $f(x_0) < f(y) + \frac{\varepsilon}{2}$ for all $y \in M$ with $\|y - x_0\| < \delta$ ([88]). Since $f(z_n) \to 0$, there exists a natural number N such that $f(z_n) < \frac{\varepsilon}{2}$ for all $n \geq N$. We denote again the sequence $\{z_n\}_{n \geq N}$ by $\{z_n\}_{n \geq 1}$. Since $z_n \xrightarrow{w} x_0$, there exist a sequence $\{y_i\}_{i \geq 1}$ in M and a sequence $\{\alpha_{in}\}_{i,n \geq 1}$ in $[0, \infty)$ such that for each i we have $y_i = \sum_{n=1}^{\infty} \alpha_{in} z_n$, where $\sum_{n=1}^{\infty} \alpha_{in} = 1$ and only finitely many $\{\alpha_{in}\}$ are not zero, and $y_i \to x_0$ originally (Theorem 3.13 in [89]). By using Lemma 4.4, we have $f(y_i) < \frac{\varepsilon}{2}$ for all $i \geq 1$. Thus, for sufficiently large i, we obtain $f(x_0) < f(y_i) + \frac{\varepsilon}{2} < \varepsilon$. Hence, $f(x_0) = 0$ and so $x_0 \in Tx_0$.

♣

If $T : M \to CB(M)$ is an upper semi-continuous multifunction, then the function $f(x) = d(x, T(x))$ is lower semi-continuous (Proposition 4.2.6 in [7]). Note that, the function $f(x) = d(x, T(x))$ is hemi-convex whenever T so is. We say that the function $f(x) = d(x, T(x))$ has the property (B) whenever $f(x_n) \to \infty$ for all sequences $\{x_n\}$ with $\|x_n\| \to \infty$. The following example shows that weakly compactness of M is a necessary condition in Theorem 4.18.

Example 60 *Consider the multifunction $T : (0,\infty) \to 2^{(0,\infty)}$ given by*

$$T(x) = \{x + \frac{1}{x}\}.$$

It is clear that T is a hemi-convex multifunction, $\inf_{x\in(0,\infty)} d(x, T(x)) = 0$ and the function $f(x) = d(x, T(x))$ is lower semi-continuous and hemi-convex. But it is clear that T has no fixed point.

The following example shows that there are many multifunctions which satisfy in the condition $\inf_{x\in M} d(x, T(x)) = 0$.

Example 61 *Let M be a convex and bounded subset of a Banach space X, $u \in M$ a fixed element and $T : M \to 2^M$ a nonexpansive multifunction. For each $n \geq 2$ define $T_n : M \to 2^M$ by $T_n x = \frac{1}{n}u + (1 - \frac{1}{n})Tx$. Since $H(T_n x, T_n y) \leq (1 - \frac{1}{n})\|x - y\|$ for all $x, y \in M$ and $n \geq 2$, T_n is a contraction multifunction and so for each $n \geq 2$ there exists $x_n \in M$ such that $x_n \in T_n x_n$. Note that*

$$d(x_n, Tx_n) \leq d(x_n, T_n x_n) + H(Tx_n, T_n x_n) = H(Tx_n, T_n x_n),$$

for all $n \geq 2$. On the other hand, $H(Tx_n, T_n x_n) \to 0$ because M is bounded. Hence, $d(x_n, Tx_n) \to 0$ and so $\inf_{x\in M} d(x, T(x)) = 0$.

Definition 4.3 Let M be a convex subset of a Banach space X and $T_n, T : M \to 2^M$ a sequence of multifunctions. We say that $\{T_n\}$ strongly converges to T whenever for each $\varepsilon > 0$ there exists a natural number n_0 such that $H(T_n(x), T(x)) < \varepsilon$ for all $n \geq n_0$ and $x \in M$. In this case, we write $T_n \twoheadrightarrow T$.

Theorem 4.19

Let M be a weakly compact subset of X, $T : M \longrightarrow CB(M)$ a multifunction and $T_n : M \longrightarrow CB(M)$ an upper semi-continuous hemi-convex multifunction for all $n \geq 1$. If each T_n has at least one fixed point in M and $T_n \twoheadrightarrow T$, then T has a fixed point.

Proof: Since each T_n has at least one fixed point in M, $\inf_{x\in M} d(x, T_n(x)) = 0$ for all $n \geq 1$. Let $\varepsilon > 0$ be given. Choose a natural number n_0 such that $H(T_n(x), T(x)) < \varepsilon$ for all $n \geq n_0$. Since

$$d(x, T(x)) \leq d(x, T_n(x)) + H(T_n(x), T(x)) \leq d(x, T_n(x)) + \varepsilon,$$

for all $n \geq n_0$, $\inf_{x \in M} d(x, T(x)) \leq \varepsilon$. Hence, we get $\inf_{x \in M} d(x, T(x)) = 0$. Now by using Theorem 4.16, T is hemi-convex and so is the function $f(x) = d(x, T(x))$. Since T is upper semi-continuous, the function $f(x) = d(x, T(x))$ is lower semi-continuous. Now by using Theorem 4.18, T has a fixed point.

♣

The following example shows that strongly convergence of the sequence $\{T_n\}_{n \geq 1}$ is a necessary condition in Theorem 4.19.

Example 62 *Let $X = \mathbb{R}$ and $M = [0,2]$. Define $T : M \to CB(M)$ by $T(x) = \{x+1\}$ if $x < 1$, $T(x) = \{x-1\}$ if $x > 1$ and $T(x) = \{0,2\}$ if $x = 1$. Moreover for each $n \geq 2$, let $T_n : M \longrightarrow CB(M)$ be defined by $T_n(x) = T(x)$ if $x \neq \frac{1}{n}$ and $T_n(x) = [0,2]$ if $x = \frac{1}{n}$. It is easily seen that T_n is upper semi-continuous, $d(x, T_n(x)) = 1$ if $x \neq \frac{1}{n}$, $d(x, T_n(x)) = 0$ if $x = \frac{1}{n}$ and T_n has a fixed point for each $n \geq 2$. This implies that T_n is hemi-convex. Evidently $H(Tn(x), T(x)) \to 0$ for all $x \in M$, but T has no fixed point.*

Theorem 4.20

Let X be an uniformly convex Banach space and $T : X \to CB(X)$ an upper semi-continuous hemi-convex multifunction and $\inf_{x \in M} d(x, T(x)) = 0$. If the function $f(x) = d(x, T(x))$ has the property (B), then T has a fixed point.

Proof: Choose a sequence $\{x_n\}$ in X such that $f(x_{n+1}) \leq f(x_n)$ and $f(x_n) \to 0$. Now, for each $n \geq 1$ define $F_n = \{x \in X : f(x) \leq f(x_n)\}$. Since the function $f(x) = d(x, T(x))$ has the property (B), each F_n is nonempty bounded subset of X. Since T is upper semi-continuous, the function $f(x) = d(x, T(x))$ is lower semi-continuous and so each F_n is a closed subset of X. Also, each F_n is convex because T is a hemi-convex multifunction. Now by using Theorem 2.3.14 in [7], there exists $x_0 \in X$ such that $x_0 \in \bigcap_{n=1}^{\infty} F_n$. Thus, $f(x_0) \leq f(x_n)$ for all $n \geq 1$. Hence, $f(x_0) = 0$ and so $x_0 \in T(x_0)$.

♣

4.7 Fixed points of set-valued mappings on ordered generalized metric spaces

In this section, our aim is to present some fixed point results about set-valued mappings on ordered generalized metric spaces. For more variety in this section, we will use new symbols.

Let (X,ρ) be a metric space. We denote the set of all nonempty closed subsets of X by $P_{cl}(X)$. Also, we denote the set of fixed points of a set-valued map T by $Fix(T)$. Let X be a nonempty set and consider the space $\mathbb{R}^m_+$ endowed with the usual component-wise partial order. A map $d : X \times X \to \mathbb{R}^m_+$ which satisfies all the usual axioms of the metric is called a generalized metric in the sense of Perov ([25]). If $v, r \in \mathbb{R}^m$, $v := (v_1, v_2, \cdots, v_m)$ and $r := (r_1, r_2, \cdots, r_m)$, then by $v \leq r$ we mean that $v_i \leq r_i$ for all $i \in \{1,2,\cdots,m\}$ while $v < r$ stands for $v_i < r_i$ for all $i \in \{1,2,\cdots,m\}$. Also, $|v| := (|v_1|, |v_2|, \cdots, |v_m|)$, $\max(v,r) := (\max(v_1,r_1), \cdots, \max(v_m,r_m))$ and if c is a real number, then $v \leq c$ means $v_i \leq c$ for all $i \in \{1,2,\cdots,m\}$. In a generalized metric space in the sense of Perov, the concepts of Cauchy sequence, convergent sequence and completeness are similarly defined as those in a metric space. We denote by $M_{m,m}(\mathbb{R}_+)$ the set of all $m \times m$ matrices with positive elements and by I the identity $m \times m$ matrix. A matrix $A \in M_{m,m}(\mathbb{R}_+)$ is said to be convergent to zero whenever $A^n \to 0$. We need next result which has been proved in ([90]).

Theorem 4.21

Let $A \in M_{m,m}(\mathbb{R}_+)$. The following are equivalents:
(i) $A^n \to 0$;
(ii) The eigenvalues of A are in the open unit disc, i.e. $|\lambda| < 1$, for all $\lambda \in C$ with $det(A - \lambda I) = 0$;
(iii) The matrix $I - A$ is non-singular and $(I-A)^{-1} = I + A + \cdots + A^n + \cdots$;
(iv) The matrix $I - A$ is non-singular and $(I-A)^{-1}$ has nonnegative elements;
(v) $A^n q \to 0$ and $qA^n \to 0$, for all $q \in \mathbb{R}^m$.

If $(X, \leq)$ is a partially ordered set, then we define

$$X_{\leq} = \{(x,y) \in X \times X : x \leq y \text{ or } y \leq x\}.$$

Let X be a nonempty set and $T : X \to 2^X$ be a set-valued operator. We set

$$(T \times T)(x,y) = \{(u,v) : u \in Tx,\ v \in Ty\}$$

for all $x, y \in X$. Note that, for each $x \in X$ there exists $b_x \in \mathbb{R}^m_+$ such that $b_x \leq d(x,y)$ for all $y \in Tx$. At least, we can set $b_x = 0$. Now, for each $x \in X$ we denote largest of these vectors by $d(x,Tx)$, that is, $d(x,Tx)$ is a vector in $\mathbb{R}^m_+$ such that $d(x,Tx) \leq d(x,y)$ for all $y \in Tx$ and $b_x \leq d(x,Tx)$ for all $b_x \in \mathbb{R}^m_+$ with $b_x \leq d(x,y)$ for all $y \in Tx$.

We say that $(X,d,\leq)$ is an ordered generalized metric space whenever (X,d) is a generalized metric space in the Perov sense and $(X,\leq)$ is a partially ordered set.

Theorem 4.22

Let $(X,d,\leq)$ be a complete ordered generalized metric space, A a matrix in $M_{m,m}(\mathbb{R}_+)$ convergent to zero and $T: X \to P_{cl}(X)$ a set-valued operator. Suppose that $(T\times T)(X_{\leq}) \subseteq X_{\leq}$ and
(i) For each $(x,y) \in X_{\leq}$ and $u \in T(x)$ there exist $v \in T(y)$ and $L(x,y) \in \mathcal{A}_{x,y}$ such that $d(u,v) \leq A\,L(x,y)$, where $\mathcal{A}_{x,y} = \{d(x,y), d(x,Tx), d(y,Ty)\}$,
(ii) For each sequence $\{x_n\}_{n\geq 1}$ in X with $x_n \to x$, there exists a subsequence $\{x_{n_k}\}_{k\geq 1}$ of $\{x_n\}_{n\geq 1}$ such that $(x_{n_k}, x) \in X_{\leq}$ for all $k \geq 1$,
(iii) There exist $x_0, x_1 \in X$ such that $(x_0,x_1) \in X_{\leq}$ and $x_1 \in Tx_0$.
Then T has a fixed point.

Proof: If $x_0 = x_1$, then x_0 is a fixed point of T. Let $x_1 \neq x_0$. By (i), there exist $x_2 \in Tx_1$ and $L(x_0,x_1) \in \mathcal{A}_{x_0,x_1}$ such that $d(x_1,x_2) \leq AL(x_0,x_1)$, where

$$\mathcal{A}_{x_0,x_1} = \{d(x_0,x_1), d(x_0,Tx_0), d(x_1,Tx_1)\}.$$

If $L(x_0,x_1) = d(x_1,Tx_1)$, then $d(x_1,x_2) \leq Ad(x_1,Tx_1) \leq Ad(x_1,x_2)$ and so

$$(I-A)d(x_1,x_2) \leq 0 \Rightarrow d(x_1,x_2) = 0 \Rightarrow x_1 = x_2.$$

If $L(x_0,x_1) = d(x_0,x_1)$ or $L(x_0,x_1) = d(x_0,Tx_0)$, then

$$d(x_1,x_2) \leq Ad(x_0,x_1). \tag{4.18}$$

Since $(x_0,x_1) \in X_{\leq}$, $x_1 \in Tx_0$, $(T\times T)(X_{\leq}) \subseteq X_{\leq}$ and $x_2 \in Tx_1$, $(x_1,x_2) \in X_{\leq}$. Now, by using (i) there exist $x_3 \in Tx_2$ and $L(x_1,x_2) \in \mathcal{A}_{x_1,x_2}$ such that

$$d(x_2,x_3) \leq AL(x_1,x_2).$$

If $L(x_1,x_2) = d(x_2,Tx_2)$, then

$$d(x_2,x_3) \leq Ad(x_2,Tx_2) \leq Ad(x_2,x_3) \Rightarrow x_2 = x_3.$$

If $L(x_1,x_2) = d(x_1,x_2)$ or $L(x_0,x_1) = d(x_1,Tx_1)$, then by using (4.18) we have

$$d(x_2,x_3) \leq Ad(x_1,x_2) \leq A^2 d(x_0,x_1).$$

Now by induction, we construct a sequence $\{x_n\}_{n\geq 0}$ in X which has these properties:
(a) $x_{n+1} \in Tx_n$ for all $n \geq 0$,
(b) $(x_n,x_{n+1}) \in X_{\leq}$ for all $n \geq 0$,
(c) $d(x_n,x_{n+1}) \leq A^n d(x_0,x_1)$ for all $n \geq 0$.
Now by using these properties and Theorem 4.21, we obtain

$$d(x_n,x_{n+p}) \leq d(x_n,x_{n+1}) + d(x_{n+1},x_{n+2}) + \cdots + d(x_{n+p-1},x_{n+p})$$

$$\leq A^n d(x_0,x_1) + A^{n+1} d(x_0,x_1) + \cdots + A^{n+p-1} d(x_0,x_1)$$
$$\leq A^n (I + A + A^2 + \cdots + A^{p-1}) d(x_0,x_1)$$
$$\leq A^n (I - A)^{-1} d(x_0,x_1) \to 0 \ \ (n \to \infty).$$

Hence, $\{x_n\}_{n\geq 0}$ is a Cauchy sequence in the complete metric space (X,d). Choose $x^* \in X$ such that $x_n \to x^*$. By (ii), there exists a subsequence $\{x_{n_k}\}_{k\geq 1}$ of $\{x_n\}_{n\geq 0}$ such that $(x_{n_k}, x^*) \in X_{\leq}$ for all $k \geq 1$. But, $x_{n_k} \in Tx_{n_k-1}$, $(x_{n_k-1}, x^*) \in X_{\leq}$ for all $n \geq 1$. Thus by using (i), for each $k \geq 1$ there exist $v_{n_k} \in Tx^*$ and $L(x_{n_k-1}, x^*) \in \mathcal{A}_{x_{n_k-1}, x^*}$ such that

$$d(v_{n_k}, x_{n_k}) \leq AL(x_{n_k-1}, x^*).$$

If $L(x_{n_k-1}, x^*) = d(x_{n_k-1}, x^*)$, then $d(v_{n_k}, x_{n_k}) \leq Ad(x_{n_k-1}, x^*)$. Hence,

$$d(v_{n_k}, x^*) \leq d(v_{n_k}, x_{n_k}) + d(x_{n_k}, x^*) \leq Ad(x_{n_k-1}, x^*) + d(x_{n_k-1}, x^*) \to 0 \ (k \to \infty).$$

If $L(x_{n_k-1}, x^*) = d(x_{n_k-1}, Tx_{n_k-1})$, then $d(v_{n_k}, x_{n_k}) \leq Ad(x_{n_k-1}, x_{n_k})$. Hence,

$$d(v_{n_k}, x^*) \leq d(v_{n_k}, x_{n_k}) + d(x_{n_k}, x^*) \leq Ad(x_{n_k-1}, x_{n_k}) + d(x_{n_k}, x^*) \to 0 \ (k \to \infty).$$

If $L(x_{n_k-1}, x^*) = d(x^*, Tx^*)$, then $d(v_{n_k}, x_{n_k}) \leq Ad(v_{n_k}, x^*)$. Hence,

$$d(v_{n_k}, x^*) \leq d(v_{n_k}, x_{n_k}) + d(x_{n_k}, x^*) \leq Ad(v_{n_k}, x^*) + d(x_{n_k}, x^*)$$
$$\Rightarrow (I - A) d(v_{n_k}, x^*) \leq d(x_{n_k}, x^*) \to 0 \ \ (k \to \infty).$$

Thus, $v_{n_k} \to x^*$ $(k \to \infty)$. Since $u_{n_k} \in Tx^*$ for all $k \geq 1$ and Tx^* is a closed subset of X, $x^* \in Tx^*$.

♣

Next example shows advantage of Theorem 4.22 respect to Theorem 3.3 in [92].

Example 63 *Assume that* $X = [-2,-1] \cup [1,2] \cup \{0\}$, $r = \frac{4}{5}$, $A = rI_{2\times 2}$, $k > 0$ *and the metric* $d : X \times X \to \mathbb{R}^2$ *is defined by* $d(x,y) = (|x-y|, k|x-y|)$ *for all* $x, y \in X$. *Then* (X,d) *is a generalized metric space. Define the multivalued mapping* $T : X \to X$ *by* $Tx = [-\frac{x}{4} + 2, \frac{5}{2}]$ *whenever* $x \in [-2,-1)$, $Tx = \{0\}$ *whenever* $x \in \{-1,0,1\}$ *and* $Tx = [\frac{3}{2}, -\frac{x}{4} + 2]$ *whenever* $x \in (1,2]$. *We show that* T *satisfies the assumptions of Theorem 4.22 while it does not satisfy the assumptions of Theorem 3.3 in [92]. In this way, note that if* $x \in \{-1,0,1\}$, *then* $d(x,Tx) = (|x|, k|x|)$ *and if* $x \in [-2,-1)$ *or* $x \in (1,2]$, *then* $d(x,Tx) = (|\frac{5x-8}{4}|, k|\frac{5x-8}{4}|)$. *Let* $x, y \in [-2,-1)$, $x \leq y$ *and* $u \in Tx$. *Then, for each* $v \in Ty$ *we have* $|u - v| \leq \frac{|y+2|}{4} \leq \frac{1}{5}\frac{|5y-8|}{4} \leq r\frac{|5y-8|}{4}$, *and so* $d(u,v) \leq Ad(y,Ty)$. *Suppose that* $x \in [-2,-1)$, $y \in \{-1,0,1\}$ *and* $u \in Tx$. *Then, for each* $v \in Ty$ *we have*

$$|u - v| \leq \frac{5}{2} \leq \frac{13}{5} \leq r\frac{|5x-8|}{4}$$

and so $d(u,v) \leq Ad(x,Tx)$. Let $x \in [-2,-1)$, $y \in (1,2]$ and $u \in Tx$. Then, for each $v \in Ty$ we have $|u-v| \leq 1 \leq \frac{1}{2}|x-y| \leq \frac{4}{5}|x-y| = r|x-y|$, and so $d(u,v) \leq Ad(x,y)$. Thus T satisfies the assumptions of Theorem 4.22. If $x = -\frac{3}{2}$ and $y = -1$, then $Tx = [\frac{19}{8}, \frac{5}{2}]$, $Ty = \{0\}$ and for each $u \in [\frac{19}{8}, \frac{5}{2}]$ and $v = 0$, we have $|u-v| \nleq r|x-y| \Rightarrow d(u,v) \nleq Ad(x,y)$. Hence, T does not satisfy the assumptions of Theorem 3.3 in [92].

Theorem 4.23

Assume that (X,d) is a complete generalized metric space, $\theta \in (0,1)$ and $T : X \to P_{cl}(X)$ is a set-valued operator. Let $\varphi : \mathbb{R}^m_+ \to \mathbb{R}^m_+$ be an increasing sublinear function such that $\varphi(0) = 0$, $\varphi(t) < t$ and $\sum_{n=1}^{\infty} \varphi^n(t) < \infty$ for all $t = (t_i)_{i=1}^m \in \mathbb{R}^m_{++}$. Also, suppose that for each $x,y \in X$ and $u \in T(x)$ there exist $v \in T(y)$ and $M(x,y) \in \mathcal{B}_{x,y}$ such that

$$d(u,v) \leq \varphi(M(x,y)), \tag{4.19}$$

where

$$\mathcal{B}_{x,y} = \{d(x,y), d(x,Tx), \theta d(y,Ty), \frac{d(x,Tx)+d(y,Ty)}{2}, \frac{d(x,Ty)+d(y,Tx)}{2}\}.$$

Then T has a fixed point.

Proof: Let $x_0 \in X$ be arbitrary and take $x_1 \in Tx_0$. If $x_0 = x_1$, then x_0 is a fixed point of T. Let $x_1 \neq x_0$. By using (4.19), there exist $x_2 \in Tx_1$ and $M(x_0,x_1) \in \mathcal{B}_{x_0,x_1}$ such that $d(x_1,x_2) \leq \varphi(M(x_0,x_1))$. If $x_1 = x_2$, then x_1 is a fixed point of T. Let $x_1 \neq x_2$. We show that $d(x_1,x_2) \leq \varphi(d(x_0,x_1))$. If $M(x_0,x_1) = d(x_0,x_1)$, then

$$d(x_1,x_2) \leq \varphi(d(x_0,x_1)). \tag{4.20}$$

If $M(x_0,x_1) = d(x_0,Tx_0)$, then (4.20) holds because $x_1 \in Tx_0$. We claim that $M(x_0,x_1) \neq \theta d(x_1,Tx_1)$. In fact, if $M(x_0,x_1) = \theta d(x_1,Tx_1)$, then

$$d(x_1,x_2) \leq \varphi(\theta d(x_1,Tx_1)) \leq \varphi(\theta d(x_1,x_2)) < \theta d(x_1,x_2),$$

which is a contradiction. If $M(x_0,x_1) = \frac{d(x_0,Tx_0)+d(x_1,Tx_1)}{2}$, then

$$d(x_1,x_2) \leq \varphi(\frac{d(x_0,Tx_0)+d(x_1,Tx_1)}{2}) \leq \frac{1}{2}\varphi(d(x_0,x_1)) + \frac{1}{2}\varphi(d(x_1,x_2))$$
$$< \frac{1}{2}\varphi(d(x_0,x_1)) + \frac{1}{2}d(x_1,x_2),$$

because $x_1 \in Tx_0$, $x_2 \in Tx_1$ and φ is sublinear. Hence, $d(x_1,x_2) < \varphi(d(x_0,x_1))$. If $M(x_0,x_1) = \frac{d(x_0,Tx_1)+d(x_1,Tx_0)}{2} = \frac{d(x_0,Tx_1)}{2}$, then by a similar way we obtain

$d(x_1,x_2) \leq \varphi(d(x_0,x_1))$. Thus, $d(x_1,x_2) \leq \varphi(d(x_0,x_1))$ holds. Now by using (4.19), there exists $x_3 \in Tx_2$ and $M(x_1,x_2) \in \mathcal{B}_{x_1,x_2}$ such that $d(x_2,x_3) \leq \varphi(M(x_1,x_2))$. If $x_2 = x_3$, then x_2 is a fixed point of T. Suppose that $x_2 \neq x_3$. Now, we show that $d(x_2,x_3) \leq \varphi^2(d(x_0,x_1))$. If $M(x_1,x_2) = d(x_1,x_2)$, then by using (4.20) we obtain

$$d(x_2,x_3) \leq \varphi(d(x_1,x_2)) \leq \varphi^2(d(x_0,x_1)). \tag{4.21}$$

If $M(x_1,x_2) = d(x_1,Tx_1)$, then (4.21) holds because $x_2 \in Tx_1$. We claim that $M(x_1,x_2) \neq \theta d(x_2,Tx_2)$. In fact, if $M(x_1,x_2) = \theta d(x_2,Tx_2)$, then

$$d(x_2,x_3) \leq \varphi(\theta d(x_2,Tx_2)) \leq \varphi(\theta d(x_2,x_3)) < \theta d(x_2,x_3),$$

which is a contradiction. If $M(x_1,x_2) = \frac{d(x_1,Tx_1)+d(x_2,Tx_2)}{2}$, then

$$\begin{aligned} d(x_2,x_3) &\leq \varphi(\frac{d(x_1,Tx_1)+d(x_2,Tx_2)}{2}) \leq \frac{1}{2}\varphi(d(x_1,x_2)) + \frac{1}{2}\varphi(d(x_2,x_3)) \\ &< \frac{1}{2}\varphi(d(x_1,x_2)) + \frac{1}{2}d(x_2,x_3) \end{aligned}$$

because $x_2 \in Tx_1$, $x_3 \in Tx_2$ and φ is sublinear. Hence,

$$d(x_2,x_3) < \varphi(d(x_1,x_2)) \leq \varphi^2(d(x_0,x_1)).$$

If $M(x_1,x_2) = \frac{d(x_1,Tx_2)+d(x_2,Tx_1)}{2}$, then by a similar way we obtain

$$d(x_2,x_3) \leq \varphi(d(x_1,x_2)) \leq \varphi^2(d(x_0,x_1)).$$

Thus, $d(x_2,x_3) \leq \varphi^2(d(x_0,x_1))$ holds. Now, by induction we construct a sequence $\{x_n\}_{n\geq 0}$ in X which has the following properties:
(a) $x_{n+1} \in Tx_n$ for all $n \geq 0$,
(b) $d(x_n,x_{n+1}) \leq \varphi^n(d(x_0,x_1))$ for all $n \geq 0$.
Now, for each natural number p we have

$$d(x_n,x_{n+p}) \leq d(x_n,x_{n+1}) + d(x_{n+1},x_{n+2}) + \cdots + d(x_{n+p-1},x_{n+p})$$

$$\leq \varphi^n(d(x_0,x_1)) + \varphi^{n+1}(d(x_0,x_1)) + \cdots + \varphi^{n+p-1}(d(x_0,x_1)) = \sum_{k=n}^{n+p-1} \varphi^k(d(x_0,x_1)).$$

Hence, $\{x_n\}_{n\geq 0}$ is a Cauchy sequence in the complete metric space (X,d). Choose $x^* \in X$ such that $x_n \to x^*$. Let $n \geq 1$ be given. Since $x_n \in Tx_{n-1}$, by using (4.19) there exist $u_n \in Tx^*$ and $M(x_{n-1},x^*) \in \mathcal{B}_{x_{n-1},x^*}$ such that

$$d(u_n,x_n) \leq \varphi(M(x_{n-1},x^*)).$$

If $u_n = x^*$ for some $n \geq 1$, then x^* is a fixed point of T. Suppose that $u_n \neq x^*$ for all $n \geq 1$. Now, we show that $\lim_{n\to\infty} d(u_n,x^*) = 0$. If $M(x_{n-1},x^*) = d(x_{n-1},x^*)$, then $d(u_n,x_n) \leq \varphi(d(x_{n-1},x^*))$. Since

$$d(u_n,x^*) \leq d(u_n,x_n) + d(x_n,x^*) \leq \varphi(d(x_{n-1},x^*)) + d(x_n,x^*),$$

$d(u_n,x^*) \to 0$. If $M(x_{n-1},x^*) = d(x_{n-1},Tx_{n-1})$, then

$$d(u_n,x_n) \leq \varphi(d(x_{n-1},Tx_{n-1})) \leq \varphi(d(x_{n-1},x_n)) \leq \varphi^{n-1}(d(x_0,x_1)).$$

Hence, $d(u_n,x^*) \leq \varphi^{n-1}(d(x_0,x_1)) + d(x_n,x^*)$ and so $d(u_n,x^*) \to 0$. If $M(x_{n-1},x^*) = \theta d(x^*,Tx^*)$, then

$$d(u_n,x_n) \leq \varphi(\theta d(x^*,Tx^*)) \leq \varphi(\theta d(x^*,u_n)) < \theta d(u_n,x^*).$$

Hence, $d(u_n,x^*) \leq \theta d(u_n,x^*) + d(x_n,x^*)$ and so $d(u_n,x^*) \leq (1-\theta)^{-1} d(x_n,x^*)$. Thus, $d(u_n,x^*) \to 0$. If $M(x_{n-1},x^*) = \frac{d(x_{n-1},Tx_{n-1})+d(x^*,Tx^*)}{2}$, then

$$\begin{aligned} d(u_n,x_n) \leq \varphi(\frac{d(x_{n-1},Tx_{n-1})+d(x^*,Tx^*)}{2}) &\leq \frac{1}{2}\varphi(d(x_{n-1},x_n)) + \frac{1}{2}\varphi(d(u_n,x^*)) \\ &< \frac{1}{2}\varphi(d(x_{n-1},x_n)) + \frac{1}{2}d(u_n,x^*). \end{aligned}$$

Hence,

$$d(u_n,x^*) \leq d(u_n,x_n) + d(x_n,x^*) < \frac{1}{2}\varphi(d(x_{n-1},x_n)) + \frac{1}{2}d(u_n,x^*) + d(x_n,x^*).$$

Thus, $d(u_n,x^*) < \varphi(d(x_{n-1},x_n)) + 2d(x_n,x^*)$ and so $d(u_n,x^*) \to 0$. If $M(x_{n-1},x^*) = \frac{d(x_{n-1},Tx^*)+d(x^*,Tx_{n-1})}{2}$, then

$$\begin{aligned} d(u_n,x_n) \leq \varphi(\frac{d(x_{n-1},Tx^*)+d(x^*,Tx_{n-1})}{2}) &\leq \frac{1}{2}\varphi(d(x_{n-1},u_n)) + \frac{1}{2}\varphi(d(x_n,x^*)) \\ &\leq \frac{1}{2}\varphi(d(x_{n-1},x^*)) + \frac{1}{2}\varphi(d(x^*,u_n)) + \frac{1}{2}\varphi(d(x_n,x^*)) \\ &< \frac{1}{2}\varphi(d(x_{n-1},x^*)) + \frac{1}{2}\varphi(d(x^*,x_n)) + \frac{1}{2}d(u_n,x^*). \end{aligned}$$

Hence,

$$\begin{aligned} d(u_n,x^*) &\leq d(u_n,x_n) + d(x_n,x^*) \\ &< \frac{1}{2}\varphi(d(x_{n-1},x^*)) + \frac{1}{2}\varphi(d(x_n,x^*)) + \frac{1}{2}d(u_n,x^*) + d(x_n,x^*) \end{aligned}$$

and so $d(u_n,x^*) \to 0$. Thus, we $\lim_{n\to\infty} d(u_n,x^*) = 0$. Since $u_n \in Tx^*$ for all $n \geq 1$ and Tx^* is a closed subset of X, $x^* \in Tx^*$.

♣

Corollary 4.5

Assume that (X,d) is a complete generalized metric space, $\theta,\alpha \in (0,1)$ and $T : X \to P_{cl}(X)$ is a set-valued operator. Suppose that each $x,y \in X$ and $u \in T(x)$ there exist $v \in T(y)$ and $M(x,y) \in \mathcal{B}_{x,y}$ such that $d(u,v) \le AM(x,y)$, where

$$\mathcal{B}_{x,y} = \{d(x,y), d(x,Tx), \theta d(y,Ty), \frac{d(x,Tx)+d(y,Ty)}{2}, \frac{d(x,Ty)+d(y,Tx)}{2}\},$$

and $A \in M_{m\times m}(\mathbb{R}_+)$ is defined by $A = \alpha I$. Then T has a fixed point.

Theorem 4.24

Let (X,d) be a complete generalized metric space, A a matrix in $M_{m,m}(\mathbb{R}_+)$ convergent to zero and $T : X \to P_{cl}(X)$ a multivalued operator. Suppose that for each $x,y \in X$, $(I+A)^{-1}d(x,Tx) \le (I-A)^{-1}d(x,y)$ implies that for each $u \in Tx$, there exists $v \in Ty$ such that

$$d(u,v) \le A\cdot d(x,y). \tag{4.22}$$

Then T has a fixed point.

Proof: First, note that for each $x \in X$, $(I+A)^{-1}d(x,Tx) \le (I-A)^{-1}d(x,Tx)$. Let $x_0 \in X$ and take $x_1 \in Tx_0$. If $x_0 = x_1$, then x_0 is a fixed point of T. Let $x_1 \ne x_0$. By using (4.22), there exists $x_2 \in Tx_1$ such that

$$d(x_1,x_2) \le Ad(x_0,x_1). \tag{4.23}$$

If $x_1 = x_2$, then x_1 is the fixed point of T. Let $x_1 \ne x_2$. Now by using (4.22) and (4.23), there exists $x_3 \in Tx_2$ such that

$$d(x_2,x_3) \le Ad(x_1,x_2) \le A^2 d(x_0,x_1).$$

Now by induction, we can construct a sequence $\{x_n\}_{n\ge 0}$ in X such that $x_{n+1} \in Tx_n$ and $d(x_n,x_{n+1}) \le A^n d(x_0,x_1)$ for all $n \ge 0$. Now, by using Theorem 4.21 we obtain

$$\begin{aligned} d(x_n,x_{n+p}) &\le d(x_n,x_{n+1}) + d(x_{n+1},x_{n+2}) + \cdots + d(x_{n+p-1},x_{n+p}) \\ &\le A^n d(x_0,x_1) + A^{n+1}d(x_0,x_1) + \cdots + A^{n+p-1}d(x_0,x_1) \\ &\le A^n(I + A + A^2 + \cdots + A^{p-1})d(x_0,x_1) \\ &\le A^n(I-A)^{-1}d(x_0,x_1) \to 0 \ (n\to\infty). \end{aligned}$$

Hence, $\{x_n\}_{n\ge 0}$ is a Cauchy sequence in the complete metric space (X,d). Choose $x^* \in X$ such that $x_n \to x^*$. We claim that for each $x \in X \setminus \{x^*\}$,

$$d(x^*,Tx) \le Ad(x^*,x). \tag{4.24}$$

Let $x \in X \setminus \{x^*\}$. Choose a natural number N such that $d(x_n, x^*) < \frac{1}{3}d(x, x^*)$ for all $n \geq N$. Hence, for each $n \geq N$ we have

$$d(x_n, Tx_n) \leq d(x_n, x_{n+1}) \leq d(x_n, x^*) + d(x^*, x_{n+1})$$

$$\leq \frac{2}{3}d(x, x^*) = d(x, x^*) - \frac{1}{3}d(x, x^*) \leq d(x, x^*) - d(x_n, x^*) \leq d(x_n, x).$$

Thus, $(I+A)^{-1}d(x_n, Tx_n) \leq (I-A)^{-1}d(x_n, Tx_n) \leq (I-A)^{-1}d(x_n, x)$ for $n \geq N$. Since $x_{n+1} \in Tx_n$, by using (4.22), for each $n \geq N$ there exists $u_n \in Tx$ such that

$$d(u_n, x_{n+1}) \leq Ad(x_n, x).$$

Hence, $d(x_{n+1}, Tx) \leq Ad(x_n, x)$ and so $\lim_{n\to\infty} d(x_{n+1}, Tx) \leq A \lim_{n\to\infty} d(x_n, x)$. Thus, $d(x^*, Tx) \leq Ad(x^*, x)$ for all $x \in X \setminus \{x^*\}$. Now, we show that for each $x \in X$ and $u \in Tx$, there exists $v \in Tx^*$ such that $d(u, v) \leq Ad(x, x^*)$. If $x = x^*$, we have nothing to prove. Let $x \neq x^*$. By definition of $d(x^*, Tx)$, for each $n \geq 1$ there exists $y_n \in Tx$ such that

$$d(x^*, y_n) < d(x^*, Tx) + \frac{1}{n}d(x, x^*).$$

Hence, we have

$$d(x, Tx) \leq d(x, y_n) \leq d(x, x^*) + d(x^*, y_n) \leq d(x, x^*) + d(x^*, Tx) + \frac{1}{n}d(x, x^*).$$

Now, by using (4.24) we obtain

$$(I+A)^{-1}d(x, Tx) \leq d(x, x^*) + \frac{1}{n}(I+A)^{-1}d(x, x^*)$$

for all $n \geq 1$. Thus, $(I+A)^{-1}d(x, Tx) \leq d(x, x^*) \leq (I-A)^{-1}d(x, x^*)$. Now by using (4.22), for each $u \in Tx$ there exists $v \in Tx^*$ such that $d(u, v) \leq Ad(x, x^*)$. Since $x_{n+1} \in Tx_n$ for all $n \geq 1$, there exists $v_n \in Tx^*$ such that $d(v_n, x_{n+1}) \leq Ad(x_n, x^*)$. Hence,

$$d(v_n, x^*) \leq d(v_n, x_{n+1}) + d(x_{n+1}, x^*) \leq Ad(x_n, x^*) + d(x_{n+1}, x^*)$$

for all $n \geq 1$. Thus, $v_n \to x^*$. Since $v_n \in Tx^*$ for all $n \geq 1$ and Tx^* is a closed subset of X, $x^* \in Tx^*$.

♣

Theorem 4.25

Let (X, d) be a complete generalized metric space, A a matrix in $M_{m,m}(\mathbb{R}_+)$ convergent to zero and $T : X \to P_{cl}(X)$ a multivalued operator. Suppose that for any $x, y \in X$,

$(I+A)^{-1}d(x,Tx) \leq (I-A)^{-1}d(x,y)$ implies that for each $u \in Tx$ there exist $v \in Ty$ and $L(x,y) \in \mathcal{A}_{x,y}$ such that $d(u,v) \leq A\,L(x,y)$, where

$$\mathcal{A}_{x,y} = \{d(x,y), d(x,Tx), d(y,Ty)\}.$$

Then T has a fixed point.

Proof: First, note that for each $x \in X$ and $y \in Tx$ we have

$$(I+A)^{-1}d(x,Tx) \leq (I-A)^{-1}d(x,Tx) \leq (I-A)^{-1}d(x,y).$$

Let $x_0 \in X$ and take $x_1 \in Tx_0$. If $x_0 = x_1$, then x_0 is a fixed point of T. Let $x_1 \neq x_0$. There exist $x_2 \in Tx_1$ and $L(x_0,x_1) \in \mathcal{A}_{x_0,x_1}$ such that $d(x_1,x_2) \leq AL(x_0,x_1)$. If $L(x_0,x_1) = d(x_1,Tx_1)$, then $d(x_1,x_2) \leq Ad(x_1,Tx_1) \leq Ad(x_1,x_2)$ and so

$$(I-A)d(x_1,x_2) \leq 0.$$

Hence, $x_1 = x_2$, that is, x_1 is a fixed point of T. If $L(x_0,x_1) = d(x_0,x_1)$ or $L(x_0,x_1) = d(x_0,Tx_0)$, then

$$d(x_1,x_2) \leq Ad(x_0,x_1). \tag{4.25}$$

Now by using the assumption, there exist $x_3 \in Tx_2$ and $L(x_1,x_2) \in \mathcal{A}_{x_1,x_2}$ such that $d(x_2,x_3) \leq AL(x_1,x_2)$. If $L(x_1,x_2) = d(x_2,Tx_2)$, then

$$d(x_2,x_3) \leq Ad(x_2,Tx_2) \leq Ad(x_2,x_3)$$

and so $x_2 = x_3$, that is, x_2 is a fixed point of T. If $L(x_1,x_2) = d(x_1,x_2)$ or $L(x_0,x_1) = d(x_1,Tx_1)$, then by using (4.25) we have

$$d(x_2,x_3) \leq Ad(x_1,x_2) \leq A^2d(x_0,x_1).$$

Now by continuing this process, we can construct a sequence $\{x_n\}_{n\geq 0}$ in X such that $x_{n+1} \in Tx_n$ and $d(x_n,x_{n+1}) \leq A^n d(x_0,x_1)$ for all $n \geq 0$. By using a similar proof of Theorem 4.21, we can show that $\{x_n\}_{n\geq 0}$ is a Cauchy sequence in the complete metric space (X,d). Choose $x^* \in X$ such that $x_n \to x^*$. We claim that for each $x \in X \setminus \{x^*\}$ we have

$$d(x^*,Tx) \leq 2(I-A)^{-1}Ad(x,x^*). \tag{4.26}$$

Let $x \in X \setminus \{x^*\}$. Since $x_n \to x^*$, we can choose a natural number N such that

$$d(x_n,x^*) < \frac{1}{3}d(x,x^*)$$

for all $n \geq N$. Thus, we obtain

$$(I+A)^{-1}d(x_n,Tx_n) \leq d(x_n,Tx_n) \leq d(x_n,x_{n+1}) \leq d(x_n,x^*) + d(x^*,x_{n+1})$$

$$\leq \frac{2}{3}d(x,x^*) = d(x,x^*) - \frac{1}{3}d(x,x^*) \leq d(x,x^*) - d(x_n,x^*)$$
$$\leq d(x_n,x) \leq (I-A)^{-1}d(x_n,x).$$

Since $x_{n+1} \in Tx_n$ for all $n \geq 1$, by using the assumption, for each $n \geq 1$ there exist $u_n \in Tx$ and $L(x_n,x) \in \mathcal{A}_{x_n,x}$ such that $d(u_n,x_{n+1}) \leq AL(x_n,x)$. If $L(x_n,x) = d(x_n,x)$, then $d(u_n,x_{n+1}) \leq Ad(x_n,x)$. Hence,

$$d(x_{n+1},Tx) \leq Ad(x_n,x) \leq 2(I-A)^{-1}Ad(x_n,x).$$

If $L(x_n,x) = d(x_n,Tx_n)$, then

$$d(u_n,x_{n+1}) \leq Ad(x_n,Tx_n) \leq Ad(x_n,x_{n+1}) \leq A[d(x_n,x)+d(x,x_{n+1})].$$

Hence,

$$d(x_{n+1},Tx) \leq d(x_{n+1},u_n) \leq (I-A)^{-1}A[d(x_n,x)+d(x,x_{n+1})].$$

If $L(x_n,x) = d(x,Tx)$, then

$$d(u_n,x_{n+1}) \leq Ad(x,Tx) \leq Ad(x,u_n) \leq Ad(x,x_{n+1}) + Ad(x_{n+1},u_n).$$

Hence, $(I-A)d(u_n,x_{n+1}) \leq Ad(x,x_{n+1})$ and so

$$d(x_{n+1},Tx) \leq d(x_{n+1},u_n) \leq (I-A)^{-1}Ad(x,x_{n+1}) \leq 2(I-A)^{-1}Ad(x,x_{n+1}).$$

Now by taking a limit, we obtain $d(x^*,Tx) \leq 2(I-A)^{-1}Ad(x,x^*)$. This completes the proof of (4.26). Here, we claim that for each $x \in X$ and $u \in Tx$ there exist $v \in Tx^*$ and $L(x,x^*) \in \mathcal{A}_{x,x^*}$ such that $d(u,v) \leq AL(x,x^*)$. If $x = x^*$, we have nothing to prove. Let $x \neq x^*$. For each $n \geq 1$ choose $y_n \in Tx$ such that

$$d(x^*,y_n) < d(x^*,Tx) + \frac{1}{n}d(x,x^*).$$

By using (4.26), we obtain

$$d(x,Tx) \leq d(x,y_n) \leq d(x,x^*) + d(x^*,y_n)$$
$$\leq d(x,x^*) + d(x^*,Tx) + \frac{1}{n}d(x,x^*)$$
$$\leq (1+\frac{1}{n})d(x,x^*) + 2(I-A)^{-1}Ad(x,x^*),$$

for all $n \geq 1$. Hence, $d(x,Tx) \leq d(x,x^*) + 2(I-A)^{-1}Ad(x,x^*)$ and so

$$(I+A)^{-1}d(x,Tx) \leq [(I+A)^{-1} + 2(I+A)^{-1}(I-A)^{-1}A]d(x,x^*) = (I-A)^{-1}d(x,x^*).$$

Now, by using the assumption, our claim is proved. Since $x_{n+1} \in Tx_n$ for all $n \geq 1$, by using our claim, there exist $v_n \in Tx^*$ and $L(x_n,x^*) \in \mathcal{A}_{x_n,x^*}$ such that

$$d(x_{n+1},v_n) \leq AL(x_n,x^*).$$

If $L(x_n,x^*) = d(x_n,x^*)$, then $d(x_{n+1},v_n) \le Ad(x_n,x^*)$ and so

$$d(v_n,x^*) \le d(v_n,x_{n+1}) + d(x_{n+1},x^*) \le Ad(x_n,x^*) + d(x_{n+1},x^*) \to 0.$$

If $L(x_n,x^*) = d(x_n,Tx_n)$, then $d(x_{n+1},v_n) \le Ad(x_n,Tx_n) \le Ad(x_n,x_{n+1})$ and so

$$d(v_n,x^*) \le d(v_n,x_{n+1}) + d(x_{n+1},x^*) \le Ad(x_n,x_{n+1}) + d(x_{n+1},x^*) \to 0.$$

If $L(x_n,x^*) = d(x^*,Tx^*)$, then $d(x_{n+1},v_n) \le Ad(x^*,Tx^*) \le Ad(x^*,v_n)$ and so

$$d(v_n,x^*) \le d(v_n,x_{n+1}) + d(x_{n+1},x^*) \le Ad(v_n,x^*) + d(x_{n+1},x^*).$$

Thus, $d(v_n,x^*) \le (I-A)^{-1}d(x_{n+1},x^*) \to 0$. Therefore, $v_n \to x^*$. Since $v_n \in Tx^*$ for all $n \ge 1$ and Tx^* is a closed subset of X, $x^* \in Tx^*$.

♣

Theorem 4.26

Let (X,d) be a complete generalized metric space, $T: X \to P_{cl}(X)$ a set-valued operator and $F: \mathbb{R}^m_+ \to \mathbb{R}^m_+$ an increasing sublinear continuous function such that $F(0) = 0$ and $F(t) > 0$ for all $t = (t_i)_{i=1}^m \in \mathbb{R}^m_{++}$, where

$$\mathbb{R}^m_{++} = \{(t_1,\cdots,t_m) : t_i > 0, \text{ for } i = 1,2,\cdots,m\}.$$

Also, suppose that for each $x,y \in X$, $(I+A)^{-1}F(d(x,Tx)) \le (I-A)^{-1}F(d(x,y))$ implies that for each $u \in Tx$ there exist $v \in Ty$ and $L(x,y) \in \mathcal{A}_{x,y}$ such that

$$F(d(u,v)) \le AF(L(x,y)), \tag{4.27}$$

where $\mathcal{A}_{x,y} = \{d(x,y), d(x,Tx), d(y,Ty)\}$ and $A \in M_{m\times m}(\mathbb{R}_+)$ converges to zero. Then T has a fixed point.

Proof: Note that, for each $x \in X$ and $y \in Tx$ we have

$$(I+A)^{-1}F(d(x,Tx)) \le (I-A)^{-1}F(d(x,Tx)) \le (I-A)^{-1}F(d(x,y)).$$

Let $x_0 \in X$ and take $x_1 \in Tx_0$. If $x_1 = x_0$, then x_0 is a fixed point of T. Let $x_1 \ne x_0$. By using (4.27), there exist $x_2 \in Tx_1$ and $L(x_0,x_1) \in \mathcal{A}_{x_0,x_1}$ such that

$$F(d(x_1,x_2)) \le AF(L(x_0,x_1)).$$

If $L(x_0,x_1) = d(x_1,Tx_1)$, then

$$F(d(x_1,x_2)) \le AF(d(x_1,Tx_1)) \le AF(d(x_1,x_2)).$$

Hence, $(I-A)F(d(x_1,x_2)) \le 0$ and so $F(d(x_1,x_2)) \le 0$. We claim that $x_1 = x_2$.

If $x_1 \neq x_2$, then $F(d(x_1,x_2)) > 0$ and so $0 < F(d(x_1,x_2)) \leq 0$. This contradiction shows that $x_1 = x_2$. Thus, x_1 is a fixed point of T. If $L(x_0,x_1) = d(x_0,Tx_0)$, then

$$F(d(x_1,x_2)) \leq AF(d(x_0,Tx_0)) \leq AF(d(x_0,x_1)).$$

If $L(x_0,x_1) = d(x_0,x_1)$, then

$$F(d(x_1,x_2)) \leq AF(d(x_0,x_1)).$$

Thus, there exists $x_2 \in Tx_1$ such that $F(d(x_1,x_2)) \leq AF(d(x_0,x_1))$. Since

$$(I+A)^{-1}F(d(x_1,Tx_1)) \leq (I-A)^{-1}F(d(x_1,x_2)),$$

by using (4.27) there exist $x_3 \in Tx_2$ and $L(x_1,x_2) \in \mathcal{A}_{x_1,x_2}$ such that

$$F(d(x_2,x_3)) \leq AF(L(x_1,x_2)).$$

If $L(x_1,x_2) = d(x_2,Tx_2)$, then $F(d(x_2,x_3)) \leq AF(d(x_2,Tx_2)) \leq AF(d(x_2,x_3))$. Hence, $x_2 = x_3$ and so x_2 is a fixed point of T. If $L(x_1,x_2) = d(x_1,Tx_1)$, then

$$F(d(x_2,x_3)) \leq AF(d(x_1,Tx_1)) \leq AF(d(x_1,x_2)) \leq A^2F(d(x_0,x_1)).$$

If $L(x_1,x_2) = d(x_1,x_2)$, then $F(d(x_2,x_3)) \leq AF(d(x_1,x_2)) \leq A^2F(d(x_0,x_1))$. Thus, there exists $x_3 \in Tx_2$ such that $F(d(x_2,x_3)) \leq A^2F(d(x_0,x_1))$. By continuing this process, we obtain a sequence $\{x_n\}_{n\geq 0}$ in X such that $x_{n+1} \in Tx_n$ and

$$F(d(x_n,x_{n+1})) \leq A^nF(d(x_0,x_1))$$

for all $n \geq 0$. Since A converges to zero, $F(d(x_n,x_{n+1})) \to 0$. We claim that

$$d(x_n,x_{n+1}) \to 0.$$

If $d(x_n,x_{n+1}) \nrightarrow 0$, then there exists $\gamma \in \mathbb{R}^m_{++}$ such that for each $k \geq 1$ there is an integer number $n_k \geq k$ such that $d(x_{n_k},x_{n_k+1}) \geq \gamma$. Hence, $0 < F(\gamma) \leq F(d(x_{n_k},x_{n_k+1})) \to 0$. This contradiction shows that $d(x_n,x_{n+1}) \to 0$. Since F is sublinear, for each natural numbers n and p we obtain

$$\begin{aligned} F(d(x_n,x_{n+p})) &\leq F(d(x_n,x_{n+1})) + F(d(x_{n+1},x_{n+2})) + \cdots + F(d(x_{n+p-1},x_{n+p})) \\ &\leq A^nF(d(x_0,x_1)) + A^{n+1}F(d(x_0,x_1)) + \cdots + A^{n+p-1}F(d(x_0,x_1)). \end{aligned}$$

Thus, $F(d(x_n,x_{n+p})) \to 0$ and so $d(x_n,x_{n+p}) \to 0$. Thus, $\{x_n\}_{n\geq 0}$ is a Cauchy sequence. Hence, there exists $x^* \in X$ such that $x_n \to x^*$. Since F is continuous, we have $F(d(x_n,x^*)) \to 0$. We claim that for each $x \in X \setminus \{x^*\}$,

$$F(d(x^*,Tx)) \leq 2(I-A)^{-1}AF(d(x,x^*)). \tag{4.28}$$

Let $x \in X \setminus \{x^*\}$. Choose a natural number N such that $d(x_n,x^*) < \frac{1}{3}d(x,x^*)$ for all $n \geq N$. Then, for each $n \geq N$ we have

$$d(x_n,Tx_n) \leq d(x_n,x_{n+1}) \leq d(x_n,x^*) + d(x^*,x_{n+1}) \leq \frac{2}{3}d(x,x^*)$$

$$= d(x,x^*) - \frac{1}{3}d(x,x^*) \leq d(x,x^*) - d(x_n,x^*) \leq d(x_n,x).$$

Therefore, $(I+A)^{-1}F(d(x_n,Tx_n)) \leq (I-A)^{-1}F(d(x_n,x))$ for all $n \geq N$. Since $x_{n+1} \in Tx_n$, by using (4.27), for each $n \geq N$ there exist $u_n \in Tx$ and $L(x_n,x) \in \mathcal{A}_{x_n,x}$ such that $F(d(x_{n+1},u_n)) \leq AF(L(x_n,x))$. If $L(x_n,x) = d(x_n,x)$, then we obtain $F(d(x_{n+1},u_n)) \leq AF(d(x_n,x))$. Since F is increasing and $u_n \in Tx$, we get

$$F(d(x_{n+1},Tx)) \leq AF(d(x_n,x)).$$

But, F is continuous and $x_n \to x^*$. Hence,

$$F(d(x^*,Tx)) \leq AF(d(x^*,x)) \leq 2(I-A)^{-1}AF(d(x^*,x)).$$

If $L(x_n,x) = d(x_n,Tx_n)$, then we obtain

$$F(d(x_{n+1},u_n)) \leq AF(d(x_n,Tx_n)) \leq AF(d(x_n,x_{n+1})).$$

Since F is increasing and sublinear, we have

$$F(d(x_{n+1},Tx)) \leq F(d(x_{n+1},u_n)) \leq AF(d(x_n,x)) + AF(d(x,x_{n+1})).$$

From continuity of F, we obtain

$$F(d(x^*,Tx)) \leq 2AF(d(x^*,x)) \leq 2(I-A)^{-1}AF(d(x^*,x)).$$

If $L(x_n,x) = d(x,Tx)$, then we have

$$F(d(x_{n+1},u_n)) \leq AF(d(x,Tx)) \leq AF(d(x,u_n)).$$

Since F is sublinear, we have $F(d(x_{n+1},u_n)) \leq AF(d(x,x_{n+1})) + AF(d(x_{n+1},u_n))$ and so $(I-A)F(d(x_{n+1},u_n)) \leq AF(d(x,x_{n+1}))$. Hence, we obtain

$$F(d(x_{n+1},u_n)) \leq (I-A)^{-1}AF(d(x,x_{n+1}))$$

and so $F(d(x^*,Tx)) \leq (I-A)^{-1}AF(d(x^*,x))$, because F is continuous and $x_n \to x^*$. Thus, for each $x \in X \setminus \{x^*\}$ we have

$$F(d(x^*,Tx)) \leq 2(I-A)^{-1}AF(d(x^*,x)).$$

Now, we claim that for each $x \in X$ and $u \in Tx$, there exist $v \in Tx^*$ and $L(x,x^*) \in \mathcal{A}_{x,x^*}$ such that $F(d(u,v)) \leq AF(L(x,x^*))$. If $x = x^*$, we have nothing to prove. If

$x \neq x^*$, for each $n \geq 1$ choose $y_n \in Tx$ such that $d(x^*, y_n) \leq d(x^*, Tx) + \frac{1}{n} d(x, x^*)$. Thus,

$$d(x, Tx) \leq d(x, y_n) \leq d(x, x^*) + d(x^*, y_n) \leq d(x, x^*) + d(x^*, Tx) + \frac{1}{n} d(x, x^*).$$

Since F is increasing and sublinear, we have

$$F(d(x, Tx)) \leq F(d(x, x^*)) + F(d(x^*, Tx)) + \frac{1}{n} F(d(x, x^*)).$$

Now by using (4.28), for each $n \geq 1$ we obtain

$$F(d(x, Tx)) \leq (1 + \frac{1}{n}) F(d(x, x^*)) + 2(I - A)^{-1} A F(d(x, x^*))$$

and so $F(d(x, Tx)) \leq F(d(x, x^*)) + 2(I - A)^{-1} A F(d(x, x^*))$. Thus, we obtain

$$(I + A)^{-1} F(d(x, Tx)) \leq (I + A)^{-1} [I + 2(I - A)^{-1} A] F(d(x, x^*)) = (I - A)^{-1} F(d(x, x^*)).$$

By using (4.27), for each $u \in Tx$ there exist $v \in Tx^*$ and $L(x, x^*) \in \mathcal{A}_{x,x^*}$ such that

$$F(d(u, v)) \leq A F(L(x, x^*)).$$

This completes proof of the claim. Since $x_{n+1} \in Tx_n$ for all $n \geq 1$, there exist $v_n \in Tx^*$ and $L(x_n, x^*) \in \mathcal{A}_{x_n, x^*}$ such that $F(d(x_{n+1}, v_n)) \leq A F(L(x_n, x^*))$. If $L(x_n, x^*) = d(x_n, x^*)$, then $F(d(x_{n+1}, v_n)) \leq A F(d(x_n, x^*))$. Since F is sublinear and continuous, we get
$$F(d(v_n, x^*)) \leq F(d(v_n, x_{n+1})) + F(d(x_{n+1}, x^*)) \leq A F(d(x_n, x^*)) + F(d(x_{n+1}, x^*))$$

and so $F(d(v_n, x^*)) \to 0$. If $L(x_n, x^*) = d(x_n, Tx_n)$, then we obtain

$$F(d(x_{n+1}, v_n)) \leq A F(d(x_n, Tx_n)) \leq A F(d(x_n, x_{n+1})).$$

Thus,
$$F(d(v_n, x^*)) \leq F(d(v_n, x_{n+1})) + F(d(x_{n+1}, x^*)) \leq A F(d(x_n, x_{n+1})) + F(d(x_{n+1}, x^*))$$

and so $F(d(v_n, x^*)) \to 0$. If $L(x_n, x^*) = d(x^*, Tx^*)$, then we have

$$F(d(x_{n+1}, v_n)) \leq A F(d(x^*, Tx^*)) \leq A F(d(x^*, v_n)).$$

Hence,

$$F(d(v_n, x^*)) \leq F(d(v_n, x_{n+1})) + F(d(x_{n+1}, x^*)) \leq A F(d(x^*, v_n)) + F(d(x_{n+1}, x^*))$$

and so $(I - A) F(d(v_n, x^*)) \leq F(d(x_{n+1}, x^*))$. Thus, $F(d(v_n, x^*)) \to 0$ and so $v_n \to x^*$. Since $v_n \in Tx^*$ for all $n \geq 1$ and Tx^* is a closed subset of X, $x^* \in Tx^*$.

♣

Corollary 4.6

Let (X,d) be a complete generalized metric space, $T : X \to P_{cl}(X)$ a set-valued operator and $\phi : \mathbb{R}^m_+ \to \mathbb{R}_+$ an integrable function such that

$$\int_0^{\varepsilon} \phi(t)dt > 0, \text{ for all } t = (t_i)_{i=1}^m \in \mathbb{R}^m_+ \text{ and } \varepsilon = (\varepsilon_i)_{i=1}^m \in \mathbb{R}^m_{++}.$$

Suppose that for each $x,y \in X$, $(I+A)^{-1}F(d(x,Tx)) \leq (I-A)^{-1}F(d(x,y))$ implies that for each $u \in T(x)$ there exist $v \in T(y)$ and $L(x,y) \in \mathcal{A}_{x,y}$ such that

$$\left(\int_0^{d_1(u,v)} \phi(t)dt_1, \cdots, \int_0^{d_m(u,v)} \phi(t)dt_m\right) \leq A\left(\int_0^{L_1(x,y)} \phi(t)dt_1, \cdots, \int_0^{L_m(x,y)} \phi(t)dt_m\right),$$

where $d(u,v) = (d_1(u,v), \cdots, d_m(u,v))$, $L(x,y) = (L_1(x,y), \cdots, L_m(x,y))$ and the matrix $A \in M_{m\times m}(\mathbb{R}_+)$ converges to zero. Then T has a fixed point.

4.8 Fixed points of set-valued mappings on cone metric spaces

In this section, we present some fixed point result about set-valued mappings on cone metric spaces. After reading this section, you are one step closer to the main aim of this book. In fact, we want you to find that there are some weak results in fixed point theory.

As we said before, next examples show that the category of cone metric spaces is bigger than the one metric spaces.

Example 64 *Let $q > 0$, $b > 1$, $E = \ell^q$, $P = \{\{x_n\}_{n\geq 1} \in E : x_n \geq 0, \text{for all } n\}$, (X,ρ) a metric space and $d : X \times X \to E$ defined by $d(x,y) = \{(\frac{\rho(x,y)}{b^n})^{\frac{1}{q}}\}_{n\geq 1}$. Then (X,d) is a normal cone metric space and the normal constant of P is equal to $M = 1$.*

Example 65 *Let $E = (C_{\mathbb{R}}([0,\infty)), \|.\|_{\infty})$, $P = \{f \in E : f(x) \geq 0\}$, (X,ρ) a metric space and $d : X \times X \to E$ defined by $d(x,y) = f_{x,y}$, where $f_{x,y}(t) = |x-y|t$. Then (X,d) is a normal cone metric space and the normal constant of P is equal to $M = 1$.*

Also, the following example shows that every metric space is a regular cone metric space.

Example 66 *Let $E = (L^1[0,1], \|.\|_1)$, $P = \{f \in E : f \geq 0 \text{ a.e.}\}$, (X,ρ) a metric space and $d : X \times X \to E$ defined by $d(x,y) = f_{x,y}$, where $f_{x,y}(t) = \rho(x,y)t^2$. Then (X,d) is a regular cone metric space. In fact, if $\{f_n\}_{n\geq 1}$ is an increasing sequence and there is $g \in L^1$ such that $f_1 \leq f_2 \leq \cdots \leq f_n \leq \cdots \leq g$ for all almost* *x, then $\{f_n(x)\}_{n\geq 1}$ converges for all almost x. Let $f(x) = \lim_{n\to\infty} f_n(x)$ for all almost x. Then, $f_n \leq f \leq g$ (a.e.) for all $n \geq 1$. Thus, $g - f_1 \in L^1$, $g - f_n \leq g - f_1$ for all $n \geq 1$ and $\lim_{n\to\infty} g - f_n = g - f$ (a.e.). Hence by the Lebesgue monotone convergence theorem, $f \in L^1$ and $\lim_{n\to\infty} \|f_n - f\|_1 = 0$. Therefore, the cone P is regular.*

Note that, the relations $P + intP \subseteq intP$ and $\lambda\, intP \subseteq intP$ $(\lambda > 0)$ hold. We appeal to these relations in the following.

Definition 4.4 Let (X,d) be a metric space and $f : X \to X$ a selfmap. We say that f has the property (KM) whenever for every $\varepsilon > 0$ there exists $\delta > 0$ such that $d(f(x), f(y)) < \varepsilon$ for all $x, y \in X$ with $\varepsilon \leq d(x,y) < \varepsilon + \delta$.

It is easy if f has the property (KM) if and only if or every $\varepsilon > 0$ there exists $\delta > 0$ such that $d(f(x), f(y)) < \varepsilon$ for all $x, y \in X$ with $d(x,y) < \varepsilon + \delta$.

Definition 4.5 Let (X,d) be a cone metric space and $f : X \to X$ a selfmap. We say that f has the property (KMC) whenever for every $0 \neq c \in P$ there exists $0 \ll d$ such that $d(f(x), f(y)) < c$ for all $x, y \in X$ with $d(x,y) < c + d$.

Definition 4.6 Let (X,d) be a cone metric space and $0 \ll c$. Then, (X,d) is called c-chainable if for each $x, y \in X$ there exist $x_0, x_1, \cdots, x_m$ in X such that $x_0 = x$, $x_m = y$ and $d(x_{i+1}, x_i) < c$ for all $0 \leq i \leq m-1$.

The example 66 shows that the following proposition extends a result about the fixed point of Meir-Keeler type contraction mappings on metric spaces to regular cone metric spaces.

Proposition 4.2

Let (X,d) be a complete regular cone metric space and $f : X \to X$ has the property (KMC). Then, f has a unique fixed point.

Proof: First, note that $d(f(x), f(y)) < d(x,y)$ for all $x, y \in X$ with $x \neq y$. Let $x_0 \in X$. Put $x_n = f(x_{n-1})$ for all $n \geq 1$. If there is a natural m such that $d(x_{m+1}, x_m) = 0$, then $f(x_m) = x_m$ and so f has a fixed point. If $d(x_{n+1}, x_n) \neq 0$ for all $n \geq 1$, then $d(x_{n+1}, x_n) < d(x_n, x_{n-1})$. Hence, there exists $\alpha \in P$ such that $d(x_{n+1}, x_n) \to \alpha$. We claim that $\alpha = 0$. If $\alpha \neq 0$, then there is $0 \ll d$ such that $d(f(x), f(y)) < \alpha$ for all $x, y \in X$ with $d(x,y) < \alpha + d$. Choose $r > 0$ such that $\frac{d}{2} + N_r(0) \subseteq P$ and take a natural number N such that $\|d(x_{n+1},x_n)-\alpha\| < r$ for all $n \geq N$. So, $d(x_{n+1},x_n)-\alpha \ll d$. Since f has the property (KMC), $d(x_{n+2},x_{n+1}) < \alpha$ for all $n \geq N$. This is a contradiction because $\alpha < d(x_{i+1},x_i)$ for all $i \geq 1$. Now, we show that $\{x_n\}_{n\geq 1}$ is a Cauchy sequence. If this is not, then there is a $0 \ll c$ such that for all natural number m there are $i_m, j_m > m$ so that the relation $d(x_{i_m},x_{j_m}) \ll c$ does not hold. For $0 \ll e \ll c$ there exists $0 \ll d$ such that $d(f(x),f(y)) < e$ for all $x,y \in X$ with $d(x,y) < e+d$. Choose a natural number M and such that $d(x_{i+1},x_i) \ll \frac{d}{2}$ for all $i \geq M$. Also, take $j_M \geq i_M > M$ so that the relation $d(x_{i_M},x_{j_M}) \ll c$ does not hold. Then, we have

$$d(x_{i_M-1},x_{i_M+1}) \leq d(x_{i_M-1},x_{i_M}) + d(x_{i_M},x_{i_M+1}) \ll \frac{d}{2}+\frac{d}{2} \ll d+e.$$

Hence, $d(x_{i_M},x_{i_M+2}) \ll e$. Similarly, $d(x_{i_M},x_{i_M+3}) \ll e$. Thus, $d(x_{i_M},x_{j_M}) \ll e \ll c$ which is a contradiction. Therefore, $\{x_n\}_{n\geq 1}$ is a Cauchy sequence. Since (X,d) is complete, there is $u \in X$ such that $x_n \to u$. Hence, $f(x_n) \to f(u)$. But, $f(x_n) = x_{n+1} \to u$ and the limit point is unique in cone metric spaces. Thus, f has at least one fixed point. Since $d(f(x),f(y)) < d(x,y)$ for all $x,y \in X$ with $x \neq y$, f has a unique fixed point.

♣

Definition 4.7 Let (X,d) be a cone metric space, p a natural number and $F: X \to 2^X$ a multifunction. The multifunction F is called weakly uniformly strict p-contraction if the following property hold:
$(\mathcal{F})$ for every $0 \neq e \in P$ there exists $0 \ll d$ such that for each $z,w \in X$ admitting the representation:

$$\exists z_0,\cdots,z_{p-1} \in X \text{ with } z \in F(z_{p-1}), z_{p-1} \in F(z_{p-2}),\cdots,z_1 \in F(z_0),$$

$$\exists w_0,\cdots,w_{p-1} \in X \text{ with } w \in F(w_{p-1}), w_{p-1} \in F(w_{p-2}),\cdots,w_1 \in F(w_0)$$

and $d(z_0,w_0) \leq e+d$, we have $d(z,w) < e$.

Theorem 4.27

Let (X,d) be a complete regular cone metric space, p a natural number and $F: X \to 2^X$ a weakly uniformly strict p-contraction. Then, F has a fixed point.

Proof: By the axiom of choice, F has a selection, that is, there is a function $f: X \to X$ such that $f(x) \in F(x)$ for all $x \in X$. We show that f^p has the property (KMC). Let $0 \neq c \in P$ and $0 \ll d$ be the same provided by $(\mathcal{F})$. Let us fix $z_0, w_0 \in X$ such that $d(z_0,w_0) < e+d$ and define

$$z_1 = f(z_0), z_2 = f(z_1),\cdots,z_{p-1} = f(z_{p-2}), w_1 = f(w_0),\cdots,w_{p-1} = f(w_{p-2}).$$

Then,

$$z_1 \in F(z_0), z_2 \in F(z_1), \cdots, z_{p-1} \in F(z_{p-2}), w_1 \in F(w_0), \cdots, w_{p-1} \in F(w_{p-2}).$$

By using the property ($\mathcal{F}$) and considering $z = f^p(z_0)$, $w = f^p(w_0)$ we have

$$d(f^p(z_0), f^p(w_0)) = d(z,w) < e.$$

Thus, f^p has the property (KMC). Therefore by Proposition 7.3, there is unique $x_0 \in X$ such that $f^p(x_0) = x_0$. Since $f^p(f(x_0)) = f(x_0)$, $x_0 = f(x_0) \in F(x_0)$.

♣

Corollary 4.7

Let (X,d) be a complete regular cone metric space, $f : X \to X$ a selfmap and f^m has the property (KMC) for some $m \geq 2$. Then, f has a unique fixed point.

Theorem 4.28

Let (X,d) be a complete regular cone metric space, $0 \ll c$, p a natural number and the multifunction $F : X \to 2^X$ has the following properties:
($\mathfrak{I}$) For each $z,w \in X$ with $z \neq w$, if there exist $z_0, \cdots, z_{p-1}$ and $w_0, \cdots, w_{p-1}$ in X such that

$$z \in F(z_{p-1}), z_{p-1} \in F(z_{p-2}), \cdots, z_1 \in F(z_0)$$

and

$$w \in F(w_{p-1}), w_{p-1} \in F(w_{p-2}), \cdots, w_1 \in F(w_0),$$

then $d(z,w) < d(z_0,w_0)$.
($\mathfrak{I}\mathfrak{I}$) For each $0 \neq e \in P$ there exists $0 \ll d$ such $e+d < c$ and for every $z,w,z_0,w_0 \in X$ which satisfy in the condition ($\mathfrak{I}$), if $d(z_0,w_0) < e+d$, then $d(z,w) \leq e$. Then, we have
(i) If there exist $z_0, w_0 \in X$ with $z_0 \neq w_0$, $a_1, \cdots, a_{p-1} \in X$ such that

$$z_0 \in F(a_1), a_1 \in F(a_2), \cdots, a_{p-1} \in F(w_0)$$

and $d(z_0,w_0) < c$, then F has a fixed point.
(ii) If (X,d) is c-chainable, then F has a fixed point.

Proof: (i) By the axiom of choice, F has at least a selection, namely f, such that $z_0 = f(a_1), a_1 = f(a_2), \cdots, a_{p-1} = f(w_0)$. Note that, $z_0 = f^p(w_0)$. We show that

$$d(f^p(x), f^p(y)) < d(x,y) \qquad (*)$$

for all $x,y \in X$ with $x \neq y$. Let $x,y \in X$ with $x \neq y$. Put

$$z_1 = f(x), z_2 = f(z_1), \cdots, z_p = f(z_{p-1}) = f^p(x)$$

and

$$w_1 = f(y), w_2 = f(w_1), \cdots, w_p = f(w_{p-1}) = f^p(y).$$

Then,

$$z_1 \in F(x), z_2 \in F(z_1), \cdots, f^p(x) \in F(z_{p-1})$$

and

$$w_1 \in F(y), w_2 \in F(w_1), \cdots, f^p(y) \in F(w_{p-1}).$$

Hence by ($\mathfrak{I}$), $d(f^p(x), f^p(y)) < d(x,y)$. Put, $z_1 = f^p(z_0)$ and $w_1 = z_0 = f^p(w_0)$. If $d(z_1,w_1) = d(z_1,z_0) = 0$, then $f^p(z_0) = z_0$. Since $f^p(f(z_0)) = f(z_0)$ and f^p has a unique fixed point, $z_0 = f(z_0) \in F(z_0)$ and so F has a fixed point. So, suppose that $0 < d(z_1,w_1)$. By $(*)$, $0 < d(z_1,w_1) < d(z_0,w_0) < c$. Now, put $z_2 = f^p(z_1)$ and $w_2 = z_1 = f^p(w_1)$. If $d(z_2,w_2) = 0$, then $z_1 = f(z_1) \in F(z_1)$. Also, we can suppose that $0 < d(z_2,w_2) < d(z_1,w_1) < c$. Thus, for each $n \geq 1$ define $z_n = f^p(z_{n-1})$, $w_n = z_{n-1} = f^p(w_{n-1})$ and without loss of generality suppose that

$$0 < d(z_n,w_n) < d(z_{n-1},w_{n-1}) < c$$

for all $n \geq 1$. Choose $\alpha \in P$ such that $\lim_{n\to\infty} d(z_n,w_n) = \alpha$. We claim that $\alpha = 0$. Suppose that $\alpha \neq 0$. Then, $0 < \alpha \leq d(z_n,w_n) < c$ for all $n \geq 1$. Take $0 \ll d$ so that satisfies in the condition ($\mathfrak{I}\mathfrak{I}$) (with $e = \alpha$). Choose a natural number N such that $d(z_n,w_n) - \alpha \ll \frac{c}{2}$ for all $n \geq N$. So, $d(z_n,w_n) < \alpha + d$ for all $n \geq N$. Hence by ($\mathfrak{I}\mathfrak{I}$), $d(z_{n+1},w_{n+1}) < \alpha$ for all $n \geq N$. This is a contradiction. Thus, $\lim_{n\to\infty} d(f^{np}(z_0), f^{np}(w_0)) = 0$. Now similar to the proof of Proposition 7.3, we can prove that $\{f^{np}(z_0)\}_{n\geq 1}$ is a Cauchy sequence. Since (X,d) is complete, there is a $x_0 \in X$ such that $\lim_{n\to\infty} d(f^{np}(z_0), x_0) = 0$. By $(*)$,

$$d(f^{(n+1)p}(z_0), f^p(x_0)) < d(f^{np}(z_0), x_0).$$

Hence, $\lim_{n\to\infty} d(f^{(n+1)p}(z_0), f^p(x_0)) = 0$. Since the limit point is unique in cone metric spaces, $f^p(x_0) = x_0$. So, $f^p(f(x_0)) = f(x_0)$. But, f^p has a unique fixed point. Hence, $x_0 = f(x_0) \in F(x_0)$.
(ii) Let $x,y \in X$. Choose $x_0, x_1, \cdots, x_m$ in X such that $x_0 = x$, $x_m = y$ and $d(x_{j+1},x_j) < c$ for all $0 \leq j \leq m-1$. Similar to the proof of the part (i), we can prove that

$$\lim_{n\to\infty} d(f^{np}(x_{j-1}), f^{np}(x_j)) = 0$$

for all $0 \leq j \leq m-1$. Since $d(f^{np}(x), f^{np}(y)) \leq \sum_{j=1}^{m} d(f^{np}(x_{j-1}), f^{np}(x_j))$,

$$\lim_{n\to\infty} d(f^{np}(x), f^{np}(y)) = 0.$$

If $y = f^p(x)$, then $\lim_{n\to\infty} d(f^{np}(x), f^{(n+1)p}(x)) = 0$. Now, similar to the proof of the part (i), $\{f^{np}(x)\}_{n\geq 1}$ is a Cauchy sequence. Since (X,d) is complete, there is a $x_0 \in X$ such that $\lim_{n\to\infty} d(f^{np}(x), x_0) = 0$. By $(*)$, $\lim_{n\to\infty} d(f^{(n+1)p}(x), f^p(x_0)) = 0$. So, $f^p(f(x_0)) = f(x_0)$. Since f^p has a unique fixed point, $x_0 = f(x_0) \in F(x_0)$.

♣

Remark 4.1 In Theorem 4.28, F has a unique fixed point. If x_0 and y_0 are two distinct fixed points of F, then suppose that g is a selection of F such that $g^p(x_0) = x_0$ and $g^p(y_0) = y_0$. Similar to the proof of Theorem 4.28, $d(g^p(x), g^p(y)) < d(x,y)$ for all $x, y \in X$ with $x \neq y$. Hence, $d(x_0, y_0) = d(g^p(x_0), g^p(y_0)) < d(x_0, y_0)$. This is a contradiction. ■

Corollary 4.8

Let (X,d) be a complete regular cone metric space, $0 \ll c$ and the selfmap $f : X \to X$ a contraction. Suppose that for each $0 \neq e \in P$ there exists $0 \ll d$ such that $e + d < c$ and for every $z_0, w_0 \in X$ with $d(z_0, w_0) < e + d$, we have $d(f(z_0), f(w_0)) < e$. Then
(i) If there exists $z_0 \in X$ with $d(z_0, f(z_0)) < c$, then f has a unique fixed point.
(ii) If (X,d) is c-chainable, then f has a unique fixed point.

In the following, we always suppose that E is a normed space, P is a cone in E with normal constant $M = 1$, $intP \neq \emptyset$ and $\leq$ is partial ordering with respect to P.

Example 67 *Let $E = \ell^1$, $P = \{\{x_n\}_{n \geq 1} \in E : x_n \geq 0, for\ all\ n\}$, (X, ρ) a metric space and $d : X \times X \to E$ defined by $d(x,y) = \{\frac{\rho(x,y)}{2^n}\}_{n \geq 1}$. Then (X,d) is a cone metric space and the normal constant of P is $M = 1$.*

This example shows that the category of cone metric spaces is bigger than the one metric spaces.

Definition 4.8 Let (X,d) be a cone metric space and $B \subseteq X$.
(i) $b \in B$ is called an interior point of B whenever there is $0 \ll p$ such that

$$N(b,p) \subseteq B,$$

where $N(b,p) := \{y \in X : d(y,b) \ll p\}$.
(ii) A subset $A \subseteq X$ is called open if each element of A is an interior point of A.

The family $\mathcal{B} = \{N(x,e) : x \in X, 0 \ll e\}$ is a sub-basis for a topology on X. We denote this cone topology by τ_c. The τ_c is Hausdorff and first countable. Most of familiar cones are normal with normal constant $M = 1$. But, for each $k > 1$ there are cones with normal constant $M > k$. Also, there are non-normal cones.

Lemma 4.5

Let (X,d) be a cone metric space, P a normal cone with normal constant $M = 1$ and

A a compact set in (X, τ_c). Then, for every $x \in X$ there exists $a_0 \in A$ such that

$$\|d(x,a_0)\| = \inf_{a \in A} \|d(x,a)\|.$$

Proof: Let $x \in X$ be given. Define $f_x : X \to [0,\infty)$ by $f_x(y) = \|d(x,y)\|$. Let $y \in X$ and $\varepsilon > 0$. Choose $0 \ll c$ such that $\|c\| < \varepsilon$ and suppose that $z \in N(y,c)$. Note that, by using the normality and the relations

$$d(y,x) \leq d(y,z) + d(z,x), \quad d(z,x) \leq d(z,y) + d(y,x),$$

we have

$$\|d(y,x)\| - \|d(z,x)\| \leq \|d(y,z)\| < \varepsilon, \quad \|d(z,x)\| - \|d(y,x)\| \leq \|d(z,y)\| < \varepsilon.$$

Thus, $|f_x(y) - f_x(z)| < \varepsilon$. Hence, f_x is continuous on X. Since A is compact, $f_x(A)$ is compact subset of $[0,\infty)$ and so there exists $a_0 \in A$ such that

$$\|d(x,a_0)\| = \inf_{a \in A} \|d(x,a)\|.$$

♣

Lemma 4.6

Let (X,d) be a cone metric space, P a normal cone with normal constant $M = 1$ and A, B two compact sets in (X, τ_c). Then,

$$\sup_{x \in B} d'(x,A) < \infty,$$

where $d'(x,A) = \inf_{a \in A} \|d(x,a)\|$.

Proof: Define $g_A : X \to [0,\infty)$ by $g_A(x) = \inf_{a \in A} \|d(x,a)\|$. Let $x \in X$ and $\varepsilon > 0$. Choose $0 \ll c$ such that $\|c\| < \varepsilon$ and suppose that $y \in N(x,c)$. Note that, by using the normality and the relation $d(x,a) \leq d(x,y) + d(y,a)$ we have

$$g_A(x) - g_A(y) \leq \|d(x,y)\|.$$

Thus, $|g_A(x) - g_A(y)| < \varepsilon$. Hence, g_A is continuous on X. Since B is compact, $g_A(B)$ is compact subset of $[0,\infty)$. This completes the proof.

♣

Definition 4.9 Let (X,d) be a cone metric space, P a normal cone with normal constant $M = 1$, $\mathcal{H}_c(X)$ the set of all compact subsets of (X, τ_c) and $A \in \mathcal{H}_c(X)$. By using Lemma 4.6, we can define

$$h_A : \mathcal{H}_c(X) \to [0,\infty) \text{ and } d_H : \mathcal{H}_c(X) \times \mathcal{H}_c(X) \to [0,\infty)$$

by

$$h_A(B) = \sup_{x \in A} d'(x,B) \text{ and } d_H(A,B) = \max\{h_A(B), h_B(A)\}$$

respectively.

Remark 4.2 Let (X,d) be a cone metric space with normal constant $M = 1$. Define $\rho : X \times X \to [0,\infty)$ by $\rho(x,y) = \|d(x,y)\|$. Then, (X,ρ) is a metric space. This implies that for each $A, B \in \mathcal{H}_c(X)$ and $x, y \in X$, we have the following relations
(i) $d'(x,A) \leq \|d(x,y)\| + d'(y,A)$,
(ii) $d'(x,A) \leq d'(x,B) + h_B(A)$,
(iii) $d'(x,A) \leq \|d(x,y)\| + d'(y,B) + h_B(A)$. ■

Theorem 4.29

Let (X,d) be a complete cone metric space with normal constant $M = 1$ and the multifunction $T : X \to \mathcal{H}_c(X)$ satisfies the relation

$$d_H(Tx,Ty) \leq c(d'(Tx,x) + d'(Ty,y))$$

for all $x, y \in X$, where $c \in (0, \frac{1}{2})$ is a constant. Then, T has a fixed point.

Proof: Let $x_0 \in X$ be given. Choose $x_1 \in Tx_0$ such that $d'(x_0,Tx_0) = \|d(x_0,x_1)\|$. If x_n has been given, then choose $x_{n+1} \in Tx_n$ such that $d'(x_n,Tx_n) = \|d(x_n,x_{n+1})\|$. Thus, we have

$$\|d(x_n,x_{n+1})\| = d'(x_n,Tx_n) \leq h_{Tx_{n-1}}(Tx_n) \leq d_H(Tx_{n-1},Tx_n)$$

$$\leq c(d'(Tx_{n-1},x_{n-1}) + d'(Tx_n,x_n)) = c(\|d(x_n,x_{n-1})\| + \|d(x_n,x_{n+1})\|),$$

for all $n \geq 1$. Hence,

$$\|d(x_n,x_{n+1})\| \leq \frac{c}{1-c}\|d(x_n,x_{n-1})\|,$$

for all $n \geq 1$. Put $s = \frac{c}{1-c}$. Then, for $n > m$ we have

$$\|d(x_n,x_m)\| \leq \sum_{i=m+1}^{n} \|d(x_i,x_{i-1})\|$$

$$\leq (s^{n-1} + \cdots + s^m)\|d(x_0,x_1)\| \leq \frac{s^m}{1-s}\|d(x_0,x_1)\|.$$

This implies that

$$\lim_{m,n\to\infty} \|d(x_n,x_m)\| = 0.$$

One can check that $\{x_n\}_{n\geq 1}$ is a Cauchy sequence in X. Thus, there exists $x^* \in X$ such that $x_n \to x^*$. Now by using Remark 4.2, we have

$$\begin{aligned} d'(x^*,Tx^*) &\leq d'(x^*,Tx_n) + h_{Tx_n}(Tx^*) \leq d'(x^*,Tx_n) + d_H(Tx_n,Tx^*) \\ &\leq \|d(x^*,x_{n+1})\| + c(d'(Tx_n,x_n) + d'(Tx^*,x^*)) \end{aligned}$$

for all $n \geq 1$. Hence,

$$\begin{aligned} d'(x^*,Tx^*) &\leq \frac{c}{1-c} d'(Tx_n,x_n) + \frac{1}{1-c}\|d(x^*,x_{n+1})\| \\ &= \frac{c}{1-c}\|d(x_{n+1},x_n)\| + \frac{1}{1-c}\|d(x^*,x_{n+1})\|, \end{aligned}$$

for all $n \geq 1$. Thus, $d'(x^*,Tx^*) = 0$ and so $x^* \in Tx^*$.

♣

Theorem 4.30

Let (X,d) be a complete cone metric space with normal constant $M = 1$ and the multifunction $T : X \to \mathcal{H}_c(X)$ satisfies the relation

$$d_H(Tx,Ty) \leq c(d'(Tx,y) + d'(Ty,x))$$

for all $x,y \in X$, where $c \in (0,\frac{1}{2})$ is a constant. Then, T has a fixed point.

Proof: Let $x_0 \in X$ be given. Choose $x_1 \in Tx_0$ such that $d'(x_0,Tx_0) = \|d(x_0,x_1)\|$. If x_n has been given, then choose $x_{n+1} \in Tx_n$ such that $d'(x_n,Tx_n) = \|d(x_n,x_{n+1})\|$. Thus, we have

$$\begin{aligned} \|d(x_n,x_{n+1})\| &= d'(x_n,Tx_n) \leq h_{Tx_{n-1}}(Tx_n) \leq d_H(Tx_{n-1},Tx_n) \\ &\leq c(d'(Tx_{n-1},x_n) + d'(Tx_n,x_{n-1})) \leq c(\|d(x_{n+1},x_{n-1})\| \\ &\leq c(\|d(x_{n-1},x_n)\| + \|d(x_n,x_{n+1})\|), \end{aligned}$$

for all $n \geq 1$. Hence,

$$\|d(x_n,x_{n+1})\| \leq \frac{c}{1-c}\|d(x_n,x_{n-1})\|,$$

for all $n \geq 1$. Put $s = \frac{c}{1-c}$. Then, for $n > m$ we have

$$\|d(x_n,x_m)\| \leq \sum_{i=m+1}^{n} \|d(x_i,x_{i-1})\|$$

$$\leq (s^{n-1}+\cdots+s^m)\|d(x_0,x_1)\| \leq \frac{s^m}{1-s}\|d(x_0,x_1)\|.$$

This implies that

$$\lim_{m,n\to\infty}\|d(x_n,x_m)\| = 0.$$

One can check that $\{x_n\}_{n\geq 1}$ is a Cauchy sequence in X. Thus, there exists $x^* \in X$ such that $x_n \to x^*$. Hence, we have

$$\begin{aligned} d'(x^*,Tx^*) &\leq d'(x^*,Tx_n)+h_{Tx_n}(Tx^*) \leq d'(x^*,Tx_n)+d_H(Tx_n,Tx^*) \\ &\leq \|d(x^*,x_{n+1})\|+c(d'(Tx_n,x^*)+d'(Tx^*,x_n)) \\ &\leq \|d(x^*,x_{n+1})\|+c(\|d(x^*,x_{n+1})\|+\|d(x_n,x^*)\|+d'(Tx^*,x^*)) \end{aligned}$$

for all $n \geq 1$. Hence,

$$d'(x^*,Tx^*) \leq \frac{1+c}{1-c}\|d(x^*,x_{n+1})\|+\frac{1}{1-c}\|d(x^*,x_n)\|,$$

for all $n \geq 1$. Thus, $d'(x^*,Tx^*) = 0$ and so $x^* \in Tx^*$.

♣

Here, we give an example that satisfies the assumptions of Theorem 4.30.

Example 68 *Let $X = \{a_1,a_2,a_3,\dots\}$ be a countable set, $E = (l^2,\|.\|_2)$ and $P = \{\{x_n\}_{n\geq 1} \in l^2 : x_n \geq 0\ (\forall n \geq 1)\}$. Put $x_i = \{\frac{3^i}{n}\}_{n\geq 1}$ for all $i \geq 1$ and note that $x_i \in l^2$ $(i \geq 1)$. Define the map $d : X\times X \to P$ by*

$$d(a_i,a_j) = |x_i - x_j| = \{\frac{|3^i-3^j|}{n}\}_{n\geq 1}.$$

It is easy to see that (X,d) is a cone metric space, the normal constant of P is $M = 1$ and there is no Cauchy sequence in (X,d). Hence, (X,d) is a complete cone metric space. Now, define the multifunction

$$T : X \to \mathcal{H}_c(X)$$

by $Ta_1 = \{a_1\}$ and $Ta_i = \{a_1,a_2,\dots,a_{i-1}\}$ for all $i \geq 1$. Then, we have $Ta_1 = Ta_2$ and for each $i \geq 3$

$$d_H(Ta_1,Ta_i) = \max\{d'(a_1,Ta_i),\ \sup_{b\in Ta_i} d'(Ta_1,b)\} = \|d(a_1,a_{i-1})\|_2$$

$$= \|x_1-x_{i-1}\|_2 = \{\sum_{n=1}^{\infty}|\frac{3}{n}-\frac{3^{i-1}}{n}|^2\}^{\frac{1}{2}} = (3^{i-1}-3)\sum_{n=1}^{\infty}\frac{1}{n^2}.$$

Also, we have $d'(a_1,Ta_i) = 0$ and

$$d'(a_i,Ta_1) = \|d(a_i,a_1)\|_2 = \|x_i-x_1\|_2 = (3^i-3)\sum_{n=1}^{\infty}\frac{1}{n^2}.$$

Thus, $d_H(Ta_1, Ta_i) \leq \frac{1}{3}(d'(a_1, Ta_i) + d'(a_i, Ta_1))$. *Now, suppose that* $j > i > 1$. *Then,* $Ta_i \subset Ta_j$ *and so* $\sup_{b \in Ta_i} d'(b, Ta_j) = 0$. *Hence,*

$$d_H(Ta_i, Ta_j) = \sup_{b \in Ta_j} d'(b, Ta_i) = \|d(a_{j-1}, a_{i-1})\|_2$$

$$= \|x_{i-1} - x_{j-1}\|_2 = (3^{j-1} - 3^{i-1}) \sum_{n=1}^{\infty} \frac{1}{n^2}.$$

Also, we have $d'(a_i, Ta_j) = 0$ *and*

$$d'(a_j, Ta_i) = \|d(a_j, a_{i-1})\|_2 = (3^j - 3^{i-1}) \sum_{n=1}^{\infty} \frac{1}{n^2}.$$

Thus, $d_H(Ta_i, Ta_j) \leq \frac{1}{3}(d'(a_i, Ta_j) + d'(a_j, Ta_i))$. *Thus,* T *satisfies in the assumptions of Theorem 4.30 and* a_1 *is the unique fixed point of* T.

Chapter 5

Some types of α-ψ-contractive set-valued mappings

In this chapter, we provide some recent fixed point results on new contractive type set-valued mappings. We would like readers of this book to be more familiar with different types of recent fixed point results. One of our aims in this chapter is for young researchers to learn different views with which they could work with set-valued mappings.

5.1 On α_*-ψ-contractive set-valued mappings

As we wrote, Samet et al. introduced the notion of α-ψ-contractive type mappings ([94]). Our aim is in this section is to generalize it to set-valued mappings. In this way, we introduce the notion of α_*-ψ-contractive set-valued maps. Let (X,d) be a metric space, $T : X \to 2^X$ a set-valued map, $\psi \in \Psi$ and $\alpha : X \times X \to [0,\infty)$ a function. In this case, we say that T is an α_*-ψ-contractive set-valued mapping whenever

$$\alpha_*(Tx,Ty)H(Tx,Ty) \leq \psi(d(x,y))$$

for all $x,y \in X$, where H is the Hausdorff metric and $\alpha_*(A,B) = \inf\{\alpha(a,b) : a \in A, b \in B\}$. Also, we say that T is α_*-admissible whenever $\alpha(x,y) \geq 1$ implies $\alpha_*(Tx,Ty) \geq 1$.

Example 69 *Let $X = [0,\infty)$, $d(x,y) = |x-y|$ and $\delta \in (0,1)$ a fixed number. Consider the set-valued map $T : X \to 2^X$ defined by $Tx = [0,\delta x]$ for all $x \in X$.*

Now, define the function $\alpha : X \times X \to [0,\infty)$ by $\alpha(x,y) = 1$ whenever $x,y \in [0,1]$ and $\alpha(x,y) = 0$ whenever $x \notin [0,1]$ or $y \notin [0,1]$. Now, we show that T is a α_*-admissible. If $\alpha(x,y) \geq 1$, then $x,y \in [0,1]$ and so Tx and Ty are subsets of $[0,1]$. Thus, $a,b \in [0,1]$ for all $a \in Tx$ and $b \in Ty$. Hence, $\alpha(a,b) = 1$ for all $a \in Tx$ and $b \in Ty$. This implies that

$$\alpha_*(Tx,Ty) = \inf\{\alpha(a,b) : a \in Tx, b \in Ty\} = 1.$$

Hence, T is α_*-admissible. Now, we show that T is an α_*-ψ-contractive set-valued map, where $\psi(t) = \delta t$ for all $t \geq 0$. If $x \notin [0,\frac{1}{\delta}]$ or $y \notin [0,\frac{1}{\delta}]$, then an easy calculation shows us that $\alpha_*(Tx,Ty) = 0$. If $0 \leq x,y \leq \frac{1}{\delta}$, then $\alpha_*(Tx,Ty) = 1$. By using the definition of the Hausdorff metric, it is easy to see that $H(Tx,Ty) \leq \delta d(x,y)$ for $x,y \in [0,\frac{1}{\delta}]$. Thus, $\alpha_*(Tx,Ty)H(Tx,Ty) \leq \psi(d(x,y))$ for all $x,y \in X$ and so T is an α_*-ψ-contractive set-valued map.

Let $(X,\preceq,d)$ be an ordered metric space and $A,B \subseteq X$. We say that $A \preceq B$ whenever for each $a \in A$ there exists $b \in B$ such that $a \preceq b$. Also, we say that $A \preceq_r B$ whenever for each $a \in A$ and $b \in B$ we have $a \preceq b$.

Theorem 5.1

Let (X,d) be a complete metric space, $\alpha : X \times X \to [0,\infty)$ a function, $\psi \in \Psi$ a strictly increasing map and T a closed valued, α_*-admissible and α_*-ψ-contractive set-valued map on X. Suppose that there exist $x_0 \in X$ and $x_1 \in Tx_0$ such that $\alpha(x_0,x_1) \geq 1$. Assume that if $\{x_n\}$ is a sequence in X such that $\alpha(x_n,x_{n+1}) \geq 1$ for all n and $x_n \to x$, then $\alpha(x_n,x) \geq 1$ for all n. Then T has a fixed point.

Proof: If $x_1 = x_0$, then we have nothing to prove. Let $x_1 \neq x_0$. If $x_1 \in Tx_1$, then x_1 is a fixed point of T. Let $x_1 \notin Tx_1$ and $q > 1$ be given. Then

$$0 < d(x_1,Tx_1) \leq \alpha_*(Tx_0,Tx_1)H(Tx_0,Tx_1) < q\alpha_*(Tx_0,Tx_1)H(Tx_0,Tx_1).$$

Hence, there exists $x_2 \in Tx_1$ such that

$$0 < d(x_1,x_2) < q\alpha_*(Tx_0,Tx_1)H(Tx_0,Tx_1) \leq q\psi(d(x_0,x_1)).$$

It is clear that $x_2 \neq x_1$ and $\alpha(x_1,x_2) \geq 1$. Thus, $\alpha_*(Tx_1,Tx_2) \geq 1$. Now, put $t_0 = d(x_0,x_1)$. Then, $t_0 > 0$ and $d(x_1,x_2) < q\psi(t_0)$. since ψ is strictly increasing,

$$\psi(d(x_1,x_2)) < \psi(q\psi(t_0)).$$

Put $q_1 = \frac{\psi(q\psi(t_0))}{\psi(d(x_1,x_2))}$. Then $q_1 > 1$. If $x_2 \in Tx_2$, then x_2 is a fixed point of T. Assume that $x_2 \notin Tx_2$. Then,

$$0 < d(x_2,Tx_2) \leq \alpha_*(Tx_1,Tx_2)H(Tx_1,Tx_2) < q_1\alpha_*(Tx_1,Tx_2)H(Tx_1,Tx_2).$$

Hence, there exists $x_3 \in Tx_2$ such that

$$0 < d(x_2,x_3) < q_1\alpha_*(Tx_1,Tx_2)H(Tx_1,Tx_2) \leq q_1\psi(d(x_1,x_2)) = \psi(q\psi(t_0)).$$

It is clear that $x_3 \neq x_2$, $\alpha(x_2,x_3) \geq 1$ and $\psi(d(x_2,x_3)) < \psi^2(q\psi(t_0))$. Put $q_2 = \frac{\psi^2(q\psi(t_0))}{\psi(d(x_2,x_3))}$. Then $q_2 > 1$. If $x_3 \in Tx_3$, then x_3 is a fixed point of T. Assume that $x_3 \notin Tx_3$. Then, we have $0 < d(x_3,Tx_3) \leq \alpha_*(Tx_2,Tx_3)H(Tx_2,Tx_3) < q_2\alpha_*(Tx_2,Tx_3)H(Tx_2,Tx_3)$. Thus, there exists $x_4 \in Tx_3$ such that

$$0 < d(x_3,x_4) < q_1\alpha_*(Tx_2,Tx_3)H(Tx_2,Tx_3) \leq q_2\psi(d(x_2,x_3)) = \psi^2(q\psi(t_0)).$$

By continuing this process, we obtain a sequence $\{x_n\}$ in X such that $x_n \in Tx_{n-1}$, $x_n \neq x_{n-1}$, $\alpha(x_n,x_{n+1}) \geq 1$ and $d(x_n,x_{n+1}) \leq \psi^{n-1}(q\psi(t_0))$ for all n. Now, for each $m > n$ we have $d(x_n,x_m) \leq \Sigma_{i=n}^{m-1} d(x_i,x_{i+1}) \leq \Sigma_{i=n}^{m-1} \psi^{i-1}(q\psi(t_0))$. Hence, $\{x_n\}$ is a Cauchy sequence in X. Choose $x^* \in X$ such that $x_n \to x^*$. Since $\alpha(x_n,x^*) \geq 1$ for all n and T is α_*-admissible, we get $\alpha_*(Tx_n,Tx^*) \geq 1$ for all n. Thus,

$$\begin{aligned} d(Tx^*,x^*) &\leq H(Tx^*,Tx_n) + d(x_{n+1},x^*) \leq \alpha_*(Tx_n,Tx^*)H(Tx_n,Tx^*) + d(x_{n+1},x^*) \\ &\leq \psi(d(x_n,x^*)) + d(x_{n+1},x^*) \end{aligned}$$

for all n. Thus, $d(Tx^*,x^*) = 0$ and so $x^* \in Tx^*$.

♣

Example 70 *Assume that $X = [0,\infty)$, $d(x,y) = |x-y|$ is the Euclidean metric, the set-valued map $T : X \to 2^X$ is defined by $Tx = [2x - \frac{3}{2}, \infty)$ for all $x > 1$ and $Tx = [0, \frac{x}{2}]$ for all $0 \leq x \leq 1$ and also the function $\alpha : X \times X \to [0,\infty)$ is defined by $\alpha(x,y) = 1$ whenever $x,y \in [0,1]$ and $\alpha(x,y) = 0$ whenever $x \notin [0,1]$ or $y \notin [0,1]$. Then it is easy to check that T is a α_*-admissible and α_*-ψ-contractive set-valued map, where $\psi(t) = \frac{t}{2}$ for all $t \geq 0$. Put $x_0 = 1$ and $x_1 = \frac{1}{2}$. Then, $\alpha(x_0,x_1) \geq 1$. Also if $\{x_n\}$ is a sequence in X such that $\alpha(x_n,x_{n+1}) \geq 1$ for all n and $x_n \to x$, then $\alpha(x_n,x) \geq 1$ for all n. Note that, T has infinitely many fixed points.*

Corollary 5.1

Let $(X,\preceq,d)$ be a complete ordered metric space, $\psi \in \Psi$ a strictly increasing map and T a closed valued multifunction on X such that

$$H(Tx,Ty) \leq \psi(d(x,y))$$

for all $x,y \in X$ with $x \preceq y$. Suppose that there exists $x_0 \in X$ and $x_1 \in Tx_0$ such that $x_0 \preceq x_1$. Assume that if $\{x_n\}$ is a sequence in X such that $x_n \preceq x_{n+1}$ for all n and $x_n \to x$, then $x_n \preceq x$ for all n. If $x \preceq y$ implies $Tx \preceq_r Ty$, then T has a fixed point.

Proof: Define $\alpha : X \times X \to [0,\infty)$ by $\alpha(x,y) = 1$ whenever $x \preceq y$ and $\alpha(x,y) = 0$ whenever $x \not\preceq y$. Since $x \preceq y$ implies $Tx \preceq_r Ty$, $\alpha(x,y) = 1$ implies $\alpha_*(Tx,Ty) = 1$. Thus, it is easy to check that T is a α_*-admissible and α_*-ψ-contractive set-valued map on X. Now by using Theorem 5.1, T has a fixed point.

♣

5.2 On β-ψ-ciric generalized set-valued mappings

In continuation of the previous section, we provide some new results. Let (X,d) be a metric space, $T : X \to 2^X$ a set-valued mapping, $\alpha : X \times X \to [0,\infty)$ a map, $\psi \in \Psi$ and $\beta : 2^X \times 2^X \to [0,\infty)$ a function. We say that T is an α-ψ-ξ-contractive selfmap whenever $\alpha(x,y)d(Tx,Ty) \leq \psi(h(x,y))$ for all $x,y \in X$, where

$$h(x,y) = d(x,Ty) + d(y,Tx) + d(x,y) - \xi(x,y)$$

and $\xi(x,y) = \max\{d(x,Ty), d(y,Tx)\}$. Also, we say that T is β-admissible whenever $\beta(A,B) \geq 1$ implies $\beta(Tx,Ty) \geq 1$ for all $x \in A$ and $y \in B$, where A and B are subsets of X. Also, we say that T is β-ψ-ξ-contractive set-valued map whenever

$$\beta(Tx,Ty)H(Tx,Ty) \leq \psi(d(x,Ty) + d(y,Tx) + d(x,y) - \xi(x,y)) = \psi(h(x,y))$$

for all $x,y \in X$, where H is the Hausdorff metric. Also, we say that the set-valued map T is lower semi-continuous (briefly, LSC) at $x_0 \in X$ whenever for each sequence $\{x_n\}$ with $x_n \to x_0$ and every $y \in Tx_0$, there exists a sequence $\{y_n\}$ such that $y_n \to y$ and $y_n \in Tx_n$ for all n. Let C a nonempty subset of X and $x \in X$. An element $y_0 \in C$ is said to be a best approximation of x whenever $d(x,y_0) = d(x,C) = \inf_{y \in C} d(x,y)$. The set C is said to be a proximinal whenever every $x \in X$ has at least one best approximation in C. As we mentioned before, proximinal subsets are closed. In this section, let's denote by $P(X)$ the set of all proximinal subsets of X.

Theorem 5.2

Let (X,d) be a complete metric space and T a continuous, α-admissible and α-ψ-ξ-contractive selfmap on X such that $\alpha(x_0,Tx_0) \geq 1$ for some $x_0 \in X$. Then T has a fixed point.

Proof: Let $x_0 \in X$ be such that $\alpha(x_0,Tx_0) \geq 1$. Put $x_{n+1} = Tx_n$ for all $n \geq 0$. If $x_n = x_{n+1}$ for some n, then we have nothing to prove. Assume that $x_n \neq x_{n+1}$ for all n. Since T is α-admissible, it is easy to check that $\alpha(x_n,x_{n+1}) \geq 1$ for all n.

Thus,

$$d(x_n,x_{n+1}) = d(Tx_{n-1},Tx_n) \leq \alpha(x_{n-1},x_n)d(Tx_{n-1},Tx_n) \leq \psi(h(x_{n-1},x_n))$$
$$= \psi(d(x_{n-1},Tx_n)+d(x_n,Tx_{n-1})+d(x_{n-1},x_n)-\xi(x_{n-1},x_n))$$
$$= \psi(d(x_{n-1},x_{n+1})+d(x_n,x_n)+d(x_{n-1},x_n)-max\{d(x_{n-1},x_{n+1}),d(x_n,x_n)\})$$
$$= \psi(d(x_{n-1},x_{n+1})+d(x_{n-1},x_n)-d(x_{n-1},x_{n+1})) = \psi(d(x_{n-1},x_n))$$

for all n. Hence, $d(x_{n+1},x_n) \leq \psi^n(d(x_0,x_1))$ for all n. Fix $\varepsilon > 0$. Then, there exists a natural number N_ε such that $\sum_{n\geq N_\varepsilon} \psi^n(d(x_0,x_1)) < \varepsilon$. Let $m > n \geq N_\varepsilon$. By using the triangular inequality, we obtain

$$d(x_n,x_m) \leq \sum_{k=n}^{m-1} d(x_k,x_{k+1}) \leq \sum_{k=n}^{m-1} \psi^k d(x_0,x_1) \leq \sum_{n\geq N_\varepsilon} \psi^n d(x_0,x_1) < \varepsilon.$$

Hence, $\{x_n\}$ is a Cauchy sequence. Since (X,d) is complete, there exists $x^* \in X$ such that $x_n \to x^*$. Since T is continuous, $Tx^* = x^*$. This completes the proof.

♣

Now, we give the following example to show the difference of Theorem 5.2 and the first result of [94].

Example 71 *Let $X = \mathbb{R}$ and $d(x,y) = |x-y|$. Define $Tx = \frac{4}{3}x$ for all $x \in \mathbb{R}$, $\psi(t) = \frac{3}{4}t$ for all $t \geq 0$ and $\alpha : X \times X \to [0,+\infty)$ by $\alpha(x,y) = 1$ whenever $y \leq \frac{7}{6}x$ and $\alpha(x,y) = 0$ otherwise. If $y \geq \frac{7}{6}x$, then $\alpha(x,y)d(Tx,Ty) = 0 \leq \psi(h(x,y))$. If $y \leq \frac{7}{6}x$, then*

$$\max\{|x-\frac{4}{3}y|,|\frac{4}{3}x-y|\} = |x-\frac{4}{3}y|.$$

Thus, we have

$$\alpha(x,y)d(Tx,Ty) = \frac{4}{3}|x-y| \leq \frac{3}{2}|\frac{4}{3}x-y|$$
$$= \frac{3}{4}(|\frac{4}{3}x-y|+|\frac{4}{3}x-y|) \leq \frac{3}{4}(|\frac{4}{3}x-y|+|x-y|)$$
$$= \psi(|\frac{4}{3}x-y|+|x-y|)) = \psi(d(x,Ty)+d(y,Tx)+d(x,y)-\xi(x,y)).$$

Hence, T is an α-ψ-ξ-contractive selfmap. On the other hand, for $y \leq \frac{7}{6}x$ we have

$$\alpha(x,y)d(Tx,Ty) = \frac{4}{3}|x-y| \gneqq \frac{3}{4}|x-y| = \psi(d(x,y))$$

and so T is not α-ψ-contractive.

Corollary 5.2

Let $(X,d,\leq)$ be a complete ordered metric space and T a continuous and nondecreasing selfmap on X such that $d(Tx,Ty) \leq \lambda h(x,y)$ for all $x,y \in X$ with $x \leq y$ or $y \leq x$, where λ is an element in $[0,1)$. If there exists $x_0 \in X$ such that $x_0 \leq Tx_0$ or $Tx_0 \leq x_0$, then T has a fixed point.

Corollary 5.3

Let (X,d) be a complete metric space, $\lambda \in [0,1)$ and T a continuous selfmap on X such that $T(A) \subset A$ for some subset A of X and $d(Tx,Ty) \leq \lambda h(x,y)$ for all $x,y \in A$. Then T has a fixed point.

Proof: Define the mapping $\alpha : X \times X \to [0,+\infty)$ by $\alpha(x,y) = 1$ whenever $x \in A$ or $y \in A$ and $\alpha(x,y) = 0$ otherwise. Then, we have $\alpha(x,y)d(Tx,Ty) \leq kh(x,y)$ for all $x,y \in X$. Define $\psi(t) = kt$ for all $t \geq 0$. Thus, T is an α-ψ-ξ-contractive mapping. Let $x,y \in X$ be such that $\alpha(x,y) \geq 1$. Since $T(A) \subset A$, $Tx \in A$ or $Ty \in A$ and so $\alpha(Tx,Ty) \geq 1$. Hence, T is α-admissible. Since A is nonempty, $\alpha(x_0,Tx_0) = 1$ for all $x_0 \in A$. Now by using Theorem 5.2, T has a fixed point.

♣

Now, we give the following result for proximinal valued set-valued mappings.

Theorem 5.3

Let (X,d) be a complete metric space and $T : X \to P(X)$ a LSC, β-admissible and β-ψ-ξ-contractive multifunction such that $\beta(A,Tx_0) \geq 1$ for some $A \subset X$ and $x_0 \in A$. Then T has a fixed point.

Proof: Choose $A \subset X$ and $x_0 \in A$ such that $\beta(A,Tx_0) \geq 1$. Define the sequence $\{x_n\}$ by $x_{n+1} \in Tx_n$ and $d(x_n,x_{n+1}) = d(x_n,Tx_n)$ for all $n \geq 0$. If $x_n = x_{n+1}$ for some n, then we have nothing to prove. Assume that $x_n \neq x_{n+1}$ for all n. Since T is β-admissible, $x_0 \in A$, $x_1 \in Tx_0$ and $\beta(A,Tx_0) \geq 1$, we have $\beta(Tx_0,Tx_1) \geq 1$. By continuing this process it is easy to show that $\beta(Tx_{n-1},Tx_n) \geq 1$ for all n. Thus,

$$d(x_n,x_{n+1}) = d(x_n,Tx_n) \leq H(Tx_{n-1},Tx_n)$$

$$\leq \beta(Tx_{n-1},Tx_n)H(Tx_{n-1},Tx_n) \leq \psi(h(x_{n-1},x_n))$$

$$= \psi(d(x_{n-1},Tx_n) + d(x_n,Tx_{n-1}) + d(x_{n-1},x_n) - \xi(x_{n-1},x_n)) = \psi(d(x_{n-1},x_n))$$

and so $d(x_n,x_{n+1}) \leq \psi^n(d(x_0,x_1))$ for all n. Fix $\varepsilon > 0$. Then there exists a natural number N_ε such that $\sum_{n\geq N_\varepsilon} \psi^n(t) < \varepsilon$. Let $m > n \geq N_\varepsilon$. Then,

$$d(x_n, x_m) \leq \sum_{k=n}^{m-1} d(x_k, x_{k+1}) \leq \sum_{k=n}^{m-1} \psi^k d(x_0, x_1) \leq \sum_{n\geq N_\varepsilon} \psi^n d(x_0, x_1) < \varepsilon.$$

Thus, $\{x_n\}$ is a Cauchy sequence. Choose $x^* \in X$ such that $x_n \to x^*$. Let $y \in Tx^*$. Since T is LSC, there exists a sequence $\{y_n\}$ such that $y_n \to y$ and $y_n \in Tx_n$ for all n. Hence, $d(x^*, Tx^*) \leq d(x^*, y) \leq d(x^*, x_{n+1}) + d(x_{n+1}, z) + d(z, y_n) + d(y_n, y)$ for every member $z \in Tx_n$. This implies that

$$\begin{aligned} d(x^*, Tx^*) &\leq d(x^*, x_{n+1}) + \inf_{z\in Tx_n} d(x_{n+1}, z) + \inf_{z\in Tx_n} d(z, y_n) + d(y_n, y) \\ &= d(x^*, x_{n+1}) + d(x_{n+1}, Tx_n) + d(Tx_n, y_n) + d(y_n, y) \\ &= d(x^*, x_{n+1}) + d(y_n, y), \end{aligned}$$

and so $d(x^*, Tx^*) \leq d(x^*, x_{n+1}) + d(y_n, y)$ for all n. Thus, we get $x^* \in Tx^*$.

♣

Next example shows that there are set-valued mappings which satisfy the assumptions of Theorem 5.3.

Example 72 *Let $X = [0, \infty)$, $a > 0$, $d(x, y) = |x - y|$ for all $x, y \in X$, H the Hausdorff metric, T a proximinal-valued multifunction on X defined by $Tx = [x, a]$ whenever $x \leq a$ and $Tx = [a, x]$ whenever $x > a$ and $\beta : 2^X \times 2^X \to [0, +\infty)$ a mapping defined by $\beta(C, D) = 1$ whenever $C \cap D = \{a\}$ and $\beta(C, D) = 0$ otherwise. Suppose that A and B are subsets of X such that $A \cap B = \{a\}$. Then, $\beta(Tx, Ty) = 1$ whenever $x \leq a < y$ or $y \leq a < x$. If $x \leq a < y$, then $\rho(Tx, Ty) = a - x$ and $\rho(Ty, Tx) = y - a$, where $\rho(A, B) = \sup_{a\in A} d(a, B)$. Hence, $H(Tx, Ty) = \max\{a - x, y - a\}$. If $a - x > y - a$, then*

$$\max\{a - x, y - a\} = a - x.$$

Also, we have

$$\begin{aligned} \beta(Tx, Ty)H(Tx, Ty) &= (a - x) < (y - a) + (y - a) + (a - x) \\ &= (a - x) + (y - a) + (y - x) - \max\{a - x, y - a\}. \end{aligned}$$

Now, by using the Archimedean property, there exists $k \in [0, 1)$ such that

$$(a - x) \leq k((a - x) + (y - a) + (y - x) - \max\{a - x, y - a\}).$$

If we define $\psi(t) = kt$, then

$$\beta(Tx, Ty)H(Tx, Ty) = (a - x) \leq \psi((a - x) + (y - a) + (y - x) - \max\{a - x, y - a\})$$

$$= \psi(d(x,Ty) + d(y,Tx) + d(x,y) - \xi(x,y)).$$

Therefore, by providing a similar proof for another case one can show that T is a β-ψ-ξ-contractive multifunction. It is easy to see that T is β-admissible and LSC. Let $a \leq c$ and $A = [a,c]$. Then, $Ta = \{a\}$ and $\beta(A,Ta) = 1$. Thus, the multifunction T satisfies the assumptions of Theorem 5.3. Note that, each element of the interval $[0,\infty)$ is a fixed point of T.

Corollary 5.4

Let (X,d) be a complete metric space, $\lambda \in [0,1)$, $T : X \to P(X)$ a LSC multifunction and C a nonempty subset of X such that $Tx \subset C$ for all $x \in C$. Suppose that $H(Tx,Ty) \leq \lambda h(x,y)$ for all $x,y \in C$. Then T has a fixed point.

Proof: Define $\beta : 2^X \times 2^X \to [0,+\infty)$ by $\beta(A,B) = 1$ whenever $A \subset C$ or $B \subset C$ and $\beta(A,B) = 0$ otherwise. Define $\psi(t) = kt$ for all $t \geq 0$. Then, we have

$$\beta(Tx,Ty)H(Tx,Ty) \leq \psi(h(x,y))$$

for all $x,y \in X$. Hence, T is a β-ψ-ξ-contractive set-valued map. If $A,B \subset X$ and $\beta(A,B) \geq 1$, then $A \subset C$ or $B \subset C$. Without loss of generality, suppose that $A \subset C$. Then, $Tx \subset C$ for all $x \in A$ and so $\beta(Tx,Ty) \geq 1$ for all $y \in B$. Therefore, T is β-admissible. If $x \in C$, then $Tx \subset C$ and so $\beta(C,Tx) = 1$. Now by using Theorem 5.3, T has a fixed point.

♣

Theorem 5.4

Let (X,d) be a complete metric space and $T : X \to P(X)$ a β-admissible and β-ψ-ξ-contractive set-valued mapping such that $\beta(A,Tx_0) \geq 1$ for some $A \subset X$ and $x_0 \in A$. Also, suppose that $\beta(Tx_{n-1},Tx) \geq 1$ for all n whenever $\{x_n\}$ is a sequence in X such that $\beta(Tx_{n-1},Tx_n) \geq 1$ for all n and $x_n \to x$. Then T has a fixed point.

Proof: Choose $A \subset X$ and $x_0 \in A$ such that $\beta(A,Tx_0) \geq 1$. Define the sequence $\{x_n\}$ by $x_{n+1} \in Tx_n$ and $d(x_n,x_{n+1}) = d(x_n,Tx_n)$ for all $n \geq 0$. If $x_n = x_{n+1}$ for some n, then we have nothing to prove. Assume that $x_n \neq x_{n+1}$ for all n. By using a similar technique in proof of Theorem 5.3, one can deduce that $\{x_n\}$ is a Cauchy sequence. Choose $x^* \in X$ such that $x_n \to x^*$. Since $\beta(Tx_{n-1},Tx_n) \geq 1$ for all n, by using the assumption we obtain $\beta(Tx_{n-1},Tx^*) \geq 1$ for all n. Hence,

$$d(x^*,Tx^*) \leq d(x^*,z) + d(z,Tx^*)$$

for all $z \in Tx_{n-1}$. But, we have

$$d(x^*,Tx^*) \leq d(x^*,Tx_{n-1}) + H(Tx_{n-1},Tx^*)$$
$$\leq d(x^*,x_n) + \beta(Tx_{n-1},Tx^*)H(Tx_{n-1},Tx^*) \leq d(x^*,x_n) + \psi(h(x_{n-1},x^*))$$
$$\leq d(x^*,x_n) + \psi(d(x_{n-1},Tx^*) + d(x^*,Tx_{n-1}) + d(x_{n-1},x^*) - \xi(x_{n-1},x^*))$$

for all n. If $\xi(x_{n-1},x^*) = d(x_{n-1},Tx^*)$, then we have

$$d(x^*,Tx^*) \leq d(x^*,x_n) + \psi(d(x_{n-1},x^*) + d(x_{n-1},x^*))$$

and if $\xi(x_{n-1},x^*) = d(x^*,Tx_{n-1})$, then we have

$$d(x^*,Tx^*) \leq d(x^*,x_n) + \psi(d(x_{n-1},Tx^*) + d(x_{n-1},x^*)).$$

These implies that $d(x^*,Tx^*) = 0$ and so $x^* \in Tx^*$.

♣

5.3 On α-ψ-ciric generalized set-valued mappings

Let (X,d) be a metric space, $\beta : 2^X \times 2^X \to [0,+\infty)$ and $\alpha : X \times X \to [0,+\infty)$ two functions and $\psi \in \Psi$. A set-valued operator $T : X \to 2^X$ is said to be β-ψ-contractive whenever

$$\beta(Tx,Ty)H(Tx,Ty) \leq \psi(d(x,y))$$

for all $x,y \in X$, where H is the Hausdorff metric. Also, we say that a set-valued map $T : X \to 2^X$ is generalized α-ψ-Ciric map whenever

$$\alpha(x,y)d(Tx,Ty) \leq \psi(max\{d(x,y),d(x,Tx),d(y,Ty),\frac{d(x,Ty)+d(y,Tx)}{2}\})$$

for all $x,y \in X$.

First let us to state and prove a single valued version fixed point result.

Lemma 5.1

Let (X,d) be a complete metric space, $\alpha : X \times X \to [0,\infty)$ a function, $\psi \in \Psi$ and T a selfmap on X such that

$$\alpha(x,y)d(Tx,Ty) \leq \psi(\max\{d(x,y),d(x,Tx),d(y,Ty),\frac{1}{2}[d(x,Ty)+d(y,Tx)]\})$$

for all $x,y \in X$. Suppose that T is α-admissible and there exists $x_0 \in X$ such that

$\alpha(x_0, Tx_0) \geq 1$. *Assume that if* $\{x_n\}$ *is a sequence in* X *such that* $\alpha(x_n, x_{n+1}) \geq 1$ *for all* n *and* $x_n \to x$, *then* $\alpha(x_n, x) \geq 1$ *for all* n. *Then* T *has a fixed point.*

Proof: Take $x_0 \in X$ such that $\alpha(x_0, Tx_0) \geq 1$ and define the sequence $\{x_n\}$ in X by $x_{n+1} = Tx_n$ for all $n \geq 0$. If $x_n = x_{n+1}$ for some n, then $x^* = x_n$ is a fixed point of T. Assume that $x_n \neq x_{n+1}$ for all n. Since T is α-admissible, it is easy to check that $\alpha(x_n, x_{n+1}) \geq 1$ for all natural number n. Thus, for each natural number n we have

$$d(x_n, x_{n+1}) = d(Tx_{n-1}, Tx_n) \leq \alpha(x_{n-1}, x_n) d(Tx_{n-1}, Tx_n)$$

$$\leq \psi(\max\{d(x_n, x_{n-1}), d(x_n, x_{n+1}), d(x_{n-1}, x_n), \frac{1}{2}[d(x_n, x_n) + d(x_{n-1}, x_{n+1})]\})$$

$$\leq \psi(\max\{d(x_n, x_{n-1}), d(x_n, x_{n+1}), \frac{1}{2}[d(x_n, x_{n-1}) + d(x_n, x_{n+1})]\})$$

$$= \psi(\max\{d(x_n, x_{n-1}), d(x_n, x_{n+1})\}).$$

If $\max\{d(x_n, x_{n-1}), d(x_n, x_{n+1})\} = d(x_n, x_{n+1})$, then

$$d(x_{n+1}, x_n) \leq \psi(d(x_n, x_{n+1})) < d(x_{n+1}, x_n)$$

which is contradiction. Thus, $\max\{d(x_n, x_{n-1}), d(x_n, x_{n-1})\} = d(x_n, x_{n+1})$ for all n. Hence, $d(x_{n+1}, x_n) \leq \psi(d(x_n, x_{n-1}))$ and so $d(x_{n+1}, x_n) \leq \psi^n(d(x_1, x_0))$ for all n. It is easy to check that $\{x_n\}$ is a Cauchy sequence. Thus, there exists $x^* \in X$ such that $x_n \to x^*$. By using the assumption, we have $\alpha(x_n, x^*) \geq 1$ for all n. Thus,

$$d(Tx^*, x^*) \leq d(Tx^*, Tx_n) + d(x_{n+1}, x^*) \leq \alpha(x_n, x^*) d(Tx^*, Tx_n) + d(x_{n+1}, x^*) \leq$$

$$\psi(\max\{d(x_n, x^*), d(x_n, x_{n+1}), d(x^*, Tx^*), \frac{1}{2}[d(x_n, Tx^*) + d(x^*, x_{n+1})]\}) + d(x_{n+1}, x^*)$$

$$= \psi(\max\{d(x_n, x^*), d(x_n, x_{n+1})\}) + d(x_{n+1}, x^*)$$

for all n. Hence, $d(Tx^*, x^*) = 0$ and so $Tx^* = x^*$.

♣

Here, we give the following example which shows that Lemma 5.1 holds while we can not use Theorem 1 in [30].

Example 73 *Let* $M_1 = \{\frac{m}{n} : m = 0, 1, 3, 9, \ldots$ *and* $n = 3k + 1$ $(k \geq 0)\}$, $d(x, y) = |x - y|$, $M_2 = \{\frac{m}{n} : m = 1, 3, 9, 27, \ldots$ *and* $n = 3k + 2$ $(k \geq 0)\}$ *and* $M_3 = \{2k : k \geq 1\}$. *Now, put* $M = M_1 \cup M_2 \cup M_3$ *and define the selfmap* $T : M \to M$ *by* $Tx = \frac{3x}{11}$ *whenever* $x \in M_1$, $Tx = \frac{x}{8}$ *whenever* $x \in M_2$ *and* $Tx = 2x$ *whenever* $x \in M_3$. *Put*

$$M(x, y) = max\{d(x, y), d(x, Tx), d(y, Ty), \frac{d(x, Ty) + d(y, Tx)}{2}\}$$

for all $x,y \in M$. If $x \in M_1$ and $y \in M_2$, then $d(Tx,Ty) = |\frac{3x}{11} - \frac{y}{8}| = \frac{3}{11}|x - \frac{11y}{24}|$. Now we consider two cases. If $x > \frac{11}{24}y$, then we have

$$\frac{3}{11}|x - \frac{11y}{24}| = \frac{3}{11}(x - \frac{11y}{24}) \leq \frac{3}{11}(x - \frac{y}{8}) = \frac{3}{11}d(x,Ty).$$

Hence, $d(Tx,Ty) \leq \frac{6}{11}\frac{d(x,Ty)}{2} \leq \frac{6}{11}M(x,y)$. If $x < \frac{11y}{24}$, then

$$\frac{3}{11}|x - \frac{11y}{24}| = \frac{3}{11}(\frac{11y}{24} - x) \leq \frac{3(y-x)}{11} = \frac{3}{11}d(x,y) \leq \frac{6}{11}M(x,y).$$

Hence, $d(Tx,Ty) \leq \frac{6}{11}M(x,y)$. If $x,y \in M_1$, then

$$d(Tx,Ty) = |\frac{3x}{11} - \frac{3y}{11}| = \frac{3}{11}d(x,y) \leq \frac{6}{11}M(x,y).$$

If $x,y \in M_2$, then $d(Tx,Ty) = |\frac{x}{8} - \frac{y}{8}| = \frac{1}{8}d(x,y) \leq \frac{6}{11}M(x,y)$. Define $\alpha(x,y) = 1$ whenever $x,y \in M_1 \cup M_2$ and $\alpha(x,y) = 1$ otherwise. Also, put $\psi(t) = \frac{6t}{11}$ for all $t \geq 0$. Then, it is easy to see that $\alpha(x,y)d(Tx,Ty) \leq \psi(M(x,y))$ for all $x,y \in M$. Also, $\alpha(1,T1) = \alpha(1,3/11) \geq 1$. If $\alpha(x,y) \geq 1$, then $x,y \in M_1 \cup M_2$ and so $\alpha(Tx,Ty) \geq 1$. Thus, T is α-admissible. It is easy to show that if $\{x_n\}$ is a sequence in M such that $\alpha(x_n,x_{n+1}) \geq 1$ for all n and $x_n \to x$, then $\alpha(x_n,x) \geq 1$ for all n. Now by using Lemma 5.1, T has a fixed point. Now, we show that Theorem 1 in [30] does not hold here. Put $x = 1$ and $y = 2$. Then, $d(T1,T2) = |\frac{3}{11} - 4| = \frac{41}{11}$, $d(1,2) = 1$, $d(1,T1) = d(1,\frac{3}{11}) = \frac{8}{11}$, $d(2,T2) = d(2,4) = 2$, $d(1,T2) = d(1,4) = 3$, $d(2,T1) = d(2,\frac{3}{11}) = \frac{19}{11}$. Thus, $N(1,2) = 3$ and so $d(T1,T2) = \frac{41}{11} > 3 = N(1,2)$, where

$$N(x,y) = max\{d(x,y),d(x,Tx),d(y,Ty),d(x,Ty),d(y,Tx)\}.$$

This shows that Theorem 1 in [30] does not hold.

5.4 On α-ψ-ciric generalized multifunctions

Let (X,d) be a metric space and G an undirected graph such that the set $V(G)$ of its vertices coincides with X and the set $E(G)$ of its edges contains all loops. We assume that G has no parallel edges. We denote it for abbreviation by (G,d) and you can call it a graph-metric space. We say that a self-map f on the graph G preserves edges of G whenever $(x,y) \in E(G)$ implies $(fx,fy) \in E(G)$ for all $x,y \in X$. A graph G is connected if there is a path $x_0 = x, \cdots, x_n = y$ between any two vertices x and y. In this case, we say that the length of this path is n. If there is a path between x and y, we denote it by $y \in [x]_G$. It is clear that there is path via minimum length between each two vertices when G is a finite graph. But, we deal with infinite graphs and so we assume that there is path via minimum length between each two vertices and we denote it by $o(x,y) = n$. We say that G is a (C)-graph whenever for each sequence $\{x_n\}_{n\geq 0}$ in X with $x_n \to x$ and $(x_n, x_{n+1}) \in E(G)$ for all $n \geq 0$, there is a subsequence $\{x_{n_k}\}_{k\geq 0}$ such that $(x_{n_k}, x) \in E(G)$ for all $k \geq 0$.

We say that a set-valued map $T : G \to 2^G$ has comparable approximative valued property whenever for each $x \in G$ there exists $y \in Tx$ such that $(x,y) \in E(G)$ and $d(x,Tx) = d(x,y)$. Denote by Φ the set of all continuous and increasing mappings $\phi : [0,\infty) \to [0,\infty)$ such that $\phi(t) < t$ for all $t > 0$. Let ϕ be a function in Φ. The set-valued map $T : G \to 2^G$ is called weakly generalized ϕ-quasi-contractive whenever $H(Tx,Ty) \leq \phi(M(x,y))$ for all $x,y \in G$ with $y \in [x]_G$, where

$$M(x,y) = max\{d(x,y), d(x,Tx), d(y,Ty), \frac{(d(x,Ty) + d(y,Tx))}{2}\}$$

and H is the Hausdorff metric.

Theorem 5.5

Let (G,d) be a complete graph-metric space, $\phi \in \Phi$ and $T : G \to 2^G$ a weakly generalized ϕ-quasi-contractive set-valued map which has comparable approximative valued property. If G is a (C)-graph, then T has a fixed point.

Proof: Let $x_0 \in X$. If $x_0 \in Tx_0$, then we have nothing to prove. Suppose that $x_0 \notin Tx_0$. Since T has comparable approximative valued property, there exists $x_1 \in Tx_0$ such that $(x_0,x_1) \in E(G)$ and $d(x_0,Tx_0) = d(x_0,x_1)$. It is clear that $x_1 \neq x_0$. If $x_1 \in Tx_1$, then we have nothing to prove. Suppose that $x_1 \notin Tx_1$. Now, there exists $x_2 \in Tx_1$ such that $(x_0,x_1) \in E(G)$ and $d(x_1,Tx_1) = d(x_1,x_2)$. It is clear that $x_2 \neq x_1$. By continuing this process, we obtain a sequence $\{x_n\}$ in X such that $x_n \in Gx_{n-1}$, $(x_{n-1},x_n) \in E(G)$ and $d(x_{n-1},x_n) = d(x_{n-1},Tx_{n-1})$ for all n. Now, we have

$$\begin{aligned} d(x_{n-1},x_n) &= d(x_{n-1},Tx_{n-1}) \leq H(Tx_{n-2},Tx_{n-1}) \leq \phi(M(x_{n-2},x_{n-1})) \\ &\leq \phi(max\{d(x_{n-2},x_{n-1}), d(x_{n-2},Tx_{n-2}), d(x_{n-1},Tx_{n-1}), \\ &\qquad \frac{1}{2}[d(x_{n-1},Tx_{n-2}) + d(x_{n-2},Tx_{n-1})]\}) \\ &\leq \phi(max\{d(x_{n-2},x_{n-1}), d(x_{n-1},x_n), 1/2d(x_{n-2},x_n)\}) \\ &\leq \phi(max\{d(x_{n-2},x_{n-1}), d(x_{n-1},x_n)\}) \end{aligned}$$

for all n and so we get $d(x_{n-1},x_n) \leq \phi(d(x_{n-2},x_{n-1}))$ for all n. Thus,

$$d(x_{n-1},x_n) \leq \phi(d(x_{n-2},x_{n-1})) \leq \phi^2(d(x_{n-3},x_{n-2}))... \leq \phi^{n-1}(d(x_0,x_1))$$

for all n and so $d(x_{n-1},x_n) \to 0$. If $\{x_n\}$ is not a Cauchy sequence, then there exists $\varepsilon > 0$ and subsequences $\{x_{n_i}\}$ and $\{x_{m_i}\}$ of $\{x_n\}$ with $n_i < m_i$ such that $d(x_{n_i},x_{m_i}) > \varepsilon$ for all i. For each n_i, put $k_i = \min\{m_i \mid d(x_{n_i},x_{m_i}) > \varepsilon\}$. Then, we have

$$\varepsilon < d(x_{n_i},x_{k_i}) \leq d(x_{n_i},x_{k_i-1}) + d(x_{k_i-1},x_{k_i}) \leq \varepsilon + d(x_{k_i-1},x_{k_i})$$

and so $d(x_{n_i},x_{k_i}) \to \varepsilon$. But, we have

$$d(x_{n_i},x_{k_i}) - d(x_{n_i},x_{n_i+1}) - d(x_{k_i},x_{k_i+1})$$

$$\leq d(x_{n_i+1},x_{k_i+1}) \leq d(x_{n_i+1},x_{k_i+1}) + d(x_{n_i},x_{n_i+1}) + d(x_{k_i},x_{k_i+1})$$

and so $d(x_{n_i+1},x_{k_i+1}) \to \varepsilon$. Thus, we get

$$\max\{d(x_{n_i},x_{k_i}), d(x_{n_i},Tx_{n_i}), d(x_{k_i},Tx_{k_i}), \frac{1}{2}[d(x_{n_i},Tx_{k_i}) + d(x_{k_i},Tx_{n_i})]\}$$

$$\leq d(x_{n_i},x_{k_i}) + \frac{3}{2}d(x_{k_i+1},x_{k_i}) + d(x_{n_i+1},x_{n_i})$$

and so $M(x_{n_i},x_{k_i}) \to \varepsilon$. On the other hand, we have

$$d(x_{n_i+1},x_{k_i+1}) \leq d(x_{n_i+1},Tx_{n_i}) + H(Tx_{n_i},Tx_{k_i})$$

$$= H(Tx_{n_i},Tx_{k_i}) \leq \phi(M(x_{n_i},x_{k_i}))$$

and so $\varepsilon \leq \phi(\varepsilon)$. This contradiction shows that $\{x_n\}$ is a Cauchy sequence. Since G is complete, there exists $x \in G$ such that $x_n \to x$. Since G is a (C)-graph, there is a subsequence $\{x_{n_k}\}_{k\geq 0}$ such that $(x_{n_k},x) \in E(G)$ for all $k \geq 0$. Thus, we obtain

$$d(x,Tx) = \lim_{n\to\infty} d(x_{n_k+1},Tx) \leq \lim_{n\to\infty} H(Tx_{n_k},Tx) \leq \lim_{n\to\infty} \phi(M(x_{n_k},x))$$

$$\leq \lim_{n\to\infty} \phi(\max\{d(x_{n_k},x), d(x_{n_k},Tx_{n_k}), d(x,Tx), \frac{d(x_{n_k},Tx) + d(x,Tx_{n_k})}{2}\})$$

$$\leq \lim_{n\to\infty} \phi(\max\{d(x_{n_k},x), (d(x_{n_k},x_{n_k+1}), d(x,Tx), \frac{d(x_{n_k},Tx) + d(x,x_{n_k+1})}{2}\})$$

$$= \phi(d(x,Tx))$$

and so $d(x,Tx) = 0$. Since T has comparable approximative valued property, there exists $y \in Tx$ such that $(x,y) \in E(G)$ and $d(x,Tx) = d(x,y)$. Hence, $d(x,y) = 0$ and so $x = y \in Tx$.

♣

Corollary 5.5

Let (G,d) be a complete graph-metric space, $\phi \in \Phi$ and $T : G \to 2^G$ a weakly generalized ϕ-quasi-contractive and compact valued set-valued map such that for each $x \in G$ and $y \in Tx$ we have $(x,y) \in E(G)$. If G is a (C)-graph, then T has a fixed point.

Corollary 5.6

Let (G,d) be a complete graph-metric space, $\phi \in \Phi$ and f a selfmap on G which preserves edges of G and $d(fx,fy) \leq \phi(M(x,y))$ for all $x,y \in X$ with $y \in [x]_G$. If G is a (C)-graph, then f has a fixed point.

Proof: It is sufficient we define $T : G \to 2^G$ by $Tx = \{fx\}$ for all $x \in G$.

♣

Note that, one can present more similar results via interesting corollaries by considering the above ones (see [55] for more details).

Proposition 5.1

Let $\psi : [0,\infty) \to [0,\infty)$ a lower semi-continuous function, $\eta : [0,\infty) \to [0,\infty)$ a map such that $\eta^{-1}(\{0\}) = \{0\}$ and $\liminf_{t\to\infty} \eta(t) > 0$ for all $t > 0$, (G,d) be a complete graph-metric space and $T : G \to 2^G$ a multifunction such that has comparable approximative valued property and

$$\psi(H(Tx,Ty)) \leq \psi(M(x,y)) - \eta(M(x,y))$$

for all $x,y \in X$ with $y \in [x]_G$. If G is a (C)-graph, then T has a fixed point.

Again, denote by Ψ the family of nondecreasing functions $\psi : [0,+\infty) \to [0,+\infty)$ such that $\sum_{n=1}^{+\infty} \psi^n(t) < +\infty$ for each $t > 0$. It is well known that $\psi(t) < t$ for all $t > 0$.

Proposition 5.2

Let (G,d) be a complete graph-metric space, $\psi \in \Psi$ a nondecreasing map which is continuous from right at each point, $\phi : [0,\infty) \to [0,\infty)$ a map such that $\phi(t) < t$ for all $t > 0$ and $T : G \to 2^G$ a multifunction such that has comparable approximative valued property and $\psi(H(Tx,Ty)) \leq \phi(\psi(M(x,y)))$ for all $x,y \in X$ with $y \in [x]_G$. If G is a (C)-graph, then T has a fixed point.

Let (G,d) be a graph-metric space, $\alpha : G \times G \to [0,\infty)$ a map and $T : G \to CB(G)$ a set-valued mapping. We say that T is α-admissible whenever for each $x \in G$ and $y \in Tx$ with $\alpha(x,y) \geq 1$ we have $\alpha(y,z) \geq 1$ for all $z \in Ty$. Finally, recall that T is continuous whenever $H(Tx_n, Tx) \to 0$ for all sequence $\{x_n\}$ in G with $x_n \to x$. Also, recall that G is called a (C_α)-graph whenever for each sequence $\{x_n\}_{n\geq 0}$ in X with $x_n \to x$, there is a subsequence $\{x_{n_k}\}_{k\geq 0}$ such that $\alpha(x_{n_k}, x) \geq 1$ for all $k \geq 0$.

Theorem 5.6

Let (G,d) be a complete graph-metric space, $\psi \in \Psi$ a strictly increasing map, $\alpha : G \times G \to [0,\infty)$ a function and $T : G \to CB(G)$ an α-admissible set-valued mapping such that $\alpha(x,y)H(Tx,Ty) \leq \psi(d(x,y))$ for all $x,y \in G$ and there exist $x_0 \in G$ and $x_1 \in Tx_0$ with $\alpha(x_0,x_1) \geq 1$. If T is continuous or G is a (C_α)-graph, then T has a fixed point.

Proof: If $x_1 = x_0$, then we have nothing to prove. Let $x_1 \neq x_0$. If $x_1 \in Tx_1$, then x_1 is a fixed point of T. Let $x_1 \notin Tx_1$ and $q > 1$ be given. Then

$$0 < d(x_1, Tx_1) \leq \alpha(x_0,x_1)H(Tx_0,Tx_1) < q\alpha(x_0,x_1)H(Tx_0,Tx_1).$$

Hence, there exists $x_2 \in Tx_1$ such that

$$d(x_1,x_2) < q\alpha(x_0,x_1)H(Tx_0,Tx_1) \leq q\psi(d(x_0,x_1)).$$

It is clear that $x_2 \neq x_1$. Put $t_0 = d(x_0,x_1) > 0$. Then, $d(x_1,x_2) < q\psi(t_0)$. since ψ is strictly increasing, $\psi(d(x_1,x_2)) < \psi(q\psi(t_0))$. Put $q_1 = \frac{\psi(q\psi(t_0))}{\psi(d(x_1,x_2))}$. Then $q_1 > 1$. If $x_2 \in Tx_2$, then x_2 is a fixed point of T. Assume that $x_2 \notin Tx_2$. Then,

$$0 < d(x_2,Tx_2) \leq \alpha(x_1,x_2)H(Tx_1,Tx_2) < q_1\alpha(x_1,x_2)H(Tx_1,Tx_2).$$

Hence, there exists $x_3 \in Tx_2$ such that

$$d(x_2,x_3) < q_1\alpha(x_1,x_2)H(Tx_1,Tx_2) \leq q_1\psi(d(x_1,x_2)) = \psi(q\psi(t_0)).$$

It is clear that $x_3 \neq x_2$ and $\psi(d(x_2,x_3)) < \psi^2(q\psi(t_0))$. Put $q_2 = \frac{\psi^2(q\psi(t_0))}{\psi(d(x_2,x_3))}$. Then $q_2 > 1$. If $x_3 \in Tx_3$, then x_3 is a fixed point of T. Assume that $x_3 \notin Tx_3$. Then,

$$0 < d(x_3,Tx_3) \leq \alpha(x_2,x_3)H(Tx_2,Tx_3) < q_2\alpha(x_2,x_3)H(Tx_2,Tx_3).$$

Thus, there exists $x_4 \in Tx_3$ such that

$$d(x_3,x_4) < q_1\alpha(x_2,x_3)H(Tx_2,Tx_3) \leq q_2\psi(d(x_2,x_3)) = \psi^2(q\psi(t_0)).$$

By continuing this process we obtain a sequence $\{x_n\}$ in G such that $x_n \in Tx_{n-1}$, $x_n \neq x_{n-1}$ and $d(x_n, x_{n+1}) \leq \psi^{n-1}(q\psi(t_0))$ for all n. Now for each $m > n$ we have

$$d(x_n, x_m) \leq \Sigma_{i=n}^{m-1} d(x_i, x_{i+1}) \leq \Sigma_{i=n}^{m-1} \psi^{i-1}(q\psi(t_0)).$$

Hence, $\{x_n\}$ is a Cauchy sequence in G. Since G is complete, there exists $x^\star \in G$ such that $x_n \to x^\star$. Now if T is continuous, then

$$d(x^\star, Tx^\star) = \lim_{n\to\infty} d(x_{n+1}, Tx^\star) \leq \lim_{n\to\infty} H(Tx_n, Tx^\star) = 0$$

and so $x^\star \in Tx^\star$. If G is a (C_α)-graph, then

$$d(x^\star, Tx^\star) = \lim_{k\to\infty} d(x_{n_k+1}, Tx^\star) \leq \lim_{k\to\infty} H(Tx_{n_k}, Tx^\star)$$

$$\leq \lim_{k\to\infty} \alpha(x_{n_k}, x^\star) H(Tx_{n_k}, Tx^\star) \leq \lim_{k\to\infty} \psi(d(x_{n_k}, x^\star)) = 0.$$

Hence, $x^\star \in Tx^\star$.

♣

Corollary 5.7

Let (G,d) be a complete graph-metric space, $\psi \in \Psi$ a strictly increasing map and $T : G \to CB(G)$ a set-valued mapping such that $H(Tx, Ty) \leq \psi(d(x,y))$ for all $x, y \in G$ with $o(x,y) \geq 3$. Assume that for each $x \in G$ and $y \in Tx$ with $o(x,y) \geq 3$ we have $o(y,z) \geq 3$ for all $z \in Ty$. Suppose that there exist $x_0 \in G$ and $x_1 \in Tx_0$ such that $o(x_0, x_1) \geq 3$. If T is continuous or G has this property that for each sequence $\{x_n\}_{n\geq 0}$ in X with $x_n \to x$, there is a subsequence $\{x_{n_k}\}_{k\geq 0}$ such that $o(x_{n_k}, x) \geq 3$ for all $k \geq 0$, then T has a fixed point.

5.5 Some endpoint results for set-valued mappings

Let (X,d) be a metric space, H the Hausdorff metric with respect to d and $T : X \to 2^X$ a set-valued mapping. An element $x \in X$ is called endpoint of T whenever $Tx = \{x\}$. We say that T has the approximate endpoint property whenever $\inf_{x\in X} \sup_{y\in Tx} d(x,y) = 0$. Let $f : X \to X$ be a mapping. We say that f has the approximate endpoint property whenever $\inf_{x\in X} d(x, fx) = 0$. Also, the function $g : \mathbb{R} \to \mathbb{R}$ is called upper semi-continuous whenever $\limsup_{n\to\infty} g(\lambda_n) \leq g(\lambda)$ for all sequence $\{\lambda_n\}_{n\geq 1}$ with $\lambda_n \to \lambda$. In 2010, Amini-Harandi defined the concept of approximate endpoint property for multifunctions and proved the following result ([14]).

Theorem 5.7

Let $\psi : [0,\infty) \to [0,\infty)$ be an upper semi-continuous function such that $\psi(t) < t$

and $\liminf_{t\to\infty}(t-\psi(t))>0$ *for all* $t>0$, (X,d) *a complete metric space and* $T: X\to CB(X)$ *a multifunction satisfy* $H(Tx,Ty)\leq\psi(d(x,y))$ *for all* $x,y\in X$*. Then* T *has a unique endpoint if and only if* T *has the approximate endpoint property.*

Later, generalized weak contractive set-valued mappings introduced and improved it by providing the following result ([93]).

Theorem 5.8

Let $\psi:[0,\infty)\to[0,\infty)$ *be an upper semi-continuous function such that* $\psi(t)<t$ *and* $\liminf_{t\to\infty}(t-\psi(t))>0$ *for all* $t>0$, (X,d) *a complete metric space and* $T: X\to CB(X)$ *a generalized weak contractive multifunction, that is,*

$$H(Tx,Ty)\leq\psi(N(x,y))$$

for all $x,y\in X$*, where* $N(x,y)=\max\{d(x,y),d(x,Tx),d(y,Ty),\frac{d(x,Ty)+d(y,Tx)}{2}\}$*. Then* T *has a unique endpoint if and only if* T *has the approximate endpoint property.*

Let (X,d) be a metric space and $\beta: 2^X\times 2^X\to[0,\infty)$ a mapping. A set-valued map $T: X\to 2^X$ is called β-generalized weak contraction whenever there exists a nondecreasing upper semi-continuous function $\psi:[0,+\infty)\to[0,+\infty)$ such that $\psi(t)<t$ for all $t>0$ and $\beta(Tx,Ty)H(Tx,Ty)\leq\psi(N(x,y))$ for all $x,y\in X$. We say that T is β-admissible whenever $\beta(A,B)\geq 1$ implies $\beta(Tx,Ty)\geq 1$ for all $x\in A$ and $y\in B$, where A and B are subsets of X. We say that T has the property (R) whenever for each convergent sequence $\{x_n\}$ in X with $x_n\to x$ and $\beta(Tx_{n-1},Tx_n)\geq 1$ for all $n\geq 1$, we have $\beta(Tx_n,Tx)\geq 1$. We say that T has the property (K) whenever for each sequence $\{x_n\}$ in X with $\beta(Tx_{n-1},Tx_n)\geq 1$ for all $n\geq 1$, there exists a natural number k such that $\beta(Tx_m,Tx_n)\geq 1$ for all $m>n\geq k$. Finally, we say that T has the property (H) whenever for each $\varepsilon>0$ there exists $z\in X$ such that $\sup_{a\in Tz}d(z,a)<\varepsilon$ implies that for every $x\in X$ there exists $y\in Tx$ such that $H(Tx,Ty)=\sup_{b\in Ty}d(y,b)$. A set-valued map $T: X\to 2^X$ is called lower semi-continuous at $x_0\in X$ whenever for each sequence $\{x_n\}$ in X with $x_n\to x_0$ and every $y\in Tx_0$, there exists a sequence $\{y_n\}$ in X with $y_n\in Tx_n$ for all $n\geq 1$ such that $y_n\to y$.

Theorem 5.9

Let (X,d) *be a complete metric space,* $\beta: 2^X\times 2^X\to[0,\infty)$ *a function and* $T: X\to CB(X)$ *a* β*-admissible,* β*-generalized weak contractive set-valued map which has the properties* (R)*,* (K) *and* (H)*. Suppose that there exist a subset A of X and* $x_0\in A$ *such that* $\beta(A,Tx_0)\geq 1$*. Then T has an endpoint if and only if T has the approximate endpoint property.*

Proof: It is clear that if T has an endpoint, Then T has the approximate endpoint property. Conversely, suppose that T has the approximate endpoint property. Choose $A \subset X$ and $x_0 \in A$ such that $\beta(A, Tx_0) \geq 1$. Since T has the approximate endpoint property, for each $\varepsilon > 0$ there exists $z \in X$ such that $\sup_{a \in Tz} d(z,a) < \varepsilon$. Now by using the condition (H), choose $x_1 \in Tx_0$ such that $H(Tx_0, Tx_1) = \sup_{a \in Tx_1} d(x_1, a)$. Also, choose $x_2 \in Tx_1$ such that $H(Tx_1, Tx_2) = \sup_{a \in Tx_2} d(x_2, a)$ and by continuing this process, we find a sequence $\{x_n\}$ in X such that $x_n \in Tx_{n-1}$ and

$$H(Tx_{n-1}, Tx_n) = \sup_{a \in Tx_n} d(x_n, a)$$

for all $n \geq 1$. Since $\beta(A, Tx_0) \geq 1$ and T is β-admissible, $\beta(Tx_0, Tx_1) \geq 1$. By using induction, it is easy to see that $\beta(Tx_{n-1}, Tx_n) \geq 1$ for all $n \geq 1$. Thus, we obtain

$$\begin{aligned} d(x_n, x_{n+1}) &\leq \sup_{a \in Tx_n} d(x_n, a) = H(Tx_{n-1}, Tx_n) \\ &\leq \beta(Tx_{n-1}, Tx_n) H(Tx_{n-1}, Tx_n) \leq \psi(N(x_{n-1}, x_n)) \end{aligned}$$

for all $n \geq 1$. If $N(x_{n-1}, x_n) = d(x_{n-1}, x_n)$, then

$$d(x_n, x_{n+1}) \leq \psi(d(x_{n-1}, x_n)).$$

If $N(x_{n-1}, x_n) = d(x_{n-1}, Tx_{n-1})$, then

$$d(x_n, x_{n+1}) \leq \psi(d(x_{n-1}, Tx_{n-1})) \leq \psi(d(x_{n-1}, x_n)).$$

If $N(x_{n-1}, x_n) = d(x_n, Tx_n)$, then

$$d(x_n, x_{n+1}) \leq \psi(d(x_n, Tx_n)) \leq \psi(d(x_n, x_{n+1}))$$

and so $d(x_n, x_{n+1}) = 0$. Thus, $d(x_n, x_{n+1}) \leq \psi(d(x_{n-1}, x_n))$. If

$$\begin{aligned} N(x_{n-1}, x_n) &= \frac{d(x_n, Tx_{n-1}) + d(x_{n-1}, Tx_n)}{2} \\ &= \frac{d(x_{n-1}, Tx_n)}{2} \frac{d(x_{n-1}, Tx_n)}{2} \leq \frac{d(x_{n-1}, x_{n+1})}{2} \\ &\leq \frac{d(x_{n-1}, x_n) + d(x_n, x_{n+1})}{2} \leq \max\{d(x_{n-1}, x_n), d(x_n, x_{n+1})\}, \end{aligned}$$

then $d(x_n, x_{n+1}) \leq \psi(d(x_{n-1}, x_n))$ (other case implies $d(x_n, x_{n+1}) = 0$). Thus,

$$d(x_n, x_{n+1}) \leq \psi(d(x_{n-1}, x_n))$$

for all $n \geq 1$. We claim that $\psi(0) = 0$. If $\psi(0) > 0$, then $\psi^2(0) \geq \psi(0) > 0$ because ψ is nondecreasing. On the other hand, since $\psi(t) < t$ for all $t > 0$, we have $\psi^2(0) < \psi(0)$ which is a contradiction. Hence, $\psi(0) = 0$. Put $d_n =$

$d(x_n, x_{n+1})$ for all n. If there exists a natural number n_0 such that $d_{n_0} = 0$, then it is easy to see that $d_n = 0$ for all $n \geq n_0$ and so $\lim_{n\to\infty} d_n = 0$. Now suppose that $d_n \neq 0$ for all n. In this case, we have $d_n \leq \psi(d_{n-1}) < d_{n-1}$ for all n. Hence, $\{d_n\}$ is a decreasing sequence and so there exists $d \geq 0$ such that $\lim_{n\to\infty} d_n = d$. If $d > 0$, then $d_n > 0$ for all n and so $d_n \leq \psi(d_{n-1}) < d_{n-1}$ for all n. Since ψ is upper semi-continuous, we obtain $d = \lim_{n\to\infty} d_n \leq \lim_{n\to\infty} \psi(d_{n-1}) \leq \psi(\lim_{n\to\infty} d_{n-1}) = \psi(d) < d$ which is a contradiction. Thus, $\lim_{n\to\infty} d_n = 0$. Now, we prove that $\{x_n\}$ is a Cauchy sequence. If $\{x_n\}$ is not a Cauchy sequence, then there exist $\varepsilon > 0$ and natural numbers m_k, n_k such that $m_k > n_k \geq k$ and $d(x_{m_k}, x_{n_k}) \geq \varepsilon$ for all $k \geq 1$. Also, we choose m_k as small as possible such that

$$d(x_{m_k-1}, x_{n_k}) < \varepsilon.$$

Thus, $\varepsilon \leq d(x_{m_k}, x_{n_k}) \leq d(x_{m_k}, x_{m_k-1}) + d(x_{m_k-1}, x_{n_k}) \leq d_{m_k-1} + \varepsilon$ for all k. Hence,

$$\lim_{k\to\infty} d(x_{m_k}, x_{n_k}) = \varepsilon.$$

Since T has the property (K), we obtain

$$\begin{aligned} d(x_{m_k}, x_{n_k}) &\leq d(x_{m_k}, x_{m_k+1}) + d(x_{m_k+1}, x_{n_k+1}) + d(x_{n_k+1}, x_{n_k}) \\ &\leq d_{m_k} + H(Tx_{m_k}, Tx_{n_k}) + d_{n_k} \\ &\leq d_{m_k} + \beta(Tx_{m_k}, Tx_{n_k}) H(Tx_{m_k}, Tx_{n_k}) + d_{n_k} \\ &\leq d_{m_k} + \psi(N(x_{m_k}, x_{n_k})) + d_{n_k} \quad (*) \end{aligned}$$

for all k. Since $\lim_{k\to\infty} d(x_{m_k}, x_{n_k}) = \varepsilon$, $\lim_{k\to\infty} N(x_{m_k}, x_{n_k}) = \varepsilon$. In fact,

$$\begin{aligned} d(x_{m_k}, x_{n_k}) &\leq N(x_{m_k}, x_{n_k}) \\ &= \max\{d(x_{m_k}, x_{n_k}), d(x_{m_k}, Tx_{m_k}), d(x_{n_k}, Tx_{n_k}), \\ &\qquad \frac{d(x_{m_k}, Tx_{n_k}) + d(x_{n_k}, Tx_{m_k})}{2}\} \\ &\leq \max\{d(x_{m_k}, x_{n_k}), d(x_{m_k}, x_{m_{k+1}}), d(x_{n_k}, x_{n_{k+1}}), \\ &\qquad \frac{d(x_{m_k}, x_{n_{k+1}}) + d(x_{n_k}, x_{m_{k+1}})}{2}\} \\ &\leq \max\{d(x_{m_k}, x_{n_k}), d(x_{m_k}, x_{m_{k+1}}), d(x_{n_k}, x_{n_{k+1}}), \\ &\qquad \frac{d(x_{m_k}, x_{n_k}) + d(x_{n_k}, x_{n_{k+1}}) + d(x_{n_k}, x_{m_k}) + d(x_{m_k}, x_{m_{k+1}})}{2}\} \end{aligned}$$

and so $\varepsilon = \lim_{k\to+\infty} d(x_{m_k}, x_{n_k}) \leq \lim_{k\to\infty} N(x_{m_k}, x_{n_k}) \leq \varepsilon$. Since ψ is upper semi continuous, by using $(*)$ we obtain

$$\varepsilon = \lim_{k\to\infty} d(x_{m_k}, x_{n_k}) \leq \lim_{k\to\infty} \psi(N(x_{m_k}, x_{n_k})) \leq \psi(\varepsilon) < \varepsilon$$

which is a contradiction and so $\{x_n\}$ is a Cauchy sequence. Choose $x^* \in X$ such that $x_n \to x^*$. Now, note that

$$H(\{x_n\}, Tx_n) = \max\{d(x_n, Tx_n), \sup_{y \in Tx_n} d(x_n, y)\} = H(Tx_{n-1}, Tx_n)$$

for all n and so

$$H(\{x_n\}, Tx_n) = H(Tx_{n-1}, Tx_n) \leq \beta(Tx_{n-1}, Tx_n)H(Tx_{n-1}, Tx_n)$$
$$\leq \psi(N(x_{n-1}, x_n)) \leq \psi(d(x_{n-1}, x_n)) \leq d(x_{n-1}, x_n)$$

for all n and so $\lim_{n\to\infty} H(\{x_n\}, Tx_n) = 0$. Since T has the property (R), we obtain

$$H(\{x^*\}, Tx^*) \leq d(x^*, x_n) + H(\{x_n\}, Tx_n) + H(Tx_n, Tx^*)$$
$$\leq d(x^*, x_n) + H(\{x_n\}, Tx_n) + \beta(Tx_n, Tx^*)H(Tx_n, Tx^*)$$
$$\leq d(x^*, x_n) + H(\{x_n\}, Tx_n) + \psi(N(x_n, x^*))$$

for all n. If $N(x_n, x^*) = d(x^*, Tx^*)$, then we have

$$H(\{x^*\}, Tx^*) \leq d(x^*, x_n) + H(\{x_n\}, Tx_n) + \psi(H(\{x^*\}, Tx^*))$$

for all n. This implies that $H(\{x^*\}, Tx^*) \leq \psi(H(\{x^*\}, Tx^*))$ and so

$$H(\{x^*\}, Tx^*) = 0.$$

If $N(x_n, x^*) = d(x_n, x^*)$ or $N(x_n, x^*) \leq d(x_n, x_{n+1})$, then it is easy to find $H(\{x^*\}, Tx^*) = 0$. Thus, x^* is an endpoint of T.

♣

Next example shows that a β-generalized weak contractive set-valued map is not necessarily generalized weak contractive multifunction.

Example 74 *Put $X = \mathbb{R}$. Define $T : X \to CB(X)$ by $Tx = [x, x+2]$ for all $x \in X$. Suppose that $\psi : [0, +\infty) \to [0, +\infty)$ is an arbitrary upper semi-continuous function such that $\psi(t) < t$ for all $t > 0$. If $x = 0$ and $y = 2$, then $H(Tx, Ty) = H([0,2],[2,4]) = 2$ and $N(x, y) = 2$. Hence,*

$$H(Tx, Ty) = 2 \not\leq \psi(2) = \psi(N(x, y)).$$

Thus, T is not a generalized weak contractive multifunction. Now, suppose that $\psi(t) = \frac{t}{2}$ for all $t \geq 0$ and define $\beta : 2^X \times 2^X \to [0, \infty)$ by $\beta(A, B) = \frac{1}{2}$ for all subsets A and B of X. Then, we have

$$\beta(Tx, Ty)H(Tx, Ty) = \frac{1}{2}d(x, y) = \psi(d(x, y)) = \psi(N(x, y))$$

for all $x, y \in \mathbb{R}$. Thus, T is a β-generalized weak contractive multifunction.

Next example shows that there are multifunctions which satisfy the conditions of Theorem 5.9 while they are not generalized weak contractive multifunctions.

Example 75 *Let $X = [0, \frac{9}{2}]$ and $d(x,y) = |x-y|$. Define $T : X \to CB(X)$ by*

$$Tx = \begin{cases} \{\frac{x}{2}\} & 0 \leq x \leq 1 \\ \{4x - \frac{3}{2}\} & 1 < x \leq \frac{3}{2} \\ \{0\} & \frac{3}{2} < x \leq \frac{9}{2}. \end{cases}$$

If $x = 1$ and $y = \frac{3}{2}$, then

$$H(Tx,Ty) = H(\{\frac{1}{2}\},\{\frac{9}{2}\}) = 4 > 3 = N(x,y) > \psi(N(x,y)),$$

where $\psi : [0,+\infty) \to [0,+\infty)$ is an arbitrary upper semi-continuous function such that $\psi(t) < t$ for all $t > 0$. Thus, T is not a generalized weak contractive multifunction. Now, we show that T satisfy all conditions of Theorem 5.9. In this way, define $\psi(t) = \frac{t}{2}$ and $\beta(A,B) = 1$ whenever A and B are subsets of $[0,1]$ and $\beta(A,B) = 0$ otherwise. First suppose that $x \notin [0,1]$ or $y \notin [0,1]$. If $x,y \in (\frac{3}{2},\frac{9}{2}]$, then $Tx,Ty \subset [0,1]$ and $\beta(Tx,Ty) = 1$. But, $H(Tx,Ty) = 0$ and so $\beta(Tx,Ty)H(Tx,Ty) \leq \psi(N(x,y))$. If $x \in (1,\frac{3}{2}]$ or $y \in (1,\frac{3}{2}]$, then $Tx \not\subseteq [0,1]$ or $Ty \not\subseteq [0,1]$ and so $\beta(Tx,Ty) = 0$. Hence, $\beta(Tx,Ty)H(Tx,Ty) \leq \psi(N(x,y))$. Now, suppose that $x,y \in [0,1]$. In this case, we have $\beta(Tx,Ty) \geq 1$, $H(Tx,Ty) = H(\{\frac{x}{2}\},\{\frac{y}{2}\}) = \frac{1}{2}d(x,y)$ and

$$N(x,y) = \max\{d(x,y), \frac{x}{2}, \frac{y}{2}, \frac{d(x,\frac{y}{2}) + d(y,\frac{x}{2})}{2}\}.$$

Thus, $d(x,y) \leq N(x,y)$ and so

$$\beta(Tx,Ty)H(Tx,Ty) = \frac{1}{2}d(x,y) \leq \psi(d(x,y)) \leq \psi(N(x,y)).$$

Therefore, T is a β-generalized weak contractive multifunction. Now, we show that T is β-admissible. If $\beta(A,B) \geq 1$, then $A,B \subset [0,1]$ and so $Tx = \{\frac{x}{2}\} \in [0,1]$ and $Ty = \{\frac{y}{2}\} \in [0,1]$ for all $x \in A$ and $y \in B$. Thus, $\beta(Tx,Ty) \geq 1$ for all $x \in A$ and $y \in B$. Now, put $A = [0,\frac{1}{2}]$ and $x_0 = \frac{1}{4}$. Then, $Tx_0 = \{\frac{1}{8}\} \in [0,1]$ and $[0,\frac{1}{2}] \subset [0,1]$. Hence, $\beta(A,Tx_0) \geq 1$. Now, we show that T satisfy the condition (H). First note that, for each $\varepsilon > 0$ there exists $z \in X$ such that $\sup_{a \in Tz} d(z,a) < \varepsilon$. Now, we show that for each $x \in X$ there exists $y \in Tx$ such that $H(Tx,Ty) = \sup_{b \in Ty} d(y,b)$. If $0 \leq x \leq 1$, then $Tx = \{\frac{x}{2}\}$, $T(\frac{x}{2}) = \{\frac{x}{4}\}$ and

$$H(Tx,T(\frac{x}{2})) = H(\{\frac{x}{2}\},\{\frac{x}{4}\}) = \frac{x}{4} = \sup_{b \in T(\frac{x}{2})} d(\frac{x}{2},b).$$

If $1 < x \leq \frac{3}{2}$, then $Tx = \{4x - \frac{3}{2}\}$. Since $\frac{5}{2} < 4x - \frac{3}{2} \leq \frac{9}{2}$, $T(4x - \frac{3}{2}) = \{0\}$. Thus,

$$H(Tx, T(4x - \frac{3}{2})) = H(\{4x - \frac{3}{2}\}, \{0\}) = 4x - \frac{3}{2} = \sup_{b \in T(4x - \frac{3}{2})} d(4x - \frac{3}{2}, b).$$

If $\frac{3}{2} < x \leq \frac{9}{2}$, then $Tx = \{0\}$ and $T(0) = \{0\}$. Hence,

$$H(Tx, T(0)) = H(\{0\}, \{0\}) = 0 = \sup_{b \in T(0)} d(0, b).$$

It is easy to check that T satisfy the conditions (R) and (K). Note that, 0 is the endpoint of T.

Now, we add an assumption to obtain uniqueness of endpoint. In this way, we introduce a new notion. Let X be a set and $\beta : 2^X \times 2^X \to [0, \infty)$ a map. We say that the set X has the property (G_β) whenever $\beta(A, B) \geq 1$ for all subsets A and B of X with $A \nsubseteq B$ or $B \nsubseteq A$.

Corollary 5.8

Let (X, d) be a complete metric space, $\beta : 2^X \times 2^X \to [0, \infty)$ a mapping and $T : X \to CB(X)$ a β-admissible, β-generalized weak contractive multifunction which has the properties (R), (K) and (H). Suppose that there exist a subset A of X and $x_0 \in A$ such that $\beta(A, Tx_0) \geq 1$. If T has the approximate endpoint property and X has the property (G_β), then T has a unique endpoint.

Proof: By using Theorem 5.9, T has a endpoint. If T has two distinct endpoints x^* and y^*, then $\beta(Tx^*, Ty^*) = \beta(\{x^*\}, \{y^*\}) \geq 1$ because X has the property (G_β). Hence,

$$\begin{aligned} d(x^*, y^*) &\leq H(Tx^*, Ty^*) \leq \beta(Tx^*, Ty^*) H(Tx^*, Ty^*) \\ &\leq \psi(N(x^*, y^*)) < N(x*, y*) = d(x^*, y^*) \end{aligned}$$

Which is a contradiction. Thus, T has a unique endpoint.

♣

In Example 75, T has a unique endpoint while X has not the property (G_β). Also, T has the property (R) while T is not lower semi-continuous. For this, consider the sequence $\{x_n\}$ defined by

$$x_n = \begin{cases} 1 - \frac{1}{n} & n = 2k \\ 1 + \frac{1}{n} & n = 2k - 1 \end{cases}$$

for $k \geq 1$ and put $y = \frac{1}{2}$ and $x_0 = 1$. Then $x_n \to 1$ and $y \in Tx_0 = \{\frac{1}{2}\}$. Let $\{y_n\}$ be an arbitrary sequence in X such that $y_n \in Tx_n$ for all $n \geq 1$. Then, $y_{2k-1} \in Tx_{2k-1}$ and $y_{2k} \in Tx_{2k}$ for all k. But, $y_{2k-1} = 4x_{2k-1} - \frac{3}{2}$ for sufficiently large k and $y_{2k} = \frac{x_{2k}}{2}$ for all k. Since $y_{2k-1} \to \frac{5}{2}$, $y_n \nrightarrow \frac{1}{2}$. This implies that T is not lower semi-continuous.

Corollary 5.9

Let (X,d) be a complete metric space, $\beta : 2^X \times 2^X \to [0,\infty)$ a mapping and $T : X \to CB(X)$ a β-admissible multifunction which has the properties (R), (K) and (H). Suppose that X has the property (G_β) and there exist a subset A of X, $x_0 \in A$ and $k \in [0,1)$ such that $\beta(A,Tx_0) \geq 1$ and $\beta(Tx,Ty)H(Tx,Ty) \leq kN(x,y)$ for all $x,y \in X$. Then T has a unique endpoint if and only if T has the approximate endpoint property.

Proof: It is sufficient we define $\psi(t) = kt$ for all $t \geq 0$. Then, Theorem 5.9 and Corollary 5.8 guarantee the result.

♣

One can prove that lower semi-continuity of the multifunction T and the property (R) are independent conditions. We can replace lower semi-continuity of the multifunction instead the property (R) to obtain next result. Its proof is similar to the proof of Theorem 5.9.

Theorem 5.10

Let (X,d) be a complete metric space, $\beta : 2^X \times 2^X \to [0,\infty)$ a mapping and $T : X \to CB(X)$ a lower semi-continuous, β-admissible, β-generalized weak contractive multifunction which has the properties (K) and (H). Suppose that there exist a subset A of X and $x_0 \in A$ such that $\beta(A,Tx_0) \geq 1$. Then T has the approximate endpoint property if and only if T has an endpoint.

Corollary 5.10

Let (X,d) be a complete metric space, $\beta : 2^X \times 2^X \to [0,\infty)$ a mapping and $T : X \to CB(X)$ a lower semi-continuous, β-admissible, β-generalized weak contractive multifunction which has the properties (K) and (H). Suppose that there exist a subset A of X and $x_0 \in A$ such that $\beta(A,Tx_0) \geq 1$. If T has the approximate endpoint property and X has the property (G_β), then T has a unique endpoint.

Corollary 5.11

Let (X,d) be a complete metric space, $\beta : 2^X \times 2^X \to [0,\infty)$ a mapping and $T : X \to CB(X)$ a β-admissible multifunction which has the properties (R), (K) and (H). Suppose that X has the property (G_β) and there exist a subset A of X, $x_0 \in A$ and $k \in [0,1)$ such that $\beta(A,Tx_0) \geq 1$ and $\beta(Tx,Ty)H(Tx,Ty) \leq kN(x,y)$ for all $x,y \in X$. If T has the approximate endpoint property, then $Fix(T) = End(T) = \{x\}$.

Proof: If we put $\psi(t) = kt$, then it is easy to see that T has a fixed point. Since T has the approximate endpoint property, by using Corollary 5.9, T has a unique endpoint such x. Let $y \in Fix(T)$. If $Tx = Ty$, then $y = x$. If $Tx \neq Ty$, then $\beta(Tx,Ty) \geq 1$ because X has the property (G_β). Also, we have

$$d(x,y) \leq H(\{x\},Ty) = H(Tx,Ty) \leq \beta(Tx,Ty)H(Tx,Ty) \leq kN(x,y).$$

But, $N(x,y) = max\{d(x,y), d(x,Tx), d(y,Ty), \frac{d(x,Ty)+d(y,Tx)}{2}\} = d(x,y)$. Thus, $d(x,y) = 0$ and so $Fix(T) = End(T) = \{x\}$.

♣

Next corollary shows us role of a point in the existence of endpoints.

Corollary 5.12

Let (X,d) be a complete metric space, $x^* \in X$ a fixed element and $T : X \to CB(X)$ a multifunction such that T has the property (H) and $x^* \in Tx \cap Ty$ for all subsets A and B of X with $x^* \in A \cap B$ and all $x \in A$ and $y \in B$. Assume that $H(Tx,Ty) \leq \psi(N(x,y))$ for all $x,y \in X$ with $x^* \in Tx \cap Ty$, where $\psi : [0,+\infty) \to [0,+\infty)$ is a nondecreasing upper semi-continuous function such that $\psi(t) < t$ for all $t > 0$. Suppose that there exist a subset A_0 of X and $x_0 \in A_0$ such that $x^* \in A_0 \cap Tx_0$. Assume that for each convergent sequence $\{x_n\}$ in X with $x_n \to x$ and $x^* \in Tx_{n-1} \cap Tx_n$ for all $n \geq 1$, we have $x^* \in Tx_n \cap Tx$. Also, for each sequence $\{x_n\}$ in X with $x^* \in Tx_{n-1} \cap Tx_n$ for all $n \geq 1$, there exists a natural number k such that $x^* \in Tx_m \cap Tx_n$ for all $m > n \geq k$. Then T has an endpoint if and only if T has the approximate endpoint property.

Proof: It is sufficient we define $\beta : 2^X \times 2^X \to [0,\infty)$ by $\beta(A,B) = 1$ whenever $x^* \in A \cap B$ and $\beta(A,B) = 0$ otherwise, and then we use Theorem 5.9.

♣

Corollary 5.13

Let (X,d) be a complete metric space, $x^* \in X$ a fixed element and $T : X \to CB(X)$ a lower semi-continuous multifunction such that T has the property (H) and $x^* \in$

$Tx \cap Ty$ for all subsets A and B of X with $x^* \in A \cap B$ and all $x \in A$ and $y \in B$. Assume that $H(Tx,Ty) \leq \psi(N(x,y))$ for all $x,y \in X$ with $x^* \in Tx \cap Ty$, where $\psi : [0,+\infty) \to [0,+\infty)$ is a nondecreasing upper semi-continuous function such that $\psi(t) < t$ for all $t > 0$. Suppose that there exist a subset A_0 of X and $x_0 \in A_0$ such that $x^* \in A_0 \cap Tx_0$. Assume that for each convergent sequence $\{x_n\}$ in X with $x_n \to x$ and $x^* \in Tx_{n-1} \cap Tx_n$ for all $n \geq 1$, we have $x^* \in Tx_n \cap Tx$. Then T has an endpoint if and only if T has the approximate endpoint property.

Proof: It is sufficient we define $\beta : 2^X \times 2^X \to [0,\infty)$ by $\beta(A,B) = 1$ whenever $x^* \in A \cap B$ and $\beta(A,B) = 0$ otherwise, and then we use Theorem 5.10.

♣

Let $(X,d,\leq)$ be an ordered metric space. Define the order $\preceq$ on arbitrary subsets A and B of X by $A \preceq B$ if and only if for each $a \in A$ there exists $b \in B$ such that $a \leq b$. It is easy to check that $(CB(X),\preceq)$ is a partially ordered set.

Theorem 5.11

Let $(X,d,\leq)$ be a complete ordered metric space and T a closed and bounded valued multifunction on X such that T has the property (H) and $Tx \preceq Ty$ for all subsets A and B of X with $A \preceq B$ and all $x \in A$ and $y \in B$. Assume that $H(Tx,Ty) \leq \psi(N(x,y))$ for all $x,y \in X$ with $Tx \preceq Ty$, where $\psi : [0,+\infty) \to [0,+\infty)$ is a nondecreasing upper semi-continuous function such that $\psi(t) < t$ for all $t > 0$. Suppose that there exist a subset A_0 of X and $x_0 \in A_0$ such that $A_0 \preceq Tx_0$. Assume that for each convergent sequence $\{x_n\}$ in X with $x_n \to x$ and $Tx_{n-1} \preceq Tx_n$ for all $n \geq 1$, we have $Tx_n \preceq Tx$. Also, for each sequence $\{x_n\}$ in X with $Tx_{n-1} \preceq Tx_n$ for all $n \geq 1$, there exists a natural number k such that $Tx_m \preceq Tx_n$ for all $m > n \geq k$. Then T has an endpoint if and only if T has the approximate endpoint property.

Proof: Define $\beta(A,B) = 1$ whenever $A \preceq B$ and $\beta(A,B) = 0$ otherwise, and then we use Theorem 5.9.

♣

Corollary 5.14

Let $(X,d,\leq)$ be a complete ordered metric space and T a closed and bounded valued multifunction on X such that T has the property (H) and $Tx \preceq Ty$ for all subsets A and B of X with $A \preceq B$ and all $x \in A$ and $y \in B$. Assume that $H(Tx,Ty) \leq \psi(N(x,y))$ for all $x,y \in X$ with $Tx \preceq Ty$, where $\psi : [0,+\infty) \to [0,+\infty)$ is a nondecreasing upper semi-continuous function such that $\psi(t) < t$ for all $t > 0$. Suppose that there exist a subset A_0 of X and $x_0 \in A_0$ such that $A_0 \preceq Tx_0$. Assume that for each convergent *sequence $\{x_n\}$ in X with $x_n \to x$ and $Tx_{n-1} \preceq Tx_n$ for all $n \geq 1$, we have $Tx_n \preceq Tx$. Also, for each sequence $\{x_n\}$ in X with $Tx_{n-1} \preceq Tx_n$ for all $n \geq 1$, there exists a natural number k such that $Tx_m \preceq Tx_n$ for all $m > n \geq k$. If T has the approximate endpoint property and $A \preceq B$ for all subsets A and B of X with $A \nsubseteq B$ or $B \nsubseteq A$, then T has a unique endpoint.*

Proof: Define $\beta(A,B) = 1$ whenever $A \preceq B$ and $\beta(A,B) = 0$ otherwise, and then we use Corollary 5.8.

♣

Let (X,d) be a metric space and $T : X \to 2^X$ a set-valued map. We say that T is a HS-set-valued map whenever for each $x \in X$ there exists $y \in Tx$ such that $H(Tx,Ty) = \sup_{b \in Ty} d(y,b)$. It is obvious that each HS-set-valued map is a set-valued map which has the property (H). Thus, one can conclude similar results to the above ones for HS-set-valued mappings. Here, we provide some. Although by considering HS-set-valued map, we restrict ourselves, but we obtain strange results with respect to above. One can prove the following by reading exactly the proofs of similar above results.

Theorem 5.12

Let (X,d) be a complete metric space, $\beta : 2^X \times 2^X \to [0,\infty)$ a mapping and $T : X \to CB(X)$ a β-admissible, β-generalized weak contractive HS-set-valued map which has the properties (R) and (K). Suppose that there exist a subset A of X and $x_0 \in A$ such that $\beta(A,Tx_0) \geq 1$. Then T has an endpoint and so T has the approximate endpoint property.

Theorem 5.13

Let (X,d) be a complete metric space, $\beta : 2^X \times 2^X \to [0,\infty)$ a mapping and $T : X \to CB(X)$ a lower semi-continuous, β-admissible and β-generalized weak contractive HS-set-valued map which has the property (K). Suppose that there exist a subset A of X and $x_0 \in A$ such that $\beta(A,Tx_0) \geq 1$. Then T has an endpoint and so T has the approximate endpoint property.

Next result is a consequence of Theorem 5.12.

Corollary 5.15

Let (X,d) be a complete metric space, $x^* \in X$ a fixed element and $T : X \to CB(X)$ a HS-set-valued map such that $x^* \in Tx \cap Ty$ for all subsets A and B of X with $x^* \in A \cap B$

and all $x \in A$ and $y \in B$. Assume that $H(Tx,Ty) \le \psi(N(x,y))$ for all $x,y \in X$ with $x^* \in Tx \cap Ty$, where $\psi : [0,+\infty) \to [0,+\infty)$ is a increasing upper semi-continuous function such that $\psi(t) < t$ for all $t > 0$. Suppose that there exist a subset A_0 of X and $x_0 \in A_0$ such that $x^* \in A_0 \cap Tx_0$. Assume that for each convergent sequence $\{x_n\}$ in X with $x_n \to x$ and $x^* \in Tx_{n-1} \cap Tx_n$ for all $n \ge 1$, we have $x^* \in Tx_n \cap Tx$. Also, for each sequence $\{x_n\}$ in X with $x^* \in Tx_{n-1} \cap Tx_n$ for all $n \ge 1$, there exists a natural number k such that $x^* \in Tx_m \cap Tx_n$ for all $m > n \ge k$. Then T has an endpoint and so T has the approximate endpoint property.

Next result is a consequence of Theorem 5.13.

Corollary 5.16

Let $(X,d,\le)$ be a complete ordered metric space and T a closed and bounded valued lower semi-continuous HS-set-valued map on X such that $Tx \preceq Ty$ for all subsets A and B of X with $A \preceq B$ and all $x \in A$ and $y \in B$. Assume that $H(Tx,Ty) \le \psi(N(x,y))$ for all $x,y \in X$ with $Tx \preceq Ty$, where $\psi : [0,+\infty) \to [0,+\infty)$ is a increasing upper semi-continuous function such that $\psi(t) < t$ for all $t > 0$. Suppose that there exist a subset A_0 of X and $x_0 \in A_0$ such that $A_0 \preceq Tx_0$. Assume that for each sequence $\{x_n\}$ in X with $Tx_{n-1} \preceq Tx_n$ for all $n \ge 1$, there exists a natural number k such that $Tx_m \preceq Tx_n$ for all $m > n \ge k$. Then T has an endpoint and so T has the approximate endpoint property.

5.6 Approximate fixed points of set-valued mappings

As you know, there are mappings having approximate fixed points while have no fixed points. Some researchers have been studied approximate fixed points of several classes of mappings till date.

Let X be a set, $T : X \to 2^X$ a multifunction and $\beta : 2^X \times 2^X \to [0,\infty)$ a mapping. As we introduced in last chapter, T is called β-admissible whenever $\beta(A,B) \ge 1$ implies $\beta(Tx,Ty) \ge 1$ for all $x \in A$ and $y \in B$, where A and B are subsets of X. Recall that we denote by $\mathcal{R}$ the set of all continuous mappings $g : [0,\infty)^5 \to [0,\infty)$ satisfying $g(1,1,1,2,0) = g(1,1,1,0,2) = h \in (0,1)$,

$$g(\alpha x_1, \alpha x_2, \alpha x_3, \alpha x_4, \alpha x_5) \le \alpha g(x_1,x_2,x_3,x_4,x_5)$$

for all nonnegative elements x_1,x_2,x_3,x_4,x_5 and $\alpha \ge 0$,

$$g(x_1,x_2,x_3,x_4,0) < g(y_1,y_2,y_3,y_4,0) \text{ and } g(x_1,x_2,x_3,0,x_4) < g(y_1,y_2,y_3,0,y_4)$$

for all $x_i, y_i \in [0,\infty)$ with $x_i < y_i$ for $i = 1,...,4$.

Let (X,d) be a metric space, $\beta : 2^X \times 2^X \to [0,\infty)$ a mapping and T a set-valued mapping on X with closed and bounded values. For your easy access, we recall here some notions. We say that T is a generalized β-contraction whenever there exists $g \in \mathcal{R}$ such that $\beta(Tx,Ty)H(Tx,Ty) \leq g(d(x,y),d(x,Tx),d(y,Ty),d(x,Ty), d(y,Tx))$ for all $x,y \in X$, where H is the Hausdorff metric with respect to d, that is,

$$H(A,B) = \max\{sup_{x\in A}d(x,B), sup_{y\in B}d(y,A)\}$$

for all closed and bounded subsets A and B of X. We say that T has approximate fixed points whenever $\inf_{x\in X} d(x,Tx) = 0$. Also, we say that T is lower semi-continuous at $x_0 \in X$ whenever for each sequence $\{x_n\}$ with $x_n \to x_0$ and every $y \in Tx_0$, there exists a sequence $\{y_n\}$ such that $y_n \to y$ and $y_n \in Tx_n$ for all n. Let C a nonempty subset of a metric space (X,d) and $x \in X$. We say that T is lower semi-continuous whenever T is lower semi-continuous at each element of X. An element $y_0 \in C$ is said to be a best approximation of x whenever $d(x,y_0) = d(x,C) = \inf_{y\in C} d(x,y)$. The set C is called proximinal whenever every $x \in X$ has at least one best approximation in C. Each proximinal set is closed and bounded. Denote by $P(X)$ the set of all proximinal subsets of X.

Theorem 5.14

Let (X,d) be a metric space, $\beta : 2^X \times 2^X \to [0,\infty)$ a function and $T : X \to P(X)$ a β-admissible generalized β-contractive set-valued mapping. Suppose that there exist $A \subset X$ and $x_0 \in A$ such that $\beta(A,Tx_0) \geq 1$. Then T has approximate fixed points.

Proof: Choose $A \subset X$ and $x_0 \in A$ such that $\beta(A,Tx_0) \geq 1$. Since T is proximinal valued, we can choose a sequence $\{x_n\}$ such that $x_{n+1} \in Tx_n$ and $d(x_n,x_{n+1}) = d(x_n,Tx_n)$ for all $n \geq 0$. Since T is β-admissible and $\beta(A,Tx_0) \geq 1$, it is easy to see that

$$\beta(Tx_{n-1},Tx_n) \geq 1$$

for all $n \geq 0$. There exists $g \in \mathcal{R}$ such that

$$\beta(Tx,Ty)H(Tx,Ty) \leq g(d(x,y),d(x,Tx),d(y,Ty),d(x,Ty),d(y,Tx))$$

for all $x,y \in X$. Fix $1 > r > h$, where $h = g(1,1,1,2,0)$. Then, we have

$$\begin{aligned} d(x_1,x_2) &\leq H(Tx_0,Tx_1) \leq \beta(Tx_0,Tx_1)H(Tx_0,Tx_1) \\ &\leq g(d(x_0,x_1),d(x_1,Tx_1),d(x_0,Tx_0),d(x_0,Tx_1),d(x_1,Tx_0)) \\ &\leq g(d(x_0,x_1),d(x_1,x_2),d(x_0,x_1),d(x_0,x_1)+d(x_1,x_2),0). \end{aligned}$$

By using Proposition 3.1, we obtain $d(x_1,Tx_1) \leq hd(x_0,x_1) < rd(x_0,x_1)$. On the other hand, we have

$$d(x_2,x_3) \leq H(Tx_1,Tx_2) \leq \beta(Tx_1,Tx_2)H(Tx_1,Tx_2)$$

$$\le g(d(x_1,x_2),d(x_2,Tx_2),d(x_1,Tx_1),d(x_1,Tx_2),d(x_2,Tx_1))$$
$$\le g(d(x_1,x_2),d(x_2,x_3),d(x_1,x_2),d(x_1,x_2)+d(x_2,x_3),0).$$

Again by using proposition 3.1, we get

$$d(x_2,x_3) \le hd(x_1,x_2) < rd(x_1,x_2) < r^2 d(x_0,x_1).$$

By continuing this process, we conclude that

$$d(x_n,x_{n+1}) \le hd(x_{n-1},x_n) < rd(x_{n-1},x_n) < r^n d(x_0,x_1)$$

for all $n \ge 0$. But, $d(x_n,Tx_n) \le d(x_n,x_{n+1})$ for all $n \ge 0$. This implies that

$$\inf_{x\in X} d(x,Tx) = 0$$

and so T has approximate fixed points.

♣

Corollary 5.17

Let (X,d) be a complete metric space, $\beta : 2^X \times 2^X \to [0,\infty)$ a mapping and $T : X \to P(X)$ a β-admissible lower semi-continuous generalized β-contractive set-valued mapping. If there exist $A \subset X$ and $x_0 \in A$ such that $\beta(A,Tx_0) \ge 1$, then T has a fixed point.

Proof: By using a similar argument in proof of Theorem 5.14, we obtain

$$d(x_n,x_{n+1}) \le hd(x_{n-1},x_n) < rd(x_{n-1},x_n) < r^n d(x_0,x_1).$$

Then for each natural numbers m and n with $m < n$, we have

$$d(x_m,x_n) \le (r^m + r^{m+1} + \dots + r^{n-1})d(x_0,x_1) < \frac{r^m}{1-r} d(x_0,x_1).$$

Hence, $\{x_n\}$ is a Cauchy sequence. Choose $x^* \in X$ such that $x_n \to x^*$. Since T is lower semi-continuous at x^*, for each $y \in Tx^*$ there exists a sequence $\{y_n\}$ such that $y_n \to y$ and $y_n \in Tx_n$ for all n. Let $n \ge 1$ be given. Then, for each $u \in Tx_n$ we have

$$d(x^*,Tx^*) \le d(x^*,y) \le d(x^*,x_{n+1}) + d(x_{n+1},u) + d(u,y_n) + d(y_n,y).$$

This implies that $d(x^*,Tx^*) \le d(x^*,x_{n+1}) + d(x_{n+1},Tx_n) + d(Tx_n,y_n) + d(y_n,y)$. Since $x_{n+1} \in Tx_n$ and $y_n \in Tx_n$, $d(x^*,Tx^*) \le d(x^*,x_{n+1}) + d(y_n,y)$ for all n. Hence,

$$d(x^*,Tx^*) = 0$$

and so $x^* \in Tx^*$.

♣

As stated before, Banach proved his contraction principle in 1922 and Nadler extended it to set-valued mappings in 1969. In fact, Nadler proved that if (X,d) is a complete metric space and there exists $k \in (0,1)$ such that $H(Tx,Ty) \leq kd(x,y)$ for all $x,y \in X$, then T has a fixed point. Let $\beta : 2^X \times 2^X \to [0,\infty)$ be a mapping. We say that T is a β-contraction whenever there exists $k \in (0,1)$ such that

$$\beta(Tx,Ty)H(Tx,Ty) \leq kd(x,y)$$

for all $x,y \in X$. It is clear that each Nadler type contractive set-valued map is a β-contraction. Next example shows that there exist β-contraction multifunctions which are not Nadler type contractions.

Example 76 *Let $X = \mathbb{R}$ and $d(x,y) = |x-y|$ for all $x,y \in X$. Define T on X defined by $Tx = [x,4]$ whenever $x \leq 4$ and $Tx = [4,x]$ whenever $x > 4$. Let $\lambda \in (0,1)$ be given. Put $x = 4$ and $y = 4+2\lambda$. Then, we have $H(Tx,Ty) = 2\lambda > \lambda d(x,y)$. Now, define $\beta : 2^X \times 2^X \to [0,\infty)$ by $\beta(A,B) = \frac{1}{4}$ whenever $A \subseteq (-\infty,4]$ and $B \subseteq [4,\infty)$ and $\beta(A,B) = 0$ otherwise. Then, it is easy to see that $\beta(Tx,Ty)H(Tx,Ty) \leq \frac{1}{2}d(x,y)$ for all $x,y \in X$. Hence, T is a β-contraction while is not a Nadler type contraction.*

Let (X,d) be a metric space, $\beta : 2^X \times 2^X \to [0,\infty)$ a mapping and T a set-valued map on X. We say that T is β-convergent whenever for each convergent sequence $\{x_n\}$ with $x_n \to x$, there exists a natural number N such that $\beta(Tx_n,Tx) \geq 1$ for all $n \geq N$.

Corollary 5.18

Let (X,d) be a metric space, $\beta : 2^X \times 2^X \to [0,\infty)$ a mapping and $T : X \to P(X)$ a β-admissible and β-contraction multifunction. Suppose that there exist $A \subset X$ and $x_0 \in A$ such that $\beta(A,Tx_0) \geq 1$. Then T has approximate fixed points. If (X,d) is a complete metric space and T is β-convergent, then T has a fixed point.

Proof: Choose $k \in (0,1)$ such that $\beta(Tx,Ty)H(Tx,Ty) \leq kd(x,y)$ for all $x,y \in X$. Now, define $g : [0,\infty)^5 \to [0,\infty)$ by $g(x_1,x_2,x_3,x_4,x_5) = kx_1$. Then, $g \in \mathcal{R}$ and

$$\beta(Tx,Ty)H(Tx,Ty) \leq g(d(x,y),d(x,Tx),d(y,Ty),d(x,Ty),d(y,Tx))$$

for all $x,y \in X$, that is, T is a generalized β-contractive multifunction. Now by using Theorem 5.14, T has approximate fixed points. Now, we show that T is lower semi-continuous. Let $x \in X$, $\{x_n\}$ be a sequence with $x_n \to x$ and $y \in Tx$. Choose $y_n \in Tx_n$ for all $n \geq 1$. We have to show that $y_n \to y$. Since T is β-convergent, there exists a natural number N such that $\beta(Tx_n,Tx) \geq 1$ for all $n \geq N$. Thus,

$$d(y_n,y) \leq H(Tx_n,Tx) \leq \beta(Tx_n,Tx)H(Tx_n,Tx) \leq kd(x_n,x)$$

for all $n \geq N$. Hence, $y_n \to y$ and so T is lower semi-continuous. If (X,d) is a complete metric space, then by using Corollary 5.17, T has a fixed point.

♣

As we said before, the notion of Kannan type contraction mappings were introduced in 1968. Later, some authors extended the notion to set-valued version. Let (X,d) be a metric space, $\beta : 2^X \times 2^X \to [0,\infty)$ a mapping and T a set-valued map on X. We say that T is a Kannan type contraction whenever there exists $\alpha \in (0,\frac{1}{2})$ such that

$$H(Tx,Ty) \leq \alpha(d(x,Tx)+d(y,Ty))$$

for all $x,y \in X$. Also, we say that T is a β-Kannan set-valued map whenever there exists $\alpha \in (0,\frac{1}{2})$ such that $\beta(Tx,Ty)H(Tx,Ty) \leq \alpha(d(x,Tx)+d(y,Ty))$ for all $x,y \in X$. Next example shows that there exist β-Kannan set-valued mappings which are not Kannan type contraction.

Example 77 *Let $X=[0,3]$ and $d(x,y)=|x-y|$ for all $x,y \in X$. Define T on X defined by $Tx=[\frac{x}{3},\frac{x}{2}]$ for all $x \in X$. Let $\alpha \in (0,\frac{1}{2})$ be given. Put $x=0$ and $y=6\alpha$. Then, we have $H(Tx,Ty)=3\alpha > \alpha(d(x,Tx)+d(y,Ty))=4\alpha^2$. Now, define $\beta : 2^X \times 2^X \to [0,\infty)$ by $\beta(A,B)=\frac{1}{4}$ whenever $A,B \subseteq [0,\frac{1}{2}]$ and $\beta(A,B)=0$ otherwise. Then, it is easy to see that $\beta(Tx,Ty)H(Tx,Ty) \leq \frac{1}{4}(d(x,Tx)+d(y,Ty))$ for all $x,y \in X$. Hence, T is a β-Kannan set-valued map while is not a Kannan type contraction.*

Corollary 5.19

Let (X,d) be a metric space, $\beta : 2^X \times 2^X \to [0,\infty)$ a function and $T : X \to P(X)$ a β-admissible and β-Kannan set-valued map. Suppose that there exist $A \subset X$ and $x_0 \in A$ such that $\beta(A,Tx_0) \geq 1$. Then T has approximate fixed points. If (X,d) is a complete metric space and T is lower semi-continuous, then T has a fixed point.

Proof: Choose $\alpha \in (0,\frac{1}{2})$ such that

$$\beta(Tx,Ty)H(Tx,Ty) \leq \alpha(d(x,Tx)+d(y,Ty))$$

for all $x,y \in X$. Define $g : [0,\infty)^5 \to [0,\infty)$ by $g(x_1,x_2,x_3,x_4,x_5)=\alpha x_2+\alpha x_3$. Then, $g \in \mathcal{R}$ and $\beta(Tx,Ty)H(Tx,Ty) \leq g(d(x,y),d(x,Tx),d(y,Ty),d(x,Ty),$ $d(y,Tx))$ for all $x,y \in X$, that is, T is a generalized β-contractive set-valued map. By using Theorem 5.14, T has approximate fixed points. If (X,d) is a complete metric space and T is lower semi-continuous, then by using Corollary 5.17, T has a fixed point.

♣

In 1971, the notion of Reich type contraction mappings were introduced. Later, the notion extended to set-valued maps. Let (X,d) be a metric space, $\beta : 2^X \times 2^X \to [0,\infty)$ a function and T a set-valued map on X. We say that T is a Reich type contraction whenever there exist nonnegative real numbers α, β, γ with $\alpha + \lambda + \gamma < 1$ such that

$$H(Tx,Ty) \leq \alpha d(x,y) + \lambda d(x,Tx) + \gamma d(y,Ty)$$

for all $x,y \in X$. Also, we say that T is a β-Reich set-valued map whenever there exists there exist nonnegative real numbers α, β, γ with $\alpha + \lambda + \gamma < 1$ such that

$$\beta(Tx,Ty)H(Tx,Ty) \leq \alpha d(x,y) + \lambda d(x,Tx) + \gamma d(y,Ty)$$

for all $x,y \in X$. Next example shows that there exist β-Reich set-valued mappings which are not Reich type contraction.

Example 78 *Let $X = [0,\infty)$ and $d(x,y) = |x-y|$ for all $x,y \in X$. Define T on X defined by $Tx = [\frac{x}{3}, x]$ for all $x \in X$. Let $\alpha, \beta, \gamma \in [0,\infty)$ with $\alpha + \lambda + \gamma < 1$ be given. Put $x = 0$ and $y = 2$. Then, we have*

$$H(Tx,Ty) = 2 > \alpha d(x,y) + \lambda d(x,Tx) + \gamma d(y,Ty).$$

Now, define $\beta : 2^X \times 2^X \to [0,\infty)$ by $\beta(A,B) = \frac{\alpha}{2}$ for all subsets A and B. Then, it is easy to see that

$$\beta(Tx,Ty)H(Tx,Ty) \leq \alpha d(x,y) + \lambda d(x,Tx) + \gamma d(y,Ty)$$

for all $x,y \in X$. Hence, T is a β-Reich set-valued map while is not a Reich type contraction.

Corollary 5.20

Let (X,d) be a metric space, $\beta : 2^X \times 2^X \to [0,\infty)$ a mapping and $T : X \to P(X)$ a β-admissible and β-Reich set-valued map. Suppose that there exist $A \subset X$ and $x_0 \in A$ such that $\beta(A,Tx_0) \geq 1$. Then T has approximate fixed points. If (X,d) is a complete metric space and T is lower semi-continuous, then T has a fixed point.

Proof: Choose nonnegative real numbers α, β, γ with $\alpha + \lambda + \gamma < 1$ such that

$$\beta(Tx,Ty)H(Tx,Ty) \leq \alpha d(x,y) + \lambda d(x,Tx) + \gamma d(y,Ty)$$

for all $x,y \in X$. It is sufficient we define $g : [0,\infty)^5 \to [0,\infty)$ by

$$g(x_1,x_2,x_3,x_4,x_5) = \alpha x_1 + \lambda x_2 + \gamma x_3.$$

Then by following the proof of Corollary 5.19, one can complete the proof.

♣

In 1972, the notion of Chatterjea type contraction mappings were introduced. Later, the notion extended to set-valued maps. Let (X,d) be a metric space, $\beta : 2^X \times 2^X \to [0,\infty)$ a mapping and T a multifunction on X. We say that T is a Chatterjea type contraction whenever there exists $\alpha \in (0,\frac{1}{2})$ such that $H(Tx,Ty) \le \alpha(d(x,Ty)+d(y,Tx))$ for all $x,y \in X$. Also, we say that T is a β-Chatterjea multifunction whenever there exists $\alpha \in (0,\frac{1}{2})$ such that $\beta(Tx,Ty)H(Tx,Ty) \le \alpha(d(x,Ty)+d(y,Tx))$ for all $x,y \in X$. Next example shows that there exist β-Chatterjea multifunctions which are not Chatterjea type contraction.

Example 79 *Let $X=[0,4]$ and $d(x,y)=|x-y|$ for all $x,y \in X$. Consider the set-valued map T on X defined by $Tx=[\frac{x}{2},x]$ for all $x \in X$. Let $\alpha \in (0,\frac{1}{2})$ be given. Put $x=0$ and $y=2\alpha$. Then, we have $H(Tx,Ty)=2\alpha > \alpha(d(x,Ty)+d(y,Tx)) = 3\alpha^2$. Now, define $\beta : 2^X \times 2^X \to [0,\infty)$ by $\beta(A,B)=\frac{1}{16}$ for all subsets A and B. Then, it is easy to see that $\beta(Tx,Ty)H(Tx,Ty) \le \frac{1}{4}(d(x,Ty)+d(y,Tx))$ for all $x,y \in X$. Hence, T is a β-Chatterjea multifunction while is not a Chatterjea type contraction.*

Corollary 5.21

Let (X,d) be a metric space, $\beta : 2^X \times 2^X \to [0,\infty)$ a mapping and $T : X \to P(X)$ a β-admissible and β-Chatterjea set-valued map. Suppose that there exist $A \subset X$ and $x_0 \in A$ such that $\beta(A,Tx_0) \ge 1$. Then T has approximate fixed points. If (X,d) is a complete metric space and T is lower semi-continuous, then T has a fixed point.

Proof: Choose $\alpha \in (0,\frac{1}{2})$ such that

$$\beta(Tx,Ty)H(Tx,Ty) \le \alpha(d(x,Ty)+d(y,Tx))$$

for all $x,y \in X$. It is sufficient we define $g : [0,\infty)^5 \to [0,\infty)$ by

$$g(x_1,x_2,x_3,x_4,x_5) = \alpha x_4 + \alpha x_5.$$

Then by following the proof of Corollary 5.19, we obtain the desired result.

♣

In 1972, the notion of Zamfirescu type contraction mappings were introduced. Later, the notion extended to set-valued maps. Let (X,d) be a metric space, $\beta : 2^X \times 2^X \to [0,\infty)$ a mapping and T a set-valued map on X. We say that T is a Zamfirescu type contraction whenever there exists $k \in [0,1)$ such that $H(Tx,Ty) \le kM_T(x,y)$ for all $x,y \in X$, where

$$M_T(x,y) = \max\{d(x,y), \frac{1}{2}[d(x,Ty)+d(y,Tx)], \frac{1}{2}[d(x,Tx)+d(y,Ty)]\}.$$

Also, we say that T is a β-Zamfirescu multifunction whenever there exists $k \in [0,1)$ such that $\beta(Tx,Ty)H(Tx,Ty) \leq kM_T(x,y)$ for all $x,y \in X$. Next example shows that there exist β-Zamfirescu set-valued maps which are not Zamfirescu type contraction.

Example 80 *Let $X = [0,2]$ and $d(x,y) = |x-y|$ for all $x,y \in X$. Define T on X defined by $Tx = [\frac{x}{3},x]$ for all $x \in X$. Let $k \in [0,1)$ be given. Put $x = 0$ and $y = 2k$. Then, we have $H(Tx,Ty) = 2k > kM_T(x,y) = 2k^2$. Now, define $\beta : 2^X \times 2^X \to [0,\infty)$ by $\beta(A,B) = \frac{1}{6}$ for all subsets A and B. Then, it is easy to see that $\beta(Tx,Ty)H(Tx,Ty) \leq \frac{1}{2}M_T(x,y)$ for all $x,y \in X$. Hence, T is a β-Zamfirescu multifunction while is not a Zamfirescu type contraction.*

Corollary 5.22

Let (X,d) be a metric space, $\beta : 2^X \times 2^X \to [0,\infty)$ a mapping and $T : X \to P(X)$ a β-admissible and β-Zamfirescu set-valued map. Suppose that there exist $A \subset X$ and $x_0 \in A$ such that $\beta(A,Tx_0) \geq 1$. Then T has approximate fixed points. If (X,d) is a complete metric space and T is lower semi-continuous, then T has a fixed point.

Proof: Choose $k \in [0,1)$ such that $\beta(Tx,Ty)H(Tx,Ty) \leq kM_T(x,y)$ for all $x,y \in X$. It is sufficient we define $g : [0,\infty)^5 \to [0,\infty)$ by

$$g(x_1,x_2,x_3,x_4,x_5) = k\max\{x_1, \frac{1}{2}[x_2+x_3], \frac{1}{2}[x_4+x_5]\}.$$

Then by following the proof of Corollary 5.19, one can complete the proof.

♣

In 1972, the notion of Ciric type contraction mappings were introduced. Later, the notion extended to set-valued maps. Let (X,d) be a metric space, $\beta : 2^X \times 2^X \to [0,\infty)$ a mapping and T a multifunction on X. We say that T is a Ciric type contraction whenever there exists $\lambda \in (0,1)$ such that $H(Tx,Ty) \leq \lambda N_T(x,y)$ for all $x,y \in X$, where

$$N_T(x,y) = \max\{d(x,y), d(x,Tx), d(y,Ty), \frac{1}{2}[d(x,Ty)+d(y,Tx)]\}.$$

We say that T is a β-Ciric set-valued map whenever there exists $\lambda \in (0,1)$ such that $\beta(Tx,Ty)H(Tx,Ty) \leq \lambda N_T(x,y)$ for all $x,y \in X$. Next example shows that there exist β-Ciric set-valued maps which are not Ciric type contraction.

Example 81 *Let $X = \mathbb{R}$ and $d(x,y) = |x-y|$ for all $x,y \in X$. Define T on X defined by $Tx = [\frac{x}{4},x]$ for all $x \in X$. Let $\lambda \in (0,1)$ be given. Put $x = 0$ and $y = \frac{\lambda}{2}$. Then, we have $H(Tx,Ty) = \frac{\lambda}{2} > \lambda N_T(x,y) = \lambda d(x,y) = \frac{\lambda^2}{2}$. Now, define $\beta : 2^X \times 2^X \to [0,\infty)$ by $\beta(A,B) = \frac{1}{6}$ for all subsets A and B. Then, it is easy to* *see that $\beta(Tx,Ty)H(Tx,Ty) \leq \frac{1}{2}N_T(x,y)$ for all $x,y \in X$. Hence, T is a β-Ciric set-valued map while is not a Ciric type contraction.*

Corollary 5.23

Let (X,d) be a metric space, $\beta : 2^X \times 2^X \to [0,\infty)$ a mapping and $T : X \to P(X)$ a β-admissible and β-Ciric set-valued map. Suppose that there exist $A \subset X$ and $x_0 \in A$ such that $\beta(A, Tx_0) \geq 1$. Then T has approximate fixed points. If (X,d) is a complete metric space and T is lower semi-continuous, then T has a fixed point.

Proof: Choose $\lambda \in (0,1)$ such that $\beta(Tx,Ty)H(Tx,Ty) \leq \lambda N_T(x,y)$ for all $x,y \in X$. It is sufficient we define $g : [0,\infty)^5 \to [0,\infty)$ by

$$g(x_1,x_2,x_3,x_4,x_5) = \lambda \max\{x_1,x_2,x_3,\frac{1}{2}[x_4+x_5]\}.$$

Then by following the proof of Corollary 5.19, we can obtain the desired results. ♣

In 1974, the notion of quasi-contractive mappings were introduced by Ciric. Later, the notion extended to set-valued maps. Let (X,d) be a metric space, $\beta : 2^X \times 2^X \to [0,\infty)$ a mapping and T a multifunction on X. We say that T is a quasi-contraction whenever there exists $\lambda \in (0,1)$ such that

$$H(Tx,Ty) \leq \lambda \max\{d(x,y),d(y,Ty),d(x,Tx),d(x,Ty),d(y,Tx)\}$$

for all $x,y \in X$. We say that T is a β-quasi-contraction whenever there exists $\lambda \in (0,1)$ such that

$$\beta(Tx,Ty)H(Tx,Ty) \leq \lambda \max\{d(x,y),d(y,Ty),d(x,Tx),d(x,Ty),d(y,Tx)\}$$

for all $x,y \in X$. The next example shows that there exist β-quasi-contractions which are not quasi-contraction.

Example 82 *Let $X = [0,5]$ and $d(x,y) = |x-y|$ for all $x,y \in X$. Define T on X defined by $Tx = [\frac{x}{2},x]$ for all $x \in X$. Let $\lambda \in (0,1)$ be given. Put $x = 0$ and $y = \lambda$. Then, we have*

$$H(Tx,Ty) = \lambda > \lambda \max\{d(x,y),d(y,Ty),d(x,Tx),d(x,Ty),d(y,Tx)\} = \lambda^2.$$

Now, define $\beta : 2^X \times 2^X \to [0,\infty)$ by $\beta(A,B) = \frac{1}{5}$ for all subsets A and B. Then, it is easy to see that

$$\beta(Tx,Ty)H(Tx,Ty) \leq \frac{1}{2}\max\{d(x,y),d(y,Ty),d(x,Tx),d(x,Ty),d(y,Tx)\}$$

for all $x,y \in X$. Hence, T is a β-quasi-contraction while is not a quasi-contraction.

Corollary 5.24

Let (X,d) be a metric space, $\beta : 2^X \times 2^X \to [0,\infty)$ a mapping and $T : X \to P(X)$ a β-quasi-contraction and β-admissible set-valued map. Suppose that there exist $A \subset X$ and $x_0 \in A$ such that $\beta(A, Tx_0) \geq 1$. Then T has approximate fixed points. If (X,d) is a complete metric space and T is lower semi-continuous, then T has a fixed point.

Proof: Choose $\lambda \in (0,1)$ such that

$$\beta(Tx,Ty)H(Tx,Ty) \leq \lambda \max\{d(x,y), d(y,Ty), d(x,Tx), d(x,Ty), d(y,Tx)\}$$

for all $x, y \in X$. It is sufficient we define $g : [0,\infty)^5 \to [0,\infty)$ by

$$g(x_1,x_2,x_3,x_4,x_5) = \lambda \max\{x_1,x_2,x_3,x_4,x_5\}.$$

Then by following the proof of Corollary 5.19, one can complete the proof.

♣

As aforesaid, Suzuki introduced in 2008 a new type of mappings and a generalization of the Banach contraction principle in which the completeness can be also characterized by the existence of fixed points of these mappings. Consider the non-increasing function $\theta : [0,1) \to (\frac{1}{2}, 1]$ by $\theta(r) = 1$ whenever $r \leq \frac{\sqrt{5}-1}{2}$, $\theta(r) = \frac{1-r}{r^2}$ whenever $\frac{\sqrt{5}-1}{2} < r \leq \frac{1}{\sqrt{2}}$ and $\theta(r) = \frac{1}{1+r}$ whenever $\frac{1}{\sqrt{2}} < r < 1$. Let (X,d) be a metric space, $r \in [0,1)$ and T be a mapping on X such that $\theta(r)d(x,Tx) \leq d(x,y)$ implies$d(Tx,Ty) \leq rd(x,y)$ for all $x,y \in X$. Suzuki proved that T has a unique fixed point. Later, some authors tried to generalize the results of Suzuki to set-valued maps. By using the main idea of [10], we give our last result about fixed point of β-generalized Suzuki type proximinal valued multifunctions.

Theorem 5.15

Let (X,d) be a complete metric space, α a constant in $(0,1)$, $\beta : 2^X \times 2^X \to [0,\infty)$ a mapping and $g \in \mathcal{R}$ with $\alpha(h+1) \leq 1$, where $h = g(1,1,1,2,0)$. Suppose that T is a β-admissible and β-convergent proximinal valued multifunction on X such that $\alpha d(x,Tx) \leq d(x,y)$ implies

$$\beta(Tx,Ty)H(Tx,Ty) \leq g(d(x,y), d(y,Ty), d(x,Tx), d(x,Ty), d(y,Tx))$$

for all $x,y \in X$. If there exits a subset A of X and $x_0 \in A$ such that $\beta(A,Tx_0) \geq 1$, then T has a fixed point.

Proof: Choose the subset A of X and $x_0 \in A$ such that $\beta(A,Tx_0) \geq 1$. Since T is proximinal valued, for each $n \geq 0$ there exists $x_{n+1} \in Tx_n$ such that $d(x_n, x_{n+1}) =$

$d(x_n,Tx_n)$. Since T is β-admissible and $\beta(A,Tx_0) \geq 1$, it is easy to see that $\beta(Tx_{n-1},Tx_n) \geq 1$ for all $n \geq 1$. Fix $1 > r > h$. Since $\alpha d(x_0,Tx_0) < d(x_0,x_1)$, by using the assumption we have

$$d(x_1,Tx_1) \leq H(Tx_0,Tx_1) \leq \beta(Tx_0,Tx_1)H(Tx_0,Tx_1)$$

$$\leq g(d(x_0,x_1),d(x_1,Tx_1),d(x_0,Tx_0),d(x_0,Tx_1),d(x_1,Tx_0))$$

$$\leq g(d(x_0,x_1),d(x_1,Tx_1),d(x_0,x_1),d(x_0,x_1)+d(x_1,Tx_1),0).$$

By using Proposition 3.1, we get $d(x_1,Tx_1) \leq hd(x_0,x_1) < rd(x_0,x_1)$. By continuing this process, it is easy to see that $d(x_{n+1},x_n) < r^n d(x_0,x_1)$ and $d(x_{n+1},Tx_{n+1}) \leq hd(x_{n+1},x_n)$ for all $n \geq 0$. If $x_m = x_{m+1}$ for some $m \geq 1$, then x_m is a fixed point of T. Suppose that $x_n \neq x_{n+1}$ for all $n \geq 1$. Choose $x \in X$ such that $x_n \to x$. We claim that either $\alpha d(x_n,Tx_n) \leq d(x_n,x)$ or $\alpha d(x_{n+1},Tx_{n+1}) \leq d(x_{n+1},x)$ hold for all n. If $\alpha d(x_n,Tx_n) > d(x_n,x)$ and $\alpha d(x_{n+1},Tx_{n+1}) > d(x_{n+1},x)$ for some $n \geq 1$, then

$$d(x_{n+1},x_n) \leq d(x_{n+1},x)+d(x,x_n) < \alpha d(x_{n+1},Tx_{n+1})+\alpha d(x_n,Tx_n)$$

$$\leq \alpha hd(x_n,x_{n+1})+\alpha d(x_n,x_{n+1})$$

and so $\alpha(h+1) > 1$ which is a contradiction. Thus, either

$$\beta(Tx_n,Tx)H(Tx_n,Tx) \leq g(d(x_n,x),d(x,Tx),d(x_n,Tx_n),d(x_n,Tx),d(x,Tx_n))$$

or

$$\beta(Tx_{n+1},Tx)H(Tx_{n+1},Tx)$$

$$\leq g(d(x_{n+1},x),d(x,Tx),d(x_{n+1},Tx_{n+1}),d(x_{n+1},Tx),d(x,Tx_{n+1}))$$

hold for all n. Hence, either there exists an infinite subset $I \subseteq \mathbb{N}$ such that

$$\beta(Tx_n,Tx)H(Tx_n,Tx) \leq g(d(x_n,x),d(x,Tx),d(x_n,Tx_n),d(x_n,Tx),d(x,Tx_n))$$

for all $n \in I$, or there exists an infinite subset $J \subseteq \mathbb{N}$ such that

$$\beta(Tx_{n+1},Tx)H(Tx_{n+1},Tx)$$

$$\leq g(d(x_{n+1},x),d(x,Tx),d(x_{n+1},Tx_{n+1}),d(x_{n+1},Tx),d(x,Tx_{n+1}))$$

for all $n \in J$. Since T is β-convergent, in the first case we obtain

$$d(x,Tx) \leq d(x,Tx_n)+H(Tx_n,Tx) \leq d(x,Tx_n)+\beta(Tx_n,Tx)H(Tx_n,Tx)$$

$$\leq d(x,x_{n+1})+g(d(x_n,x),d(x,Tx),d(x_n,Tx_n),d(x_n,Tx),d(x,Tx_n))$$

$$\leq d(x,x_{n+1})+g(d(x_n,x),d(x,Tx),d(x_n,x_{n+1}),d(x_n,x)+d(x,Tx),d(x,x_{n+1}))$$

for sufficiently large $n \in I$. Since g is continuous, we get

$$d(x,Tx) \leq g(0,d(x,Tx),0,0+d(x,Tx),0)$$

and so by using Proposition 3.1, we conclude that $d(x,Tx) = 0$. Since T is β-convergent, in the second case we obtain

$$d(x,Tx) \leq d(x,Tx_{n+1}) + H(Tx_{n+1},Tx) \leq d(x,Tx_{n+1}) + \beta(Tx_{n+1},Tx)H(Tx_{n+1},Tx)$$

$$\leq d(x,x_{n+2}) + g(d(x_{n+1},x),d(x,Tx),d(x_{n+1},Tx_{n+1}),d(x_{n+1},Tx),d(x,Tx_{n+1}))$$

$$d(x,x_{n+2}) + g(d(x_{n+1},x),d(x,Tx),d(x_{n+1},x_{n+2}),d(x_{n+1}x) + d(x,Tx),d(x,x_{n+2}))$$

for sufficiently large $n \in J$. Since g is continuous, we get

$$d(x,Tx) \leq g(0,d(x,Tx),0,0 + d(x,Tx),0)$$

and so by using Proposition 3.1, we obtain $d(x,Tx) = 0$. Thus, $x \in Tx$ and so T has a fixed point.

♣

Finally, we recommend young researchers to try to obtain more approximate fixed point results for single-valued or set-valued mappings. For example, we provide approximate version of Theorem 5.2 as following. The proof is similar via some little changes.

Theorem 5.16

Let (X,d) be a metric space and T a continuous, α-admissible and α-ψ-ξ-contractive selfmap on X such that $\alpha(x_0,Tx_0) \geq 1$ for some $x_0 \in X$. Then T has an approximate fixed point.

Proof: Let $x_0 \in X$ be such that $\alpha(x_0,Tx_0) \geq 1$. Put $x_{n+1} = Tx_n$ for all $n \geq 0$. If $x_n = x_{n+1}$ for some n, then we have nothing to prove. Assume that $x_n \neq x_{n+1}$ for all n. Since T is α-admissible, it is easy to check that $\alpha(x_n,x_{n+1}) \geq 1$ for all n. Thus,

$$d(x_n,x_{n+1}) = d(Tx_{n-1},Tx_n) \leq \alpha(x_{n-1},x_n)d(Tx_{n-1},Tx_n) \leq \psi(h(x_{n-1},x_n))$$

$$= \psi(d(x_{n-1},Tx_n) + d(x_n,Tx_{n-1}) + d(x_{n-1},x_n) - \xi(x_{n-1},x_n))$$

$$= \psi(d(x_{n-1},x_{n+1}) + d(x_n,x_n) + d(x_{n-1},x_n) - max\{d(x_{n-1},x_{n+1}),d(x_n,x_n)\})$$

$$= \psi(d(x_{n-1},x_{n+1}) + d(x_{n-1},x_n) - d(x_{n-1},x_{n+1})) = \psi(d(x_{n-1},x_n))$$

for all n. Hence, $d(x_{n+1},x_n) \leq \psi^n(d(x_0,x_1))$ for all n. Fix $\varepsilon > 0$. Then, there exists a natural number N_ε such that $\psi^n(d(x_0,x_1)) < \varepsilon$. Let $n \geq N_\varepsilon$. Then, we have

$$d(Tx_{n-1},Tx_n) = d(x_n,x_{n+1}) \leq \psi^n d(x_0,x_1) < \varepsilon.$$

Hence, $d(Tx_{n-1},Tx_n) \to 0$ and so T has an approximate fixed point.

♣

5.7 Absolute retract set-valued mappings

In this section, our aim is to provide some relations hip between the topological notion absolute retractivity and set-valued mappings.

Topological properties of the fixed point set of set-valued mappings have been reviewed by some researchers. In 2008, Sintamarian proved some results on absolute retractivity of the common fixed points set of two multivalued operators. On the other hand, Suzuki generalized the notion of contractive mappings in the year. In this section, we provide some new results in this area. These could help young researchers to learn how they can find interesting ideas.

We say that a topological space X is an absolute retract for metric spaces whenever for each metric space Y, $A \in P_{cl}(Y)$ and continuous function $\psi : A \to X$, there exists a continuous function $\varphi : Y \to X$ such that $\varphi|_A = \psi$. Let $\mathcal{M}$ be the set of all metric spaces, $X \in \mathcal{M}$, $\mathcal{D} \in P(\mathcal{M})$ and $F : X \to P_{b,cl}(X)$ a lower semicontinuous multifunction. We say that F has the selection property with respect to $\mathcal{D}$ if for each $Y \in \mathcal{D}$, continuous function $f : Y \to X$ and continuous functional $g : Y \to (0,\infty)$ such that $G(y) := \overline{F(f(y)) \cap B(f(y), g(y))} \neq \emptyset$ for all $y \in Y$, $A \in P_{cl}(Y)$, every continuous selection $\psi : A \to X$ of $G|_A$ admits a continuous extension $\varphi : Y \to X$, which is a selection of G. If $\mathcal{D} = \mathcal{M}$, then we say that F has the selection property and we denote this by $F \in Sp(X)$.

We say that $F : Y \to P_{b,cl}(X)$ is KS-multifunction whenever there exists $a \in [0,1)$ such that

$$\eta(a)d(x,F(x)) \leq d(x,y) \text{ implies } H(F(x),F(y)) \leq ad(x,y)$$

for all $x,y \in Y$, where $\eta(a) = \frac{1}{1+a}$. In 2008, Kikkawa and Suzuki proved that each KS-multifunction has a fixed point. Let (X,d) be a metric space, $\alpha : X \times X \to [0,\infty)$ a function, $\psi \in \Psi$ and $F : X \to 2^X$ a set-valued map. We say that F is an α-ψ-Suzuki contractive multifunction whenever there exists $a \in [0,1)$ such that

$$\eta(a)d(x,F(x)) \leq d(x,y) \text{ implies } \alpha(x,y)H(F(x),F(y)) \leq \psi(d(x,y))$$

for all $x,y \in X$, where $\eta(a) = \frac{1}{1+a}$. Recall that a multifunction F is said α-admissible whenever for each $x \in X$ and $y \in Fx$ with $\alpha(x,y) \geq 1$ we have $\alpha(y,z) \geq 1$ for all $z \in Fy$. Also, recall that F is continuous whenever $H(Fx_n, Fx) \to 0$ for all sequence $\{x_n\}$ in X with $x_n \to x$.

Theorem 5.17

Let (X,d) be a complete metric space, $\alpha : X \times X \to [0,\infty)$ a function, $\psi \in \Psi$ a *strictly increasing map and $F : X \to CB(X)$ an α-admissible multifunction such that*

$$\eta(a)d(x,F(x)) \leq d(x,y) \Longrightarrow \alpha(x,y)H(F(x),F(y)) \leq \psi(d(x,y)) \qquad (5.1)$$

for all $x,y \in X$ and there exist $x_0 \in X$ and $x_1 \in Fx_0$ with $\alpha(x_0,x_1) \geq 1$. If F is continuous, then F has a fixed point.

Proof: If $x_1 = x_0$, then we have nothing to prove. Let $x_1 \neq x_0$. If $x_1 \in Fx_1$, then x_1 is a fixed point of F. Let $x_1 \notin Fx_1$ and $q > 1$ be given. Then

$$0 < d(x_1,Fx_1) \leq \alpha(x_0,x_1)H(Fx_0,Fx_1) < q\alpha(x_0,x_1)H(Fx_0,Fx_1).$$

In (5.1) set $x = x_0,\ y = x_1$, we have,

$$\eta(a)D(x_0,Fx_0) \leq D(x_0,Fx_0) \leq d(x_0,x_1)$$

there exists $x_2 \in Fx_1$ such that

$$d(x_1,x_2) < q\alpha(x_0,x_1)H(Fx_0,Fx_1),$$

hence by (5.1),

$$d(x_1,x_2) < q\psi(d(x_0,x_1)).$$

It is clear that $x_2 \neq x_1$. Put $t_0 = d(x_0,x_1)$. Then, $t_0 > 0$ and $d(x_1,x_2) < q\psi(t_0)$. since ψ is strictly increasing, $\psi(d(x_1,x_2)) < \psi(q\psi(t_0))$. Put $q_1 = \frac{\psi(q\psi(t_0))}{\psi(d(x_1,x_2))}$. Then $q_1 > 1$. If $x_2 \in Fx_2$, then x_2 is a fixed point of F. Assume that $x_2 \notin Fx_2$. Then,

$$0 < d(x_2,Fx_2) \leq \alpha(x_1,x_2)H(Fx_1,Fx_2) < q_1\alpha(x_1,x_2)H(Fx_1,Fx_2).$$

In (5.1) set $x = x_1,\ y = x_2$, we have,

$$\eta(a)D(x_1,Fx_1) \leq D(x_1,Fx_1) \leq d(x_1,x_2)$$

there exists $x_3 \in Fx_2$ such that

$$d(x_2,x_3) < q_1\alpha(x_1,x_2)H(Fx_1,Fx_2),$$

hence by (5.1),

$$d(x_2,x_3) < q_1\psi(d(x_1,x_2)) = \psi^2(q_1\psi(t_0)).$$

It is clear that $x_3 \neq x_2$ and $\psi(d(x_2,x_3)) < \psi^2(q\psi(t_0))$. Put $q_2 = \frac{\psi^2(q\psi(t_0))}{\psi(d(x_2,x_3))}$. Then $q_2 > 1$. If $x_3 \in Fx_3$, then x_3 is a fixed point of F. Assume that $x_3 \notin Fx_3$. Then,

$$0 < d(x_3,Fx_3) \leq \alpha(x_2,x_3)H(Fx_2,Fx_3) < q_2\alpha(x_2,x_3)H(Fx_2,Fx_3).$$

In (5.1) set $x = x_2,\ y = x_3$, we have,

$$\eta(a)D(x_2,Fx_2) \leq D(x_2,Fx_2) \leq d(x_2,x_3)$$

there exists $x_4 \in Fx_3$ such that

$$d(x_3,x_4) < q_2\alpha(x_2,x_3)H(Fx_2,Fx_3),$$

hence by (5.1),

$$d(x_3,x_4) < q_2\psi(d(x_2,x_3)) = \psi(q_2\psi(t_0)).$$

Thus, there exists $x_4 \in Fx_3$ such that

$$d(x_3,x_4) < q_1\alpha(x_2,x_3)H(Fx_2,Fx_3) \leq q_2\psi(d(x_2,x_3)) = \psi^2(q\psi(t_0)).$$

By continuing this process we obtain a sequence $\{x_n\}$ in X such that $x_n \in Fx_{n-1}$, $x_n \neq x_{n-1}$ and $d(x_n,x_{n+1}) \leq \psi^{n-1}(q\psi(t_0))$ for all n. Now for each $m > n$ we have

$$d(x_n,x_m) \leq \Sigma_{i=n}^{m-1} d(x_i,x_{i+1}) \leq \Sigma_{i=n}^{m-1} \psi^{i-1}(q\psi(t_0)).$$

Hence, $\{x_n\}$ is a Cauchy sequence in X. Since X is complete, there exists $x^\star \in X$ such that $x_n \to x^\star$. Now if F is continuous, then

$$d(x^\star,Fx^\star) = \lim_{n\to\infty} d(x_{n+1},Fx^\star) \leq \lim_{n\to\infty} H(Fx_n,Fx^\star) = 0$$

and so $x^\star \in Fx^\star$.

♣

Example 83 *Put* $X = \{1\} \cup \{m + \frac{1}{n+2} : m,n \in \mathbb{N}\}$ *and define a metric* d *on* X *by*

$$d(x,y) = \mid x - y \mid .$$

Define a mapping F *on* X *by*

$$F(x) = \begin{cases} 1 & x = 1, \\ 7m + \frac{1}{n+2} & x = m + \frac{1}{n}. \end{cases}$$

Then F *satisfies in the assumption of Theorem 5.17.*

Proof: It is obviously that (X,d) is a complete metric space and 1 is a unique fixed point of F. if $n < m$, we have

$$\eta(a)d(m + \frac{1}{n+2}, F(m + \frac{1}{n+2})) < d(m + \frac{1}{n+2}, n + \frac{1}{n+2})$$

$$\Updownarrow$$

$$\eta(a)d(m + \frac{1}{n+2}, 7m + \frac{1}{n+2}) < d(m + \frac{1}{n+2}, n + \frac{1}{n+2})$$

$$\Updownarrow$$

$$\eta(a)\mid m+\frac{1}{n+2}-7m-\frac{1}{n+2}\mid<\mid m+\frac{1}{n+2}-n-\frac{1}{n+2}\mid$$
$$\Updownarrow$$
$$\frac{1}{2}\mid -6m\mid\leq\eta(a)\mid -6m\mid<\mid m+\frac{1}{n+2}-n-\frac{1}{n+2}\mid.$$
$$\Updownarrow$$
$$3m<m-n<m.$$

This is a contradiction. Therefore F satisfies in the assumption of Theorem 5.17.

♣

Theorem 5.18

Let (X,d) be a complete metric space and absolute retract for metric spaces, $F: X\to P_{b,cl}(X)$ a α-ψ-Suzuki contractive multifunction and $F\in Sp(X)$. If $\alpha(x,y)\geq 1$ for all $x\in X$ and $y\in F(x)$, then $\mathcal{F}_F=\{x\in X: x\in F(x)\}$ is an absolute retract for metric spaces.

Proof: Let Y be a metric space, $A\in P_{cl}(Y)$, $q>1$ and $\xi: A\to\mathcal{F}_F$ a continuous function. Since X is an absolute retract for metric spaces, there exists a continuous function $\varphi_0: Y\to X$ such that $\varphi_0|_A=\xi$. Define the function $g_0: Y\to(0,\infty)$ by

$$g_0(y)=\sup\{d(\varphi_0(y),z)|z\in F(\varphi_0(y))\}+1$$

for all $y\in Y$. It is not difficult to see that g_0 is continuous and

$$F(\varphi_0(y))\cap B(\varphi_0(y),g_0(y))=F(\varphi_0(y))$$

for all $y\in A$. Also we observe that the function $\xi: A\to\mathcal{F}_F$ has the property

$$\xi(y)\in F(\varphi_0(y))\ (y\in A)$$

so is a continuous selection of the multivalued mapping. Since $F\in Sp(X)$, there exists a continuous function $\varphi_1: Y\to X$ such that $\varphi_1|_A=\xi$ and $\varphi_1(y)\in F(\varphi_0(y))$ for all $y\in Y$. Since $\eta(a)d(\varphi_0(y),F(\varphi_0(y)))\leq\eta(a)d(\varphi_0(y),\varphi_1(y))\leq d(\varphi_0(y),\varphi_1(y))$, we have

$$\alpha(\varphi_0(y),\varphi_1(y))H(F(\varphi_0(y),F(\varphi_1(y))))\leq\psi(d(\varphi_0(y),\varphi_1(y))).$$

But, $\alpha(\varphi_0(y),\varphi_1(y))\geq 1$. Thus,

$$d(\varphi_1(y),F(\varphi_1(y)))\leq H(F(\varphi_0(y)),F(\varphi_1(y)))$$

$$\leq \alpha(\varphi_0(y),\varphi_1(y))H(F(\varphi_0(y)),F(\varphi_1(y)))$$
$$\leq \psi(d(\varphi_0(y),\varphi_1(y))) < \psi(d(\varphi_0(y),\varphi_1(y))) + q^{-1}.$$

Hence, $G_2(y) := F(\varphi_1(y)) \cap B(\varphi_1(y), \psi(d(\varphi_0(y),\varphi_1(y))) + q^{-1}) \neq \emptyset$ for all $y \in Y$. Since we know that $F \in Sp(X)$, there exists a continuous function $\varphi_2 : Y \to X$ such that $\varphi_2|_A = \xi$ and $\varphi_2(y) \in \overline{G_2(y)}$ for all $y \in Y$. Thus, $\varphi_2|_A = \xi$, $\varphi_2(y) \in F(\varphi_1(y))$ for all $y \in Y$ and $d(\varphi_1(y),\varphi_2(y)) < \psi(d(\varphi_0(y),\varphi_1(y))) + q^{-1}$. Now, note that

$$\eta(a)d(\varphi_1(y),F(\varphi_1(y))) \leq \eta(a)d(\varphi_1(y),\varphi_2(y)) \leq d(\varphi_1(y),\varphi_2(y)).$$

Thus,

$$d(\varphi_2(y),F(\varphi_2(y))) \leq H(F(\varphi_1(y)),F(\varphi_2(y))) \leq \alpha(\varphi_0(y),\varphi_1(y))H(F(\varphi_0(y)),$$

$$F(\varphi_1(y))) \leq \psi(d(\varphi_1(y),\varphi_2(y))) \leq \psi^2(d(\varphi_1(y),\varphi_2(y))) < \psi^2(d(\varphi_1(y),\varphi_2(y)))+q^{-2}.$$

Hence, $G_3(y) := F(\varphi_2(y)) \cap B(\varphi_2(y)), \psi^2(d(\varphi_0(y),\varphi_1(y)) + q^{-2}) \neq \emptyset$. Since $F \in Sp(X)$, there exists a continuous function $\varphi_3 : Y \to X$ such that $\varphi_3|_A = \xi$ and $\varphi_3(y) \in \overline{G_3(y)}$ for all $y \in Y$. Also, we have $d(\varphi_2(y),\varphi_3(y)) < \psi^2 d(\varphi_0(y),\varphi_1(y)) + q^{-2}$ and $\varphi_3(y) \in F(\varphi_2(y))$ for all $y \in Y$. By continuing this process, we obtain $\{\varphi_n : Y \to X\}_{n \geq 0}$ a sequence of continuous functions such that $\varphi_n|_A = \xi$ and $d(\varphi_{n-1}(y),\varphi_n(y)) < \psi^{n-1}d(\varphi_0(y),\varphi_1(y)) + q^{-(n-1)}$ and $\varphi_n(y) \in F(\varphi_{n-1}(y))$ for all $y \in Y$ and $n \geq 1$. Let $\varepsilon > 0$ be given and $t \geq 0$. Choose a natural number $n(\varepsilon)$ such that $\sum_{n \geq n(\varepsilon)} \psi^n(t) < \varepsilon$. Put

$$Y_\varepsilon := \{y \in Y | \sum_{n \geq n(\varepsilon)} \psi^n(d(\varphi_0(y),\varphi_1(y))) < \varepsilon\}$$

for all $\varepsilon > 0$. Then the family $\{Y_\varepsilon | \varepsilon > 0\}$ is an open cover for Y. Let $m > n > n(\varepsilon)$. By using the triangular inequality, we have

$$d(\varphi_n(y),\varphi_m(y)) \leq \sum_{k=n}^{m-1} d(\varphi_k(y),\varphi_{k+1}(y))$$

$$\leq \sum_{k=n}^{m-1} \psi^k(d(\varphi_0(y),\varphi_1(y)) \leq \sum_{n \geq n(\varepsilon)} \psi^n(d(\varphi_0(y),\varphi_1(y))) < \varepsilon$$

for all $y \in Y_\varepsilon$, the sequence $(\varphi_n)_{n \geq 0}$ converges uniformly on Y_ε for all $\varepsilon > 0$. Let $\varphi : Y \to X$ be the point-wise limit of $\{\varphi_n\}_{n \geq 0}$. Then φ is continuous and $\varphi|_A = \xi$ because $\varphi_n|_A = \xi$ for all $n \geq 0$. Since F is lower semi-continuous, $\varphi(y) \in F(\varphi(y))$ for all $y \in Y$. Thus, $\varphi : Y \to \mathcal{F}_F$ is a continuous extension of ξ. Therefore, $\mathcal{F}_F$ is an absolute retract for metric spaces.

♣

The following example shows that there are α-ψ-Suzuki contractive multifunctions which do not satisfy the conditions of Sintamarian or Suzuki.

Example 84 *Let* $X=\{(2,0),(-4,2),(2,3),(-3,2)\}$ *and its metric is defined by*

$$d((x_1,x_2),(y_1,y_2))=|x_1-y_1|+|x_2-y_2|.$$

Define the multifunction F *on* X *and the mapping* $\alpha: X\times X\to[0,\infty)$ *by*

$$F(x_1,x_2)=\begin{cases}\{(-4,2)\} & x_1=2\\ \{(2,0)\} & (x_1,x_2)=(-3,2)\\ \{(2,3)\} & (x_1,x_2)=(-4,2)\end{cases}$$

and

$$\alpha((x_1,x_2),(y_1,y_2))=\begin{cases}1 & (x_1,x_2),(y_1,y_2)\in\{(2,0),(2,3)\}\\ 0 & else.\end{cases}$$

Let $\psi\in\Psi$ *be arbitrary and* $a=0$. *It is easy to check that* F *is an* α-ψ-*Suzuki contractive multifunction. Now, we show that* F *does not satisfy the condition of Sintamarian. If*

$$\begin{aligned}H(F(2,3),F(-3,2))&\le a_1d((2,3),(-3,2))+a_2d((2,3),F(2,3))\\ +a_3d((-3,2),F(-3,2))&+a_4d((2,3),F(-3,2))+a_5d((-3,2),F(2,3))\\ &=6a_1+a_2d((2,3),(-4,2))+a_3d((-3,2),(2,0))\\ &\quad+a_4d((2,3),(2,0))+a_5d((-3,2),(-4,2))\\ &=6a_1+7a_2+7a_3+3a_4+a_5,\end{aligned}$$

then $8\le 6a_1+7a_2+7a_3+3a_4+a_5\le 7a_1+7a_2+7a_3+3a_4+a_5$ *and so*

$$\frac{8}{7}\le a_1+a_2+a_3+a_4+\frac{1}{7}a_5\le a_1+a_2+a_3+a_4+\frac{1}{7}.$$

Thus, $1\le a_1+a_2+a_3+a_4+a_5$ *which is a contradiction. Hence,* F *does not satisfy the condition of Sintamarian. Now, we show that* F *does not satisfy the condition of Suzuki. In this way, put* $x=(2,3)$ *and* $y=(-3,2)$. *Then*

$$d(x,F(x))=d((2,3),(2,0))<d((2,3),(-3,2))$$

while

$$H(F(2,3),F(-3,2))=d((-4,2),(2,0))>d((2,3),(2,0)).$$

Theorem 5.19

Let (X,d) *be a complete metric space and absolute retract for metric spaces,* $F:$

$X \to P_{b,cl}(X)$ a multifunction and $F \in Sp(X)$. Suppose that there exist $\alpha \in (0,\infty)$ and $g \in \mathcal{R}$ such that $\alpha(h+1) \leq 1$ and $\alpha d(x,Fx) \leq d(x,y)$ implies

$$H(Fx,Fy) \leq g(d(x,y),d(x,Fx),d(y,Fy),d(x,Fy),d(y,Fx))$$

for all $x,y \in X$. Then $\mathcal{F}_F$ is an absolute retract for metric spaces.

Proof: Let Y be a metric space, $A \in P_{cl}(Y)$, $q > 1$ and $\xi : A \to \mathcal{F}_F$ a continuous function. Since X is an absolute retract for metric spaces, there exists a continuous function $\varphi_0 : Y \to X$ such that $\varphi_0|_A = \xi$. Define the function $g_0 : Y \to (0,\infty)$ by

$$g_0(y) = \sup\{d(\varphi_0(y),z)|z \in F(\varphi_0(y))\} + 1$$

for all $y \in Y$. It is not difficult to see that g_0 is continuous and

$$F(\varphi_0(y)) \cap B(\varphi_0(y), g_0(y)) = F(\varphi_0(y))$$

for all $y \in A$. Also we observe that the function $\xi : A \to \mathcal{F}_F$ has the property

$$\xi(y) \in F(\varphi_0(y)) \ \ (y \in A)$$

so is a continuous selection of the set-valued mapping. Since $F \in Sp(X)$, there exists a continuous function $\varphi_1 : Y \to X$ such that $\varphi_1|_A = \xi$ and $\varphi_1(y) \in F(\varphi_0(y))$ for all $y \in Y$. Since $\alpha d(\varphi_0(y), F(\varphi_0(y))) \leq \alpha d(\varphi_0(y), \varphi_1(y)) \leq d(\varphi_0(y), \varphi_1(y))$, we have

$$\begin{aligned} &H(F(\varphi_0(y)),F(\varphi_1(y))) \\ &\leq g(d(\varphi_0(y),\varphi_1(y)),d(\varphi_0(y),F(\varphi_0(y))),d(\varphi_1(y),F(\varphi_1(y))), \\ &\quad d(\varphi_0(y),F(\varphi_1(y))),d(\varphi_1(y),F(\varphi_0(y)))). \end{aligned}$$

Hence

$$\begin{aligned} d(\varphi_1(y),F(\varphi_1(y))) &\leq H(F(\varphi_0(y)),F(\varphi_1(y))) \\ &\leq g(d(\varphi_0(y),\varphi_1(y)),d(\varphi_0(y),F(\varphi_0(y))),d(\varphi_1(y),F(\varphi_1(y))), \\ &\quad d(\varphi_0(y),F(\varphi_1(y))),d(\varphi_1(y),F(\varphi_0(y)))) \\ &\leq g(d(\varphi_0(y),\varphi_1(y)),d(\varphi_0(y),\varphi_1(y)),d(\varphi_1(y),F(\varphi_1(y))), \\ &\quad d(\varphi_0(y),\varphi_1(y))+d(\varphi_1(y),F(\varphi_1(y))),0). \end{aligned}$$

Fix $h < r < 1$. By using Proposition 3.1, we obtain

$$d(\varphi_1(y),F(\varphi_1(y))) \leq hd(\varphi_0(y),\varphi_1(y)) < rd(\varphi_0(y),\varphi_1(y)) < rd(\varphi_0(y),\varphi_1(y))+q^{-1}.$$

Thus, $G_2(y) := F(\varphi_1(y)) \cap B(\varphi_1(y), rd(\varphi_0(y),\varphi_1(y)) + q^{-1}) \neq \emptyset$ for all $y \in Y$. Since $F \in Sp(X)$, there exists a continuous function $\varphi_2 : Y \to X$ such that

$$\varphi_2|_A = \xi \text{ and } \varphi_2(y) \in \overline{G_2(y)}$$

for all $y \in Y$. But, note that

$$\alpha d(\varphi_1(y), F(\varphi_1(y))) \leq \alpha d(\varphi_1(y), \varphi_2(y)) \leq d(\varphi_1(y), \varphi_2(y))$$

hence by assumption of theorem we have

$$\begin{aligned} d(\varphi_2(y), F(\varphi_2(y))) &\leq H(F(\varphi_1(y)), F(\varphi_2(y))) \leq \\ g(d(\varphi_1(y), \varphi_2(y)), &d(\varphi_1(y), F(\varphi_1(y))), d(\varphi_2(y), F(\varphi_2(y))), \\ &d(\varphi_1(y), F(\varphi_2(y))), d(\varphi_2(y), F(\varphi_1(y)))) \\ \leq g(d(\varphi_1(y), \varphi_2(y)), &d(\varphi_1(y), \varphi_2(y)), d(\varphi_2(y), F(\varphi_2(y))), \\ &d(\varphi_1(y), \varphi_2(y)) + d(\varphi_2(y), F(\varphi_2(y))), 0). \end{aligned}$$

Again, by using Proposition 3.1, we obtain

$$\begin{aligned} d(\varphi_2(y), F(\varphi_2(y))) &\leq hd(\varphi_1(y), \varphi_2(y)) < rd(\varphi_1(y), \varphi_2(y)) \\ \leq r(rd(\varphi_0(y), \varphi_1(y))) &= r^2 d(\varphi_0(y), \varphi_1(y)) < r^2 d(\varphi_0(y), \varphi_1(y)) + q^{-2} \end{aligned}$$

and so $G_3(y) := F(\varphi_2(y)) \cap B(\varphi_2(y)), r^2(d(\varphi_0(y), \varphi_1(y)) + q^{-2}) \neq \emptyset$. Since $F \in Sp(X)$, there exists a continuous function $\varphi_3 : Y \to X$ such that $\varphi_3(y) \in \overline{G_3(y)}$ for all $y \in Y$ and $\varphi_3|_A = \xi$. Also, we have $d(\varphi_2(y), \varphi_3(y)) < r^2 d(\varphi_0(y), \varphi_1(y)) + q^{-2}$ and $\varphi_3(y) \in F(\varphi_2(y))$ for all $y \in Y$. By continuing this process, we obtain $\{\varphi_n : Y \to X\}_{n \geq 0}$ a sequence of continuous functions such that $\varphi_n(y) \in F(\varphi_{n-1}(y))$, $\varphi_n|_A = \xi$ and

$$d(\varphi_{n-1}(y), \varphi_n(y)) < r^n d(\varphi_0(y), \varphi_1(y)) + q^{-(n-1)}$$

for all $y \in Y$ and $n \geq 1$. Now for each $\lambda > 0$, put $Y_\lambda := \{y \in Y | d(\varphi_0(y), \varphi_1(y)) < \lambda\}$. Then the family $\{Y_\lambda | \lambda > 0\}$ is an open covering for Y. Since

$$d(\varphi_{n-1}(y), \varphi_n(y)) < r^{n-1} d(\varphi_0(y), \varphi_1(y)) + q^{-(n-1)}$$

for all $y \in Y$ and $n \geq 1$, $r < 1$ and X is complete, the sequence $\{\varphi_n\}_{n \geq 0}$ converges uniformly on Y_λ for all $\lambda > 0$. Let $\varphi : Y \to X$ be the point-wise limit of $\{\varphi_n\}_{n \geq 0}$. Then φ is continuous and $\varphi|_A = \xi$ because $\varphi_n|_A = \xi$ for all $n \geq 0$. Since F is lower semi-continuous, $\varphi(y) \in F(\varphi(y))$ for all $y \in Y$. Thus, $\varphi : Y \to \mathcal{F}_F$ is a continuous extension of ξ. Therefore, $\mathcal{F}_F$ is an absolute retract for metric spaces.

♣

Corollary 5.25

Let (X, d) be a complete metric space and absolute retract for metric spaces, F :

$X \to P_{b,cl}(X)$ a set-valued map and $F \in Sp(X)$. Assume that there exists $r \in [2^{\frac{-1}{2}}, 1)$ such that $\theta(r)d(x,Fx) \le d(x,y)$ implies

$$H(Fx,Fy) \le r\max\{d(x,y), d(x,Fx), d(y,Fy)\}$$

for all $x, y \in X$. Then $\mathcal{F}_F$ is an absolute retract for metric spaces.

Proof: Define $g \in \mathcal{R}$ by $g(x_1,x_2,x_3,x_4,x_5) = r\max\{x_1,x_2,x_3\}$ and $\alpha(x,y) = \theta(r)$. Since $h = r$ and $\alpha(1+h) \le 1$, by using Theorem 5.19, $\mathcal{F}_F$ is an absolute retract for metric spaces.

♣

Corollary 5.26

Let (X,d) be a complete metric space and absolute retract for metric spaces, $F : X \to P_{b,cl}(X)$ a multifunction and $F \in Sp(X)$. Assume that there exists $\beta, \gamma \in [0,1)$ such that $\frac{1}{2\beta+\gamma+1}d(x,Fx) \le d(x,y)$ implies $H(Fx,Fy) \le \gamma(x,y) + \beta d(x,Fx) + \beta d(y,Fy)$ for all $x, y \in X$. Then $\mathcal{F}_F$ is an absolute retract for metric spaces.

Proof: Define $g \in \mathcal{R}$ by $g(x_1,x_2,x_3,x_4,x_5) = \gamma x_1 + \beta x_2 + \beta x_3$ and $\alpha(x,y) = \frac{1}{2\beta+\gamma+1}$. Since $h = 2\beta + \gamma$ and $\alpha(1+h) \le 1$, by using Theorem 5.19, $\mathcal{F}_F$ is an absolute retract for metric spaces.

♣

5.8 Absolute retraction Suzuki type set-valued mappings

Let (X,d) be a metric space and $Y \in P(X)$. As we said in last section, the set-valued mapping $F : Y \to P(X)$ is called a KS-multifunction whenever there exists $a \in [0,1)$ such that

$$\eta(a)D(x,F(x)) \le d(x,y) \text{ implies } H(F(x),F(y)) \le ad(x,y)$$

for all $x, y \in Y$, where $\eta(a) = \frac{1}{1+a}$. In 2008, Kikawa and Suzuki proved that if (X,d) is a complete metric space, then each KS-multifunction $F : Y \to P_{b,cl}(X)$ has a fixed point. In this section, by using the Suzuki's method we present some results on absolute retractivity of the fixed points set of KS-multifunctions. In 2009, Sintamarian proved the next result.

Theorem 5.20

Let (X,d) be a complete metric space and absolute retract for metric spaces and $F \in SP(X)$ with the property that there exist $a_1,\dots,a_5 \in (0,\infty)$ such that $a_1+a_2+a_3+2a_4 < 1$ and

$$H(F(x),F(y)) \leq a_1 d(x,y) + a_2 D(x,F(x))$$

$$+a_3 D(y,F(y)) + a_4 D(x,F(y)) + a_5 D(y,F(x))$$

for all $x,y \in X$. Then $\mathcal{F}_F$ is an absolute retract for metric spaces.

Theorem 5.21

Let (X,d) be a complete metric space and absolute retract for metric spaces, $F : X \to P_{b,cl}(X)$ a KS-multifunction and $F \in SP(X)$. Then $\mathcal{F}_F$ is an absolute retract for metric spaces.

Proof: Let Y be a metric space, $A \in P_{cl}(Y)$ and $\psi : A \to \mathcal{F}_F$ a continuous function. Since X is an absolute retract for metric spaces, there exists a continuous function $\varphi_0 : Y \to X$ such that $\varphi_0|_A = \psi$. Define the function $g_0 : Y \to (0,\infty)$ by

$$g_0(y) = \sup\{d(\varphi_0(y),z) | z \in F(\varphi_0(y))\} + 1$$

for all $y \in Y$. It is not difficult to see that g_0 is continuous and

$$F(\varphi_0(y)) \cap B(\varphi_0(y), g_0(y)) = F(\varphi_0(y))$$

for all $y \in Y$. Also we observe that the function $\psi : A \to \mathcal{F}_F$ is a continuous selection of the multivalued mapping $A \ni y \vdash F(\varphi_0(y))$. Since $F \in SP(X)$, there exists a continuous function $\varphi_1 : Y \to X$ such that $\varphi_1|_A = \psi$ and $\varphi_1(y) \in F(\varphi_0(y))$ for all $y \in Y$. Suppose that $a \in (0,1)$ and $q \in (1, a^{-1})$. Note that, $\eta(a) = \frac{1}{1+a}$. Since $\eta(a) D(\varphi_0(y), F(\varphi_0(y))) \leq \eta(a) d(\varphi_0(y), \varphi_1(y)) \leq d(\varphi_0(y), \varphi_1(y))$, by using the hypothesis we have $H(F(\varphi_0(y)), F(\varphi_1(y))) \leq ad(\varphi_0(y), \varphi_1(y))$ and so

$$D(\varphi_1(y), F(\varphi_1(y))) \leq H(F(\varphi_0(y)), F(\varphi_1(y)))$$

$$\leq ad(\varphi_0(y), \varphi_1(y)) < ad(\varphi_0(y), \varphi_1(y)) + a < ad(\varphi_0(y), \varphi_1(y)) + q^{-1}.$$

Thus,

$$G_2(y) := F(\varphi_1(y)) \cap B(\varphi_1(y), ad(\varphi_0(y), \varphi_1(y)) + q^{-1}) \neq \emptyset$$

for all $y \in Y$. Since $F \in SP(X)$, there exists a continuous function $\varphi_2 : Y \to X$ such that $\varphi_2|_A = \psi$ and $\varphi_2(y) \in G_2(y)$ for all $y \in Y$. Thus, $\varphi_2|_A = \psi$, $\varphi_2(y) \in F(\varphi_1(y))$ for all $y \in Y$ and $d(\varphi_1(y), \varphi_2(y)) \leq ad(\varphi_0(y), \varphi_1(y)) + q^{-1}$. Now, note that

$$\eta(a) D(\varphi_1(y), F(\varphi_1(y))) \leq \eta(a) d(\varphi_1(y), \varphi_2(y)) \leq d(\varphi_1(y), \varphi_2(y))$$

and so

$$D(\varphi_2(y),F(\varphi_2(y))) \leq H(F(\varphi_1(y)),F(\varphi_2(y)))$$
$$\leq ad(\varphi_1(y),\varphi_2(y)) < a^2d(\varphi_0(y),\varphi_1(y)) + aq^{-1} < a^2d(\varphi_0(y),\varphi_1(y)) + q^{-2}.$$

Thus,

$$G_3(y) := F(\varphi_2(y)) \cap B(\varphi_2(y), a^2d(\varphi_0(y),\varphi_1(y)) + q^{-2}) \neq \emptyset.$$

Since $F \in SP(X)$, there exists a continuous function $\varphi_3 : Y \to X$ such that $\varphi_3(y) \in \overline{G_3(y)}$ for all $y \in Y$ and $\varphi_3|_A = \psi$. Hence, $d(\varphi_2(y),\varphi_3(y)) \leq a^2d(\varphi_0(y),\varphi_1(y)) + q^{-2}$ and $\varphi_3(y) \in F(\varphi_2(y))$ for all $y \in Y$. By continuing this process, we obtain a sequence $(\varphi_n)_{n\geq 0}$, where $\varphi_n : Y \to X$ is a continuous function for all $n \geq 0$, such that $\varphi_n(y) \in F(\varphi_{n-1}(y))$ for all $y \in Y$, $\varphi_n|_A = \psi$ and

$$d(\varphi_{n-1}(y),\varphi_n(y)) \leq a^{n-1}d(\varphi_0(y),\varphi_1(y)) + q^{-(n-1)}$$

for all $y \in Y$. Now for each $\lambda > 0$, we put $Y_\lambda := \{y \in Y | d(\varphi_0(y),\varphi_1(y)) < \lambda\}$. The family of sets $\{Y_\lambda | \lambda > 0\}$ is an open covering of Y. Since

$$d(\varphi_{n-1}(y),\varphi_n(y)) \leq a^{n-1}d(\varphi_0(y),\varphi_1(y)) + q^{-(n-1)}$$

for all $y \in Y$ and $n \geq 0$, $a \in (0,1)$ and $q > 1$ are arbitrary and X is complete, the sequence $(\varphi_n)_{n\geq 0}$ converges uniformly on Y_λ for all $\lambda > 0$. Let $\varphi : Y \to X$ be the point wise limit of $(\varphi_n)_{n\geq 0}$ and note that φ is continuous and $\varphi|_A = \psi$ because $\varphi_n|_A = \psi$ for all $n \geq 0$. Since F is lower semi-continuous, we get that $\varphi(y) \in F(\varphi(y))$ for all $y \in Y$. Thus, $\varphi : Y \to \mathcal{F}_F$ is a continuous extension of ψ, that is, $\mathcal{F}_F$ is an absolute retract for metric spaces.

♣

The following example shows that there are KS-multifunctions which do not satisfy the condition of Sintamarian.

Example 85 *Let* $X = \{(0,0),(4.5,0),(0,4.5),(2,4.5),(4.5,2),(2,5.1)\}$ *and define the metric d on X by* $d((x_1,x_2),(y_1,y_2)) = |x_1 - y_1| + |x_2 - y_2|$. *Now, define the set-valued map F on X by*

$$F(x) = \begin{cases} \{(4.5,0)\} & (x_1,x_2) = (2,4.5) \\ \{(0,4.5)\} & (x_1,x_2) = (4.5,2) \\ \{(4.5,2)\} & (x_1,x_2) = (2,5.1) \\ \{(0,0)\} & x_1 = 0 \text{ or } x_2 = 0 \end{cases}$$

It is easy to check that F is a KS-multifunction. Now, we show that F does not satisfy the condition of Sintamarian. If F satisfy the condition of Sintamarian, then

$$H(F(2,5.1),F(4.5,2)) \leq a_1d((2,5.1),(4.5,2)) + a_2D((2,5.1),F(2,5.1))$$

$$+a_3D((4.5,2),F(4.5,2))+a_4D((2,5.1),F(4.5,2))+a_5D((4.5,2),F(2,5.1))$$
$$=5.6a_1+5.6a_2+7a_3+2.6a_4.$$

Hence, $7\leq 5.6a_1+5.6a_2+7a_3+2.6a_4<7a_1+7a_2+7a_3+14a_4$. *This implies that* $1<a_1+a_2+a_3+2a_4$ *which is a contradiction.*

Proposition 5.3

Let (X,d) *be a complete metric space and absolute retract for metric spaces and* $\eta:[0,1)\to(\frac{1}{2},1]$ *defined by*

$$\eta(r)=\begin{cases}1 & if\, 0\leq r<\frac{1}{\sqrt{2}}\\ \frac{1}{1+r} & if\, \frac{1}{\sqrt{2}}\leq r<1.\end{cases}$$

Suppose that $F:X\to P_{b,cl}(X)$ *is a set-valued map such that* $F\in SP(X)$ *and there exists* $\alpha\in[0,\frac{1}{2})$ *such that for* $r=\frac{\alpha}{1-\alpha}$

$$\eta(r)D(x,F(x))\leq d(x,y) \text{ implies } H(F(x),F(y))\leq \alpha D(x,F(x))+\alpha D(y,F(x))$$

for all $x,y\in X$. *Then* $\mathcal{F}_F$ *is an absolute retract for metric spaces.*

Proof: By using similar terminologies as in Theorem 5.21, we have

$$\eta(r)D(\varphi_0(y),F(\varphi_0(y)))\leq\eta(r)d(\varphi_0(y),\varphi_1(y))\leq d(\varphi_0(y),\varphi_1(y)).$$

Hence

$$H(F(\varphi_0(y)),F(\varphi_1(y)))\leq\alpha D(\varphi_0(y),F(\varphi_0(y)))+\alpha D(\varphi_1(y),F(\varphi_0(y)))$$
$$=\alpha D(\varphi_0(y),F(\varphi_0(y))).$$

Therefore, $H(F(\varphi_0(y)),F(\varphi_1(y)))\leq\frac{\alpha}{1-\alpha}d(\varphi_0(y),\varphi_1(y))=rd(\varphi_0(y),\varphi_1(y))$. Now by using a similar proof as in Theorem 5.21, we can complete the proof.

♣

Proposition 5.4

Let (X,d) *be a complete metric space and absolute retract for metric spaces. Suppose that* $F:X\to P_{b,cl}(X)$ *is a set-valued map such that* $F\in SP(X)$ *and there exists* $\beta\in[0,\frac{1}{3})$ *such that for* $r=\frac{2\beta}{1-\beta}$

$$\eta(r)D(x,F(x))\leq d(x,y) \text{ implies } H(F(x),F(y))\leq\beta d(x,y)+\beta D(x,F(x))+\beta D(y,F(x))$$

for all $x,y\in X$, *where* $\eta(r)=\frac{1}{1+r}$. *Then* $\mathcal{F}_F$ *is an absolute retract for metric spaces.*

Proof: By using similar terminologies as in Theorem 5.21, we have

$$\eta(r)D(\varphi_0(y),F(\varphi_0(y))) \leq \eta(r)d(\varphi_0(y),\varphi_1(y)) \leq d(\varphi_0(y),\varphi_1(y)).$$

Hence

$$H(F(\varphi_0(y)),F(\varphi_1(y))) \leq \beta d(\varphi_0(y),\varphi_1(y)) + \beta D(\varphi_0(y),F(\varphi_0(y))) + \beta D(\varphi_1(y),F(\varphi_0(y)))$$

$$\leq \beta d(\varphi_0(y),\varphi_1(y)) + \beta d(\varphi_0(y),\varphi_1(y)) \leq \frac{2\beta}{1-\beta} d(\varphi_0(y),\varphi_1(y)) = rd(\varphi_0(y),\varphi_1(y)).$$

Now by using a similar proof as in Theorem 5.21, we can complete the proof.

♣

Proposition 5.5

Let (X,d) be a complete metric space and absolute retract for metric spaces and $F : X \to P_{b,cl}(X)$ a set-valued map such that $F \in SP(X)$ and $\mathcal{F}_F \neq \emptyset$. Suppose that there exists $\beta \in [0,\frac{1}{3})$ such that for $r = \frac{2\beta}{1-\beta}$ and a natural number n such that

$$\eta(r)D(x,F(x)) \leq d(x,y) \text{ implies } [H(F(x),F(y))]^n \leq \beta[d(x,y)]^n + \beta[D(x,F(x))]^n + \beta[D(y,F(x))]^n$$

for all $x,y \in X$, where $\eta(r) = \frac{1}{1+r}$. Then $\mathcal{F}_F$ is an absolute retract for metric spaces.

Proof: Again by using similar terminologies as in Theorem 5.21 and a similar proof in Proposition 5.4, we have

$$\eta(r)D(\varphi_0(y),F(\varphi_0(y))) \leq \eta(r)d(\varphi_0(y),\varphi_1(y)) \leq d(\varphi_0(y),\varphi_1(y)).$$

Hence,

$$[H(F(\varphi_0(y)),F(\varphi_1(y)))]^n \leq \beta[d(\varphi_0(y),\varphi_1(y))]^n + \beta[D(\varphi_0(y),F(\varphi_0(y)))]^n$$

$$\leq \frac{2\beta}{1-\beta}[d(\varphi_0(y),\varphi_1(y))]^n = r[d(\varphi_0(y),\varphi_1(y))]^n.$$

Since $r \in [0,1)$, by putting $r^{\frac{1}{n}} = a$ we have $a \in [0,1)$ and

$$H(F(\varphi_0(y)),F(\varphi_1(y))) \leq ad(\varphi_0(y),\varphi_1(y)).$$

Now by using a similar proof as in Theorem 5.21, we can complete the proof.

♣

Proposition 5.6

Let (X,d) be a complete metric space and absolute retract for metric spaces. Suppose that $F : X \to P_{b,cl}(X)$ is a set-valued map such that $F \in SP(X)$ and there exist $a,b,c \in (0,1)$ with $a+2b+2c < 1$ such that $D(x,F(x)) \le d(x,y)$ implies

$$H(F(x),F(y)) \le ad(x,y)+b[D(x,F(x))+D(y,F(y))]+c[D(x,F(y))+D(y,F(x))]$$

for all $x,y \in X$. Then $\mathcal{F}_F$ is an absolute retract for metric spaces.

Proof: By using similar terminologies as in Theorem 5.21, we have

$$D(\varphi_0(y),F(\varphi_0(y))) \le d(\varphi_0(y),\varphi_1(y)).$$

Since $\varphi_1(y) \in F(\varphi_0(y))$, we get

$$\begin{aligned} D(\varphi_1(y),F(\varphi_1(y))) \le H(F(\varphi_0(y)),F(\varphi_1(y))) &\le ad(\varphi_0(y),\varphi_1(y)) \\ &+b[D(\varphi_0(y),F(\varphi_0(y)))+D(\varphi_1(y),F(\varphi_1(y)))] \\ &+c[D(\varphi_0(y),F(\varphi_1(y)))+D(\varphi_1(y),F(\varphi_0(y))) \\ \le ad(\varphi_0(y),\varphi_1(y))&+bd(\varphi_0(y),\varphi_1(y))+bD(\varphi_1(y),F(\varphi_1(y))) \\ &+cD(\varphi_1(y),F(\varphi_1(y)))+cd(\varphi_0(y),\varphi_1(y)). \end{aligned}$$

Therefore,

$$D(\varphi_1(y),F(\varphi_1(y))) \le \frac{a+b+c}{1-b-c}d(\varphi_0(y),\varphi_1(y)).$$

Since $\frac{a+b+c}{1-b-c} < 1$, this converts to first stage of Theorem 5.21. Now by using a similar proof, we can complete the proof.

♣

Chapter 6

Quasi-contractions and T-stability

6.1 Introduction

This chapter includes some technical fixed point results about quasi-contractive mappings and T-stability. These could extend our sight about this field.

6.2 Quasi-contractions

Let (X,d) be a metric space. A selfmap $T : X \to X$ is called a quasi-contraction whenever there exists $\lambda \in (0,1)$ such that

$$d(Tx,Ty) \leq \lambda \max\{d(x,y),d(x,Tx),d(y,Ty),d(x,Ty),d(y,Tx)\}$$

for all $x,y \in X$. This notion has been generalized to cone metric spaces. Suppose that (X,d) is a cone metric space (with values of the metric in a Banach space $(E,\|.\|)$). A selfmap $T : X \to X$ is called a quasi-contraction whenever there exists $\lambda \in (0,1)$ such that for every $x,y \in X$ there exists

$$u \in \{d(x,y),d(x,Tx),d(y,Ty),d(x,Ty),d(y,Tx)\}$$

with $d(Tx,Ty) \leq \lambda u$. For $x \in X$, put $O(x;\infty) = \{x,Tx,T^2x,\cdots,T^nx,\cdots\}$. The set $O(x;\infty)$ is called orbit of x with respect to T. We say that a subset $A \subseteq X$ is bounded whenever $diam(A) = \sup\{\|d(a,b)\| : a,b \in A\} < \infty$. A space (X,d) is called T-orbitally complete whenever every Cauchy sequence $\{T^{n_i}x\}_{i\geq 1}$ $(x \in X)$ has a limit point in X. Also, a map $f : X \to E$ is called lower semicontinuous at $x \in X$ if for any sequence $\{x_n\}$ in X with $x_n \to x$, we have $f(x) \leq$

$\liminf_{n\to\infty} f(x_n)$. It is obvious that each complete cone metric space is a T-orbitally complete space but it has been proved that the converse does not hold (Example 3.1 in [80]).

Theorem 6.1

Let (X,d) be a complete cone metric space, $x \in X$ and $T : X \to X$ a quasi-contraction map. Then, the following statements hold.
(i) The orbit $O(x;\infty)$ is bounded,
(ii) $\lim_{n\to\infty} T^n x = \bar{x}$ exists and $T(\bar{x}) = \bar{x}$,
(iii) $\bar{x}$ is a unique fixed point of T whenever $\bigcap_{n=0}^{\infty} T^n(X)$ is a singleton set.

Proof: (i) Put $\delta_1(x) = d(x,Tx)$. Since $d(x,T^2x) \leq d(x,Tx) + d(Tx,T^2x)$ and T is quasi-contraction, we have two cases:

$$d(Tx,T^2x) \leq \lambda d(x,Tx) \text{ or } d(Tx,T^2x) \leq \lambda d(x,T^2x).$$

If $d(Tx,T^2x) \leq \lambda d(x,Tx)$, then

$$d(x,T^2x) \leq (1+\lambda)d(x,Tx)$$

. In this case, put $\delta_2(x) = (1+\lambda)d(x,Tx)$. If $d(Tx,T^2x) \leq \lambda d(x,T^2x)$, then $d(x,T^2x) \leq \frac{1}{1-\lambda} d(x,Tx)$. In this case, put $\delta_2(x) = \frac{1}{1-\lambda} d(x,Tx)$. Note that, in both cases we have $\delta_1(x) \leq \delta_2(x)$ and $d(x,T^2x) \leq \delta_2(x)$. Suppose that $\delta_{n-1}(x)$ has been choose. Now by induction, we show that for each $n \geq 2$ there exists $1 \leq m \leq n$ such that

$$d(T^n x, T^{n-1}x) \leq \lambda^{n-1} d(x,T^m x). \qquad (*)$$

If $n = 2$, then $d(Tx,T^2x) \leq \lambda u$, where $u \in \{d(x,Tx), d(Tx,T^2x), d(x,T^2x)\}$. Thus, $(*)$ holds for $n = 2$. Suppose that $(*)$ holds for each $k < n$. We show that $(*)$ holds for $k = n$. In this case we have $d(T^n x, T^{n-1}x) \leq \lambda u$, where

$$u \in \{d(T^{n-1}x, T^{n-2}x), d(T^n x, T^{n-1}x), d(T^n x, T^{n-2}x)\}.$$

It is trivial that $(*)$ holds if $u = d(T^{n-1}x, T^{n-2}x)$ or $u = d(T^n x, T^{n-1}x)$. Now suppose that $u = d(T^n x, T^{n-2}x)$. In this case we have $d(T^n x, T^{n-2}x) \leq \lambda u_1$, where

$$u_1 \in \{d(T^{n-1}x, T^{n-3}x), d(T^{n-1}x, T^{n-2}x), d(T^n x, T^{n-3}x),$$
$$d(T^{n-2}x, T^{n-3}x), d(T^n x, T^{n-1}x)\}.$$

Again, it is trivial that $(*)$ holds if

$$u_1 = d(T^n x, T^{n-1}x) \text{ or } u_1 = d(T^{n-2}x, T^{n-3}x).$$

If $u_1 = d(T^{n-1}x, T^{n-2}x)$, then $d(T^n x, T^{n-1}x) \leq \lambda^2 d(T^{n-1}x, T^{n-2}x)$. By assumption of the induction, there exists $1 \leq m \leq n-1$ such that $d(T^{n-1}x, T^{n-2}x) \leq \lambda^{n-2} d(x, T^m x)$. Hence, $d(T^n x, T^{n-1}x) \leq \lambda^n d(x, T^m x) \leq \lambda^{n-1} d(x, T^m x)$. If $u_1 = d(T^{n-1}x, T^{n-3}x)$, then

$$d(T^n x, T^{n-1}x) \leq \lambda^2 d(T^{n-3}x, T^{n-1}x).$$

If $u_1 = d(T^n x, T^{n-3}x)$, then $d(T^n x, T^{n-1}x) \leq \lambda^2 d(T^n x, T^{n-3}x)$. By continuing this process, we see that $(*)$ holds for each $n \geq 2$. Now, we have

$$d(x, T^n x) \leq d(x, T^{n-1}x) + d(T^{n-1}x, T^n x) \leq d(x, T^{n-1}x) + \lambda^{n-1} d(x, T^m x)$$

for some $1 \leq m \leq n$. If $n > m$, then

$$d(x, T^n x) \leq d(x, T^{n-1}x) + \lambda^{n-1} d(x, T^m x) \leq (1 + \lambda^{n-1})\delta_{n-1}(x).$$

In this case, put $\delta_n(x) = (1 + \lambda^{n-1})\delta_{n-1}(x)$. If $n = m$, then

$$d(x, T^n x) \leq \frac{1}{1 - \lambda^{n-1}} d(x, T^{n-1}x) \leq (\frac{1}{1 - \lambda^{n-1}})\delta_{n-1}(x).$$

In this case, put $\delta_n(x) = \frac{1}{1-\lambda^{n-1}}\delta_{n-1}(x)$. Note that, in both cases we have

$$\delta_{n-1}(x) \leq \delta_n(x) \text{ and } d(x, T^n x) \leq \delta_n(x).$$

Thus, $\{\delta_n(x)\}_{n \geq 1}$ is an increasing sequence and $\delta_n(x)$ is a bound for the set

$$\{x, Tx, T^2x, \cdots, T^n x\}$$

for all $n \geq 1$. Note that, for each $m \geq 2$ there exists a subset I of $\{1, 2, \cdots, m-1\}$ such that $\delta_m(x) = \frac{\prod_{k \in I}(1+\lambda^k)}{\prod_{k \in \{1,2,\cdots,m-1\} \setminus I}(1-\lambda^k)}\delta_1(x)$. Put $u_m = \prod_{i=1}^m (1 + \lambda^i)$ for all $m \geq 1$. Since $\ln u_m = \sum_{i=1}^m \ln(1 + \lambda^i)$ and the series $\sum_{i=1}^\infty \ln(1 + \lambda^i)$ converges, the sequence $\{u_m\}_{m \geq 1}$ is bounded and increasing. Thus, $\prod_{i=1}^\infty (1 + \lambda^i)$ converges. Similarly, one can prove that $\prod_{i=1}^\infty \frac{1}{1-\lambda^i}$ converges. Now, note that $\delta_m(x) \leq \frac{\prod_{i=1}^\infty (1+\lambda^i)}{\prod_{i=1}^\infty (1-\lambda^i)}\delta_1(x)$ for all $m \geq 1$. If we put $\delta(x) = \frac{\prod_{i=1}^\infty (1+\lambda^i)}{\prod_{i=1}^\infty (1-\lambda^i)}\delta_1(x)$, then $\delta(x)$ is a bound for the orbit $O(x; \infty)$.

(ii) For each $m > n \geq 1$ we have

$$d(T^n x, T^m x) \leq d(T^n x, T^{n+1}x) + d(T^{n+1}x, T^m x) \leq \lambda^n d(x, T^s x) + \lambda^{n+1} d(x, T^p x)$$

for some $s, p \leq m$. Thus, $d(T^n x, T^m x) \leq (\lambda^n + \lambda^{n+1})\delta(x)$ for all $m > n \geq 1$. Hence, $\{T^n x\}_{n \geq 1}$ is a Cauchy sequence. Since (X, d) is a complete cone metric space, there exists $\bar{x} \in X$ such that $\lim_{n \to \infty} T^n x = \bar{x}$. For each $n \geq 1$ there exists

$$u_n \in \{d(\bar{x}, T^n x), d(\bar{x}, T\bar{x}), d(T^n x, T^{n+1}x), d(\bar{x}, T^{n+1}x), d(T^n x, T\bar{x})\}$$

such that $d(T\bar{x}, T^{n+1}x) \leq \lambda u_n$. Let $0 \ll c$ be given. Choose a natural number N such that $d(T^n x, \bar{x}) \ll c$ for all $n \geq N$. Then, we have $d(\bar{x}, T\bar{x}) \ll \frac{2\lambda+1}{(1-\lambda)}c$ for all $n \geq N$. Since c was arbitrary, we obtain $T\bar{x} = \bar{x}$.

(iii) If $\bigcap_{n=0}^{\infty} T^n(X) = \{w\}$ and y is a fixed point of T, then $y \in \bigcap_{n=0}^{\infty} T^n(X)$. Hence, $y = w$ and so T has a unique fixed point.

♣

Theorem 6.2

Let (X,d) be a cone metric space and $T : X \to X$ a quasi-contraction. Then, T has the property (P).

Proof: It is clear that $F(T) \subseteq F(T^n)$ for all $n \geq 1$. Let $n \geq 1$ and $u \in F(T^n)$. Then, $T^n u = u$ and so $T^{kn}u = u$ for all $k \geq 1$. Hence, for every $k \geq 1$ we have

$$d(u,Tu) = d(T^{kn}u, T^{kn+1}u) \leq \lambda^{kn} d(u, T^s u) \leq \lambda^{kn} \delta(u),$$

where $s \leq kn+1$ and $\delta(u)$ is defined as in Theorem 6.1. If $k \to \infty$, then we obtain $d(u,Tu) = 0$. Thus, $Tu = u$ and so T has the property P.

♣

Theorem 6.3

Let (X,d) be a cone metric space and $T : X \to X$ a selfmap. Suppose that (X,d) is T-orbitally complete and there exists $x \in X$ and $\lambda_1, \lambda_2 \in [0,1)$ such that $\lambda_1 + 2\lambda_2 < 1$ and $d(Ty, T^2y) \leq \lambda_1 d(y,Ty) + \lambda_2 d(y, T^2y)$ for all $y \in O(x;\infty)$. Then, the following statements hold.
(i) $\lim_{n\to\infty} T^n x = \bar{x}$ exists,
(ii) $T(\bar{x}) = \bar{x}$ if the mapping $X \ni z \mapsto \|d(z,Tz)\|$ is lower semicontinuous at x,
(iii) $\bar{x}$ is a unique fixed point of T whenever $\bigcap_{n=0}^{\infty} T^n(X)$ is a singleton set.

Proof: (i) First, note that $d(Ty,T^2y) \leq \lambda_1 d(y,Ty) + \lambda_2(d(y,Ty) + d(Ty,T^2y))$. Hence, $d(Ty,T^2y) \leq \frac{\lambda_1+\lambda_2}{1-\lambda_2} d(y,Ty)$. Since $\lambda_1 + 2\lambda_2 < 1$, $\mu = \frac{\lambda_1+\lambda_2}{1-\lambda_2} < 1$. Thus,

$$d(T^n x, T^{n+1}x) \leq \mu^n d(x,Tx)$$

for all $n \geq 1$. Hence, for each $m > n \geq 1$ we have

$$d(T^n x, T^m x) \leq (\mu^n + \mu^{n+1} + \cdots + \mu^m) d(x,Tx).$$

Thus, $\{T^n x\}_{n\geq 1}$ is a Cauchy sequence. Since (X,d) is T-orbitally complete, there exists $\bar{x} \in X$ such that $\lim_{n\to\infty} T^n x = \bar{x}$.

(ii) Since the mapping $z \mapsto \|d(z,Tz)\|$ is lower semicontinuous at x, we have

$$\|d(\bar{x},T\bar{x})\| \leq \liminf_{n\to\infty} \|d(T^n x, T^{n+1}x)\| = 0.$$

Thus, $T(\bar{x}) = \bar{x}$.

(iii) It is obvious.

♣

Now, let (X,d) be a metric space. The set-valued $T : X \to CB(X)$ is said to be a set-valued quasi-contractive map whenever there exists $\lambda \in (0,1)$ such that

$$H(Tx,Ty) \leq \lambda \max\{d(x,y), d(x,Tx), d(y,Ty), d(x,Ty), d(y,Tx)\}$$

for all $x,y \in X$. In 2011, Amini-Harandi proved next result on the existence of fixed points of set-valued quasi-contractions ([15]) by using the technique of Theorem 6.1.

Theorem 6.4

Let (X,d) be a complete metric space and $T : X \to CB(X)$ a set-valued quasi-contraction for some $\lambda \in (0, \frac{1}{2})$. Then T has a fixed point.

He raised immediately the following question.

Question. Does the conclusion of Theorem 6.4 remains true for any $\lambda \in [\frac{1}{2}, 1)$?

A set-valued $T : X \to CB(X)$ is said to be a set-valued quasi-contraction type whenever there exists $\lambda \in (0,1)$ such that

$$H(Tx,Ty) \leq \lambda \max\{d(x,Tx), d(y,Ty), d(x,Ty), d(y,Tx)\}$$

for all $x,y \in X$. It is obvious that, each quasi-contraction type set-valued map is quasi-contraction. Finding a quasi-contraction set-valued map which is not quasi-contraction type, is an interesting issue. We reprove Theorem 6.4 by using a simple method. Also, we give a positive answer to the above question for quasi-contraction type multifunctions.

Theorem 6.5

Let (X,d) be a complete metric space and $T : X \to CB(X)$ a set-valued quasi-contraction for some $\lambda \in (0, \frac{1}{2})$. Then T has a fixed point.

Proof: First, note that

$$H(Tx,Ty) \leq (2\lambda)\max\{d(x,y), d(x,Tx), d(y,Ty), \frac{d(x,Ty)}{2}, \frac{d(y,Tx)}{2}\}$$

for all $x, y \in X$. Choose $q > 1$ such that $2\lambda q < 1$, $x_0 \in X$ and $x_1 \in Tx_0$. If $x_1 = x_0$, then x_0 is a fixed point of T. Let $x_1 \neq x_0$. Take $x_2 \in Tx_1$ such that $d(x_2,x_1) \leq qH(Tx_1,Tx_0)$. If $x_2 = x_1$, then x_1 is a fixed point of T. Let $x_2 \neq x_1$. If x_{n-1} has been chosen, then choose $x_n \in Tx_{n-1}$ such that $d(x_n,x_{n-1}) \leq qH(Tx_{n-1},Tx_{n-2})$ for all $n \geq 2$. If $x_n = x_{n-1}$, then x_{n-1} is a fixed point of T. Thus, we suppose that $x_n \neq x_{n-1}$ for all $n \geq 1$. Hence, we have

$$\begin{aligned} d(x_n,x_{n-1}) &\leq (2\lambda q)\max\{d(x_{n-1},x_{n-2}), d(x_{n-1},Tx_{n-1}), \\ &\quad d(x_{n-2},Tx_{n-2}), \frac{d(x_{n-1},Tx_{n-2})}{2}, \frac{d(x_{n-2},Tx_{n-1})}{2}\} \\ &\leq (2\lambda q)\max\{d(x_{n-1},x_{n-2}), d(x_{n-1},x_n), \frac{d(x_{n-2},x_n)}{2}\} \end{aligned}$$

for all $n \geq 2$. On the other hand, we have

$$\frac{d(x_{n-2},x_n)}{2} \leq \frac{d(x_{n-2},x_{n-1}) + d(x_{n-1},x_n)}{2} \leq \max\{d(x_{n-2},x_{n-1}), d(x_{n-1},x_n)\}.$$

Note that, we have $\max\{d(x_{n-2},x_{n-1}), d(x_{n-1},x_n)\} = d(x_{n-2},x_{n-1})$ for all $n \geq 2$. Thus, $d(x_{n-1},x_n) \leq (2\lambda q)d(x_{n-2},x_{n-1})$ for all $n \geq 2$. This implies that $\{x_n\}_{n\geq 1}$ is a Cauchy sequence. Since (X,d) is a complete metric space, there is $x^* \in X$ such that $x_n \to x^*$. Now, we get

$$\begin{aligned} d(x^*,Tx^*) &= \lim_{n\to\infty} d(x_{n+1},Tx^*) \leq \lim_{n\to\infty} H(Tx_n,Tx^*) \\ &\leq \lim_{n\to\infty}(2\lambda)\max\{d(x_n,x^*), d(x_n,Tx_n), d(x^*,Tx^*), \frac{d(x_n,Tx^*)}{2}, \frac{d(x^*,Tx_n)}{2}\} \\ &\leq \lim_{n\to\infty}(2\lambda)\max\{d(x_n,x^*), d(x_n,x_{n+1}), d(x^*,Tx^*), \frac{d(x_n,Tx^*)}{2}, \frac{d(x^*,x_{n+1})}{2}\} \\ &= (2\lambda)d(x^*,Tx^*). \end{aligned}$$

Hence, $d(x^*,Tx^*) = 0$ and so $x^* \in Tx^*$.

♣

Now, we give a positive answer to the question for quasi-contraction type set-valued maps.

Theorem 6.6

Let (X,d) be a complete metric space and $T : X \to CB(X)$ a quasi-contraction type set-valued map. Then T has a fixed point.

Proof: Choose $q > 1$ such that $q\lambda < 1$, $x_0 \in X$ and $x_1 \in Tx_0$. Take $x_2 \in Tx_1$ such that $d(x_2,x_1) \leq qd(x_1,Tx_1) \leq qH(Tx_1,Tx_0)$. If $x_1 = x_0$ or $x_2 = x_1$, then we have nothing to prove. If $x_0 \neq x_1$ and $x_2 \neq x_1$, then we have

$$\begin{aligned} d(x_2,x_1) &\leq qH(Tx_1,Tx_0) \\ &\leq (q\lambda)\max\{d(x_0,Tx_0),d(x_1,Tx_1),d(x_0,Tx_1),d(x_1,Tx_0)\} \\ &\leq (q\lambda)\max\{d(x_1,x_0),d(x_1,x_2),d(x_0,x_2)\} \\ &= (q\lambda)\max\{d(x_1,x_0),d(x_0,x_2)\}. \end{aligned} \tag{6.1}$$

Now, choose $x_3 \in Tx_2$ such that $d(x_2,x_3) \leq qd(x_2,Tx_2) \leq qH(Tx_1,Tx_2)$. If $x_3 = x_2$, then we have nothing to prove. If $x_3 \neq x_2$, then we have

$$\begin{aligned} d(x_3,x_2) &\leq qH(Tx_1,Tx_2) \\ &\leq (q\lambda)\max\{d(x_1,Tx_1),d(x_2,Tx_2),d(x_1,Tx_2),d(x_2,Tx_1)\} \\ &\leq (q\lambda)\max\{d(x_1,x_2),d(x_2,x_3),d(x_1,Tx_2)\} \\ &= (q\lambda)\max\{d(x_1,x_2),d(x_1,Tx_2)\}. \end{aligned} \tag{6.2}$$

Note that,

$$\begin{aligned} d(x_1,Tx_2) &\leq H(Tx_0,Tx_2) \\ &\leq \lambda\max\{d(x_0,Tx_0),d(x_2,Tx_2),d(x_0,Tx_2),d(x_2,Tx_0)\} \\ &\leq (q\lambda)\max\{d(x_0,x_1),d(x_2,x_3),d(x_0,x_3),d(x_2,x_1)\}. \end{aligned} \tag{6.3}$$

Now, (6.1), (6.2) and (6.3) imply that $d(x_3,x_2) \leq (q\lambda)^2 d(x_0,x_m)$ for some $1 \leq m \leq 3$. Now, we claim that for each natural number $n \geq 3$ there exists an element $x_n \in Tx_{n-1}$ such that $d(x_{n-1},x_n) \leq (q\lambda)^{n-1} d(x_0,x_m)$ for some $1 \leq m \leq n$. Let $x_1,x_2,\cdots,x_{n-1}$ have been chosen. Take $x_n \in Tx_{n-1}$ such that

$$d(x_{n-1},x_n) \leq qd(x_{n-1},Tx_{n-1}) \leq qH(Tx_{n-2},Tx_{n-1}).$$

If $x_n = x_{n-1}$, then we have nothing to prove. If $x_n \neq x_{n-1}$, then we have

$$d(x_{n-1},x_n) \leq qH(Tx_{n-2},Tx_{n-1})$$

$$\leq (q\lambda)\max\{d(x_{n-2},Tx_{n-2}),d(x_{n-1},Tx_{n-1}),d(x_{n-2},Tx_{n-1}),d(x_{n-1},Tx_{n-2})\}$$

$$\leq (q\lambda)\max\{d(x_{n-2},x_{n-1}),d(x_{n-1},x_n),d(x_{n-2},Tx_{n-1})\}$$
$$= (q\lambda)\max\{d(x_{n-2},x_{n-1}),d(x_{n-2},Tx_{n-1})\}. \quad (4)$$

But, we have

$$d(x_{n-2},Tx_{n-1}) \leq H(Tx_{n-3},Tx_{n-1})$$
$$\leq \lambda\max\{d(x_{n-3},Tx_{n-3}),d(x_{n-1},Tx_{n-1}),d(x_{n-1},Tx_{n-3}),d(x_{n-3},Tx_{n-1})\}$$
$$\leq (q\lambda)\max\{d(x_{n-3},x_{n-2}),d(x_{n-1},x_n),d(x_{n-1},x_{n-2}),d(x_{n-3},Tx_{n-1})\}. \quad (5)$$

Now, (4) and (5) imply that

$$d(x_{n-1},x_n)$$
$$\leq \max\{(q\lambda)d(x_{n-2},x_{n-1}),(q\lambda)^2d(x_{n-3},x_{n-2}),(q\lambda)^2d(x_{n-3},Tx_{n-1})\}.$$

If $d(x_{n-1},x_n) \leq (q\lambda)d(x_{n-2},x_{n-1})$ or $d(x_{n-1},x_n) \leq (q\lambda)^2d(x_{n-3},x_{n-2})$, then by using the assumption of our induction, we get that our claim is proved. On the other hand, if $d(x_{n-1},x_n) \leq (q\lambda)^2d(x_{n-3},Tx_{n-1})$, then

$$d(x_{n-3},Tx_{n-1}) \leq H(Tx_{n-4},Tx_{n-1})$$
$$\leq \lambda\max\{d(x_{n-4},Tx_{n-4}),d(x_{n-1},Tx_{n-1}),d(x_{n-1},Tx_{n-4}),d(x_{n-4},Tx_{n-1})\}$$
$$\leq (q\lambda)\max\{d(x_{n-4},x_{n-3}),d(x_{n-1},x_n),d(x_{n-1},Tx_{n-4}),d(x_{n-4},Tx_{n-1})\}$$

and so in this case we get that

$$d(x_{n-1},x_n) \leq$$
$$\max\{(q\lambda)^3d(x_{n-4},x_{n-3}),(q\lambda)^3d(x_{n-1},Tx_{n-4}),(q\lambda)^3d(x_{n-4},Tx_{n-1})\}.$$

As we stated above, it is sufficient we consider only the cases

$$d(x_{n-1},x_n) \leq (q\lambda)^3d(x_{n-1},Tx_{n-4})$$

and

$$d(x_{n-1},x_n) \leq (q\lambda)^3d(x_{n-4},Tx_{n-1}).$$

Now, note that

$$d(x_{n-4},Tx_{n-1}) \leq H(Tx_{n-5},Tx_{n-1})$$
$$\leq \lambda\max\{d(x_{n-5},Tx_{n-5}),d(x_{n-1},Tx_{n-1}),d(x_{n-1},Tx_{n-5}),d(x_{n-5},Tx_{n-1})\}$$
$$\leq (q\lambda)\max\{d(x_{n-5},x_{n-4}),d(x_{n-1},x_n),d(x_{n-1},x_{n-4}),d(x_{n-5},Tx_{n-1})\}$$

and

$$d(x_{n-1},Tx_{n-4})$$
$$\leq H(Tx_{n-2},Tx_{n-4})$$
$$\leq \lambda\max\{d(x_{n-2},Tx_{n-2}),d(x_{n-4},Tx_{n-4}),d(x_{n-2},Tx_{n-4}),d(x_{n-4},Tx_{n-2})\}$$
$$\leq (q\lambda)\max\{d(x_{n-3},x_{n-2}),d(x_{n-4},x_{n-3}),d(x_{n-4},Tx_{n-2})\}.$$

By continuing this process, these relations lead us to $d(x_0, Tx_m)$ or $d(x_s, Tx_0)$ for some $1 \leq m \leq n-1$ and $1 \leq s \leq n-1$, after some steps. Note that, it is sufficient we consider these cases because another cases prove that the claim holds. First, suppose that we get the first case $d(x_{n-1}, x_n) \leq (q\lambda)^{n-1} d(x_0, Tx_m)$ for some $1 \leq m \leq n-1$. Then, we obtain $d(x_{n-1}, x_n) \leq (q\lambda)^{n-1} d(x_0, x_m)$ for some $1 \leq m \leq n$. If we get that the second case $d(x_{n-1}, x_n) \leq (q\lambda)^{n-1} d(x_s, Tx_0)$ holds for some $1 \leq s \leq n-1$, then

$$d(x_{n-1}, x_n) \leq (q\lambda)^{n-1} d(x_s, Tx_0) \leq (q\lambda)^{n-1} H(Tx_{s-1}, Tx_0)$$
$$\leq (q\lambda)^n \max\{d(x_{s-1}, Tx_{s-1}), d(x_0, Tx_0), d(x_{s-1}, Tx_0), d(x_0, Tx_{s-1})\}$$
$$\leq (q\lambda)^{n-1} \max\{d(x_{s-1}, x_s), d(x_0, x_1), d(x_{s-1}, Tx_0), d(x_0, x_s)\}.$$

It is sufficient we suppose that $d(x_{n-1}, x_n) \leq (q\lambda)^{n-1} d(x_{s-1}, Tx_0)$. By continuing this process, we get that $d(x_{n-1}, x_n) \leq (q\lambda)^{n-1} d(x_0, Tx_0) \leq (q\lambda)^{n-1} d(x_0, x_1)$. Thus, our claim is proved, that is, for each natural number $n \geq 3$ there exists an element $x_n \in Tx_{n-1}$ such that $d(x_{n-1}, x_n) \leq (q\lambda)^{n-1} d(x_0, x_m)$ for some $1 \leq m \leq n$.

Now, first put $\delta_1 = d(x_0, x_1)$. If $\max\{d(x_0, x_1), d(x_0, x_2)\} = d(x_0, x_1)$, then we put $\delta_2 = (1 + q\lambda)\delta_1$. If $\max\{d(x_0, x_1), d(x_0, x_2)\} = d(x_0, x_2)$, then we put $\delta_2 = \frac{1}{1-q\lambda}\delta_1$. Note that, $\delta_2 \geq \delta_1$ and $\delta_2 \geq d(x_0, x_2)$. Note that, $d(x_0, x_3) \leq d(x_0, x_2) + (q\lambda)^2 u$, where $u \in \{d(x_0, x_1), d(x_0, x_2), d(x_0, x_3)\}$. If

$$\max\{d(x_0, x_1), d(x_0, x_2), d(x_0, x_3)\} = d(x_0, x_1),$$

or

$$\max\{d(x_0, x_1), d(x_0, x_2), d(x_0, x_3)\} = d(x_0, x_2),$$

then put $\delta_3 = (1 + (q\lambda)^2)\delta_2$. If $\max\{d(x_0, x_1), d(x_0, x_2), d(x_0, x_3)\} = d(x_0, x_3)$, then we put $\delta_3 = \frac{1}{1-(q\lambda)^2}\delta_2$. In this case, we have $\delta_3 \geq \delta_2$ and $\delta_3 \geq d(x_0, x_3)$. Suppose that δ_{n-1} has been chosen such that $\delta_{n-2} \leq \delta_{n-1}$ and $d(x_0, x_{n-1}) \leq \delta_{n-1}$. On the other hand, we have $d(x_0, x_n) \leq d(x_0, x_{n-1}) + (q\lambda)^{n-1} d(x_0, x_m)$ for some $1 \leq m \leq n$. If $m < n$, then put $\delta_n = (1 + (q\lambda)^{n-1})\delta_{n-1}$. If $m = n$, then put $\delta_n = \frac{1}{1-(q\lambda)^{n-1}}\delta_{n-1}$. It is obvious that $\delta_n \geq \delta_{n-1}$ and $\delta_n \geq d(x_0, x_n)$ for all $n \geq 2$. Put $u_m = \prod_{i=1}^{m}(1 + (q\lambda)^i)$ for all $m \geq 1$. Since $\ln u_m = \sum_{i=1}^{m} \ln(1 + (q\lambda)^i)$ and the series $\sum_{i=1}^{\infty} \ln(1 + (q\lambda)^i)$ converges, the sequence $\{u_m\}_{m\geq 1}$ is bounded and increasing. Thus, $\prod_{i=1}^{\infty}(1 + (q\lambda)^i)$ converges. Similarly, one can prove that $\prod_{i=1}^{\infty} \frac{1}{1-(q\lambda)^i}$ converges. If we put $\delta = \frac{\prod_{i=1}^{\infty}(1+(q\lambda)^i)}{\prod_{i=1}^{\infty}(1-(q\lambda)^i)}$, then $\delta < \infty$ and $\delta_n \leq \delta$ for all $n \geq 2$. Now, we show that $\{x_n\}_{n\geq 1}$ is a Cauchy sequence. For see this, note that

$$d(x_n, x_m) \leq \sum_{i=m+1}^{n} d(x_i, x_{i-1}) \leq \sum_{i=m+1}^{n} (q\lambda)^{i-1}\delta$$

for all $n > m$. Thus, $\{x_n\}_{n\geq 1}$ is a Cauchy sequence because $\sum_{i=1}^{\infty}(q\lambda)^{i-1} < \infty$.

Since (X,d) is a complete metric space, there exists $x^* \in X$ such that $x_n \to x^*$. Hence,

$$d(x^*,Tx^*) = \lim_{n\to\infty} d(x_{n+1},Tx^*) \le \lim_{n\to\infty} H(Tx_n,Tx^*)$$

$$\le \lim_{n\to\infty} \lambda \max\{d(x_n,Tx_n),d(x^*,Tx^*),d(x_n,Tx^*),d(x^*,Tx_n)\}$$

$$\le \lim_{n\to\infty} \lambda \max\{d(x_n,x_{n+1}),d(x^*,Tx^*),d(x_n,Tx^*),d(x^*,x_{n+1})\}$$

$$= \lambda d(x^*,Tx^*).$$

Thus, $d(x^*,Tx^*) = 0$ and so $x^* \in Tx^*$.

♣

Corollary 6.1

Let (X,d) be a complete metric space, $T : X \to CB(X)$ a set-valued mapping and f a selfmap on X with $T(X) \subseteq f(X)$. Also, suppose that there exists $\lambda \in (0,1)$ such that

$$H(Tx,Ty) \le \lambda \max\{d(fx,Tx),d(fy,Ty),d(fx,Ty),d(fy,Tx)\}$$

for all $x,y \in X$. Then f and T have a coincidence point.

Proof: Define $g : f(X) \to CB(f(X))$ by $g(f(x)) = T(f(x))$. Then, g is a quasi-contraction type set-valued map. By using Theorem 6.6, there is $x_0 \in X$ such that

$$f(x_0) \in g(f(x_0)) = T(f(x_0)).$$

Hence, f and T have a coincidence point.

♣

Now by providing next example, we give a positive answer to the question.

Example 86 *Let $X = [0,2]$ and $d(x,y) = |x-y|$ be the Euclidean metric. Consider the set-valued map $T : X \to CB(X)$ defined by $Tx = [1,\frac{5}{4}]$ whenever $x \in [0,3/4]$, $Tx = [\frac{7}{8},\frac{9}{8}]$ whenever $x \in (\frac{3}{4},\frac{5}{4})$ and $Tx = [\frac{3}{4},1]$ whenever $x \in [\frac{5}{4},2]$. We show that T is a quasi-contraction while is not a quasi-contractive type set-valued map. Put*

$$M(x,y) = max\{d(x,y),d(x,Tx),d(y,Ty),d(x,Ty),d(y,Tx)\}$$

for all $x,y \in X$. First, we show that T is a quasi-contraction. It is easy to check that $H(Tx,Ty) \le \frac{1}{2}M(x,y)$ whenever $x,y \in [0,\frac{3}{4}]$ or $x,y \in (\frac{3}{4},\frac{5}{4})$ or $x,y \in [\frac{5}{4},2]$. If $x \in [0,\frac{3}{4}]$ and $y \in [\frac{5}{4},2]$, then $H(Tx,Ty) = \frac{1}{4}$ and $d(x,y) \ge \frac{5}{4} - \frac{3}{4} = \frac{1}{2}$.

Hence, $H(Tx,Ty) = \frac{1}{4} \leq \frac{1}{2}d(x,y) \leq \frac{1}{2}M(x,y)$. *If* $x \in [0,\frac{3}{4}]$ *and* $y \in (\frac{3}{4},\frac{5}{4})$, *then* $H(Tx,Ty) = \frac{1}{8}$ *and*

$$d(x,Tx) \geq \frac{1}{4}.$$

Hence, $H(Tx,Ty) = \frac{1}{8} \leq \frac{1}{2}d(x,Tx) \leq \frac{1}{2}M(x,y)$. *If* $x \in [\frac{5}{4},2]$ *and* $y \in (\frac{3}{4},\frac{5}{4})$, *then*

$$H(Tx,Ty) = H([\frac{3}{4},1],[\frac{7}{8},\frac{9}{8}]) = \frac{1}{8}$$

and $d(x,Tx) \geq \frac{1}{4}$. *Hence,* $H(Tx,Ty) = \frac{1}{8} \leq \frac{1}{2}d(x,Tx) \leq \frac{1}{2}M(x,y)$. *Thus, T is a quasi-contraction with* $\lambda = \frac{1}{2}$. *Now, put* $x = \frac{3}{4}$ *and* $y = \frac{5}{4}$. *Then, we have*

$$d(\frac{3}{4},T\frac{3}{4}) = d(\frac{3}{4},[1,\frac{5}{4}]) = \frac{1}{4},$$

$d(\frac{5}{4},T\frac{5}{4}) = d(\frac{5}{4},[\frac{3}{4},1]) = \frac{1}{4}$, $d(\frac{3}{4},T\frac{5}{4}) = d(\frac{3}{4},[\frac{3}{4},1]) = 0$. *And*

$$d(\frac{5}{4},T\frac{3}{4}) = d(\frac{5}{4},[1,\frac{5}{4}]) = 0.$$

Hence, $M(\frac{3}{4},\frac{5}{4}) = \max\{d(\frac{3}{4},T\frac{3}{4}),d(\frac{5}{4},T\frac{5}{4}),d(\frac{3}{4},T\frac{5}{4}),d(\frac{5}{4},T\frac{3}{4})\} = \max\{\frac{1}{4},0\} = \frac{1}{4}$. *Also,* $H(T\frac{3}{4},T\frac{5}{4}) = H([1,\frac{5}{4}],[\frac{3}{4},1]) = \frac{1}{4}$. *Thus,* $H(T\frac{3}{4},T\frac{5}{4}) > \lambda M(\frac{3}{4},\frac{5}{4})$ *for all* $\lambda \in (0,1)$. *This shows that T is not a quasi-contractive type set-valued map.*

This helps us to improve our insight into set-valued mappings and study the existing concepts about those more carefully.

6.3 On T-stability of the Picard iteration for generalized φ-contractive mappings

Let (X,d) be a metric space, $T : X \to X$ a map and $x_{n+1} = f(T,x_n)$ an iteration procedure. Suppose that T has at least one fixed point and that the sequence $\{x_n\}$ converges to a fixed point $x^* \in X$. We denote the set of fixed points of the mapping T by F_T. Let $\{y_n\}$ be an arbitrary sequence in X and $\varepsilon_n = d(y_{n+1}, f(T,y_n))$. If $\lim_{n\to\infty} \varepsilon_n = 0$ implies that $\lim_{n\to\infty} y_n = x^*$, then the iteration procedure $x_{n+1} = f(T,x_n)$ is said to be T-stable. If $\{y_n\}$ is a bounded sequence and $\lim_{n\to\infty} \varepsilon_n = d(y_{n+1}, f(T,y_n)) = 0$ implies that $\lim_{n\to\infty} y_n = x^*$, then the iteration procedure $x_{n+1} = f(T,x_n)$ is said to be boundedly T-stable. In most papers of T-stability, some authors consider the notion of boundedly T-stability instead of T-stability. Recall now the Picard iteration method. Let $x_0 \in X$. The Picard iteration is given by $x_{n+1} = Tx_n$. The following example shows that the notion of T-Stability is different from the notion of boundedly T-stability.

Example 87 *Consider the mapping* $T : [0,\infty) \to [0,\infty)$ *given by* $Tx = \frac{x+1}{2}$ *whenever* $x \in [0,1]$ *and* $Tx = x+1$ *whenever* $x > 1$. *Put* $y_n = n + \frac{1}{n}$ *for all* $n \geq 1$. *Note that,* $\{y_n\}$ *is unbounded while* $\lim_{n\to\infty} |y_{n+1} - Ty_n| = 0$.

We first study the T-stability of the Picard iteration for φ-contraction and generalized φ-contraction mappings on metric spaces. A function $\varphi : [0,\infty) \to [0,\infty)$ is said to be comparison if φ is increasing and $\varphi^n(t)$ converges to 0 for all $t \geq 0$. If φ is comparison, then one can check that $\varphi(t) < t$ for all $t > 0$ and $\varphi(0) = 0$. Let (X,d) be a metric space and $\varphi : [0,\infty) \to [0,\infty)$ a comparison function. A mapping $T : X \to X$ is called φ-contraction whenever $d(Tx,Ty) \leq \varphi(d(x,y))$ for all $x,y \in X$.

We say that $\varphi : [0,\infty) \to [0,\infty)$ is a sub-additive comparison function whenever φ is comparison and $\varphi(t+s) \leq \varphi(t) + \varphi(s)$ for all $t,s \in [0,\infty)$. There are many sub-additive comparison mappings. For example, if $\lambda < 1$ and $g : [0,\infty) \to [0,\lambda)$ is a decreasing function, then $\varphi(t) = \int_0^t g(x)dx$ is a comparison function. In fact, φ is increasing because $g > 0$. Also, $\varphi(t) < \min\{t,\lambda\}$. Hence, $\varphi^n(t)$ converges to 0 for all $t \geq 0$. Since g is decreasing, we have

$$\varphi(u+v) = \int_0^{u+v} g(x)dx = \int_0^u g(x)dx + \int_u^{u+v} g(x)dx$$

$$\leq \int_0^u g(x)dx + \int_0^v g(x)dx = \varphi(u) + \varphi(v)$$

for all $u,v \geq 0$. In particular case, by considering $g(t) = \lambda e^{-t}$, and

$$\varphi(t) = \int_0^t g(x)dx$$

is a sub-additive comparison function.

Theorem 6.7

Let (X,d) *be a complete metric space,* $\varphi : [0,\infty) \to [0,\infty)$ *a sub-additive comparison function and* $T : X \to X$ *a* φ*-contraction. Then, the Picard iteration is T-stable.*

Proof: By using Theorem 2.7 in [21], T has a unique fixed point q. Let $\{y_n\}$ ba a sequence in X with $\lim_{n\to\infty} d(y_{n+1}, Ty_n) = 0$. First, we show that $\{y_n\}$ is bounded. If $\{y_n\}$ is not bounded, then there exist subsequence $\{z_n\}$ of $\{y_n\}$ for which

$d(z_n,q) \geq n$. Since $\lim_{n\to\infty} d(y_{n+1},Ty_n) = 0$, we can take a subsequence $\{x_n\}$ of $\{z_n\}$ such that $d(x_{n+1},Tx_n) \leq \frac{1}{n^2}$. Now, we have

$$\begin{aligned} d(Tx_n,q) &\leq \varphi(d(x_n,q)) \leq \varphi(d(x_n,Tx_{n-1})) + \varphi^2(d(x_{n-1}),q)) \\ &\leq 1 + \frac{1}{2^2} + \cdots + \frac{1}{n^2} + \varphi^n d(x_1,q) \leq \sum_{i=1}^{\infty} \frac{1}{i^2} + d(x_1,q). \end{aligned}$$

Thus, $\{Tx_n\}$ is bounded and so is $\{x_n\}$. This is a contradiction and so $\{y_n\}$ is bounded. Now, choose $M > 0$ such that $d(y_n,q) < M$ for all $n \geq 1$. For each $\varepsilon > 0$ there exist natural numbers p_0 and N such that

$$\varphi^{p_0}(M) < \varepsilon \text{ and } d(y_{n+1},Ty_n) < \varepsilon,$$

for all $n \geq N$. But, we have

$$d(y_{n+2},q) \leq d(y_{n+2},Ty_{n+1}) + d(Ty_{n+1},q) \leq d(y_{n+2},Ty_{n+1}) + \varphi(d(y_{n+1},q))$$

and

$$\begin{aligned} d(y_{n+3},q) &\leq d(y_{n+3},Ty_{n+2}) + d(Ty_{n+2},q) \leq d(y_{n+3},Ty_{n+2}) + \varphi(d(y_{n+2},q)) \\ &\leq d(y_{n+3},Ty_{n+2}) + \varphi(d(y_{n+2},Ty_{n+1}) + \varphi^2(d(y_{n+1},q)). \end{aligned}$$

By continuing this process, we obtain

$$\begin{aligned} d(y_{n+p_0+1},q) &\leq d(y_{n+p_0+1},Ty_{n+p_0}) + \varphi(d(y_{n+p_0},Ty_{n+p_0-1})) \\ &\quad + \cdots + \varphi^{p_0-1}(d(y_{n+1},Ty_n)) + \varphi^{p_0}(d(y_n,q)) \\ &< d(y_{n+p_0+1},Ty_{n+p_0}) + \varphi(d(y_{n+p_0},Ty_{n+p_0-1}) \\ &\quad + \cdots + \varphi^{p_0-1}(d(y_{n+1},Ty_n)) + \varepsilon. \end{aligned}$$

Hence, $\limsup_{n\to\infty} d(y_n,q) \leq \varepsilon$. Since $\varepsilon > 0$ was arbitrary, $\lim_{n\to\infty} d(y_n,q) = 0$.

♣

A function $\varphi : \mathbb{R}_+{}^5 \to \mathbb{R}_+$ is called (5-dimensional) comparison function whenever $\varphi(u) \leq \varphi(v)$ for each $u,v \in \mathbb{R}^5_+$ with $u \leq v$ and the function $\psi : \mathbb{R}_+ \to \mathbb{R}_+$ defined by

$$\psi(t) = \varphi(t,t,t,t,t) \qquad (**)$$

satisfies $\lim_{n\to\infty} \psi^n(t) = 0$ for all $t > 0$. Note that, ψ is a comparison function.

Example 88 *[21] The followings are 5-dimensional comparison functions:*
(i) $\varphi(t) = a\max\{t_1,t_2,t_3,t_4,t_5\}$ *for each* $t = (t_1,t_2,t_3,t_4,t_5)$, *where* $a \in [0,1)$.
(ii) $\varphi(t) = a\max\{t_1,t_2,t_3,t_4,\frac{t_4+t_5}{2}\}$, $a \in [0,1)$.
(iii) $\varphi(t) = at_1 + b(t_2+t_3)$, $a,b \in \mathbb{R}_+$ *with* $a+2b < 1$.
(iv) $\varphi(t) = a\max\{t_2,t_3\}$, $a \in (0,1)$.

Note that in above examples, the function ψ given by (**) is a sub-additive comparison function. Let (X,d) be a metric space and $\varphi : \mathbb{R}_+{}^5 \to \mathbb{R}_+$ a 5-dimensional comparison function. A mapping $T : X \to X$ is called generalized φ-contraction whenever

$$d(Tx,Ty) \leq \varphi(d(x,y),d(x,Tx),d(y,Ty),d(x,Ty),d(y,Tx)) \qquad (***)$$

for all x,y in X.

In the sequel, we shall use the functions φ such that the function ψ is sub-additive.

Lemma 6.1

Let (X,d) be a metric space and $T : X \to X$ a generalized φ-contraction map. Suppose $\{y_n\}$ be a bounded sequence in X such that $\lim_{n\to\infty} d(y_{n+1},Ty_n) = 0$. Let p_n be the diameter of the set $A_n = \{y_i\}_{i\geq n}\bigcup\{Ty_i\}_{i\geq n}$. Then, $\lim_{n\to\infty} p_n = 0$. In particular, $\lim_{n\to\infty} d(y_n,Ty_n) = 0$.

Proof: By using definition of T, For each n and $i,j \geq n$ we have

$$d(Ty_i,Ty_j) \leq \varphi(d(y_i,y_j),(d(y_i,Ty_i),(d(y_j,Ty_j),(d(y_i,Ty_j),(d(y_j,Ty_i)) \leq \psi(p_n).$$

Let $\varepsilon_i = d(y_{i+1},Ty_i)$. Then

$$\begin{aligned} d(y_i,y_j) &\leq d(y_i,Ty_{i-1}) + d(Ty_{i-1},Ty_{j-1}) + d(Ty_{j-1},y_i) \\ &\leq \varepsilon_{i-1} + \psi(p_{n-1}) + \varepsilon_{j-1}, \end{aligned}$$

and

$$d(y_i,Ty_j) \leq d(y_i,Ty_{i-1}) + d(Ty_{i-1},Ty_j) \leq \varepsilon_{i-1} + \psi(p_{n-1}).$$

Let $a_n = \sup_{i\geq n} 2\varepsilon_i$. It is easy to see that $\lim_{n\to\infty} a_n = 0$ and we have

$$p_n \leq a_n + \psi(p_{n-1}). \qquad (V)$$

By using (V), we observe that

$$\psi(p_n) \leq \psi(a_n) + \psi^2(p_{n-1}).$$

Since $\{y_n\}$ is bounded, $\{Ty_n\}$ so is. Choose $M > 0$ such that $p_n \leq M$ for all $n \geq 1$. Since ψ is comparison, for each $\varepsilon > 0$ there exists a natural number k_o such that $\psi^{k_0}(M) < \frac{\varepsilon}{2}$. But, for each $n \geq 1$ we obtain

$$\psi(p_{n+1}) \leq \psi(a_{n+1}) + \psi^2(p_n) \leq \psi(a_{n+1}) + \psi^2(a_n) + \psi^3(p_{n-1}).$$

Hence,

$$\psi(p_{n+2}) \le \psi(a_{n+2}) + \psi^2(a_{n+1}) + \psi^3(a_n) + \psi^4(p_{n-1}).$$

Since $\psi(t) < t$ for all $t > 0$ and ψ is increasing, $\sum_{i=1}^{k+1} \psi^i(a_{n-i+3}) \to 0$ for all natural number k. Thus by continuing these relations, for each $k \ge k_0$ we have

$$\psi(p_{n+k}) \le \sum_{i=1}^{k+1} \psi^i(a_{n-i+3}) + \psi^{k+2}(p_{n-1}) \le \sum_{i=1}^{k+1} \psi^i(a_{n-i+3}) + \varepsilon.$$

It implies that $\limsup_{n\to\infty} \psi(p_n) \le \varepsilon$. Since $\varepsilon > 0$ was arbitrary, $\limsup_{n\to\infty} \psi(p_n) = 0$. Therefore by using (V), $\lim_{n\to\infty} p_n = 0$.

♣

Theorem 6.8

Let (X,d) be a metric space, $T : X \to X$ a generalized φ-contraction map and $F_T = \{q\}$. Then the Picard iteration is boundedly T-stable.

Proof: Let $\{y_n\}$ be a bounded sequence in X such that $\lim_{n\to\infty} d(y_{n+1}, Ty_n) = 0$. Choose $M > 0$ such that $d(y_n, q) < M$ for all $n \ge 1$. Observe that

$$d(Ty_n, q) \le \varphi(d(y_n,q), d(y_n,Ty_n), 0, d(y_n,q), d(q,Ty_n))$$

$$\le \psi(\max\{d(y_n,q), d(y_n,Ty_n), d(q,Ty_n)\}).$$

If $\max\{d(y_n,q), d(y_n,Ty_n), d(q,Ty_n)\} = d(Ty_n,q)$, then $d(Ty_n,q) = 0$. Without loss of generality, suppose that the last equality does not hold. Therefore, we get

$$d(Ty_n, q) \le \psi(\max\{d(y_n,q), d(y_n,Ty_n)\})$$

$$\le \psi(d(y_n,q) + d(y_n,Ty_n)) \le \psi(d(y_n,q)) + \psi(d(y_n,Ty_n)).$$

For any given $\varepsilon > 0$, choose $p_0 \in \mathbb{N}$ such that

$$\psi^{p_0}(M) < \varepsilon.$$

Now, for each $n \ge 1$ we have

$$d(Ty_{n+1}, q) \le \psi(d(y_{n+1}, Ty_{n+1})) + \psi(d(y_{n+1}, q))$$

$$\le \psi(d(y_{n+1}, Ty_{n+1})) + \psi(d(y_{n+1}, Ty_n)) + d(Ty_n, q))$$

$$\le \psi(d(y_{n+1}, Ty_{n+1})) + \psi(d(y_{n+1}, Ty_n)) + \psi^2(d(y_n, Ty_n)) + \psi^2(d(y_n, q)).$$

Similarly

$$
\begin{aligned}
d(Ty_{n+2},q) &\le \psi(d(y_{n+2},Ty_{n+2}))+\psi(d(y_{n+2},q))\\
&\le \psi(d(y_{n+2},Ty_{n+2}))+\psi(d(y_{n+2},Ty_{n+1}))+d(Ty_{n+1},q))\\
&\le \psi(d(y_{n+2},Ty_{n+2}))+\psi(d(y_{n+2},Ty_{n+1}))+\psi^2(d(y_{n+1},Ty_{n+1}))\\
&\quad +\psi^2(d(y_{n+1},Ty_n))+\psi^3(d(y_n,Ty_n))+\psi^3(d(y_n,q)).
\end{aligned}
$$

Now for each $p \ge p_0$ we have

$$
\begin{aligned}
d(Ty_{n+p},q) &\le \sum_{i=1}^{p}\psi^{p-i+1}(d(y_{n+i},Ty_{n+i}))+\sum_{i=1}^{p}\psi^{p-i+1}(d(y_{n+i},Ty_{n+i-1}))\\
&\quad +\psi^{p+1}(d(y_n,q))+\psi^{p+1}(d(y_n,Ty_n))\\
&\le \sum_{i=1}^{p}\psi^{p-i+1}(d(y_{n+i},Ty_{n+i}))+\sum_{i=1}^{p}\psi^{p-i+1}(d(y_{n+i},Ty_{n+i-1}))\\
&\quad +\psi^{p+1}(d(y_n,Ty_n))+\varepsilon.
\end{aligned}
$$

If $n\to\infty$, then by a similar method in Lemma 6.1, $\limsup_{n\to\infty} d(Ty_n,q)\le\varepsilon$. Since $\varepsilon>0$ is arbitrary, $\lim_{n\to\infty} d(Ty_n,q)=0$. Finally note that the inequality

$$d(y_n,q)\le d(y_n,Ty_n)+d(Ty_n,q)$$

implies that $\lim_{n\to\infty} d(y_n,q)=0$.

♣

6.4 T-stability and almost T-stability

Let (X,d) be a complete metric space, $T: X\to X$ a map and $x_{n+1}=f(T,x_n)$ an iteration procedure. Suppose that T has at least one fixed point and that the sequence $\{x_n\}$ converges to a fixed point $q\in X$. Let $\{y_n\}$ be an arbitrary sequence in X and define $\varepsilon_n=d(y_{n+1},f(T,y_n))$. If $\lim_{n\to\infty}\varepsilon_n=0$ implies that $\lim_{n\to\infty} y_n=q$, then as we said in last section, the iteration procedure $x_{n+1}=f(T,x_n)$ is called T-stable. If the convergence of the series $\sum_{i=1}^{\infty}\varepsilon_i$ implies that $\lim_{n\to\infty} y_n=q$, then the iteration procedure is said to be almost T-stable. In this section, we suppose that X is a normed space. Let us mention three iteration methods here. For $x_1,v_1,s_1\in X$, the Picard iteration is given by

$$x_{n+1}=Tx_n,$$

the Mann iteration is given by

$$u_{n+1} = (1-\alpha_n)u_n + \alpha_n T u_n,$$

and the Ishikawa iteration given by

$$s_{n+1} = (1-\alpha_n)s_n + \alpha_n T t_n$$

$$t_n = (1-\beta_n)s_n + \beta_n T s_n,$$

where $\{\alpha_n\}_{n\geq 1}$ and $\{\beta_n\}_{n\geq 1}$ are sequences in $[0,1]$ and satisfy $\lim_{n\to\infty} \alpha_n = 0$ and $\sum_{i=1}^{\infty} \alpha_i = \infty$.

Property (A_1)**:** We say that a mapping φ satisfies property (A_1) whenever φ is a convex comparison function and

$$\varphi(u+v) \leq u + \varphi(v)$$

for all $u, v \in [0,\infty)$.

There are many mappings which satisfy property (A_1). For example, if $g : [0,\infty) \to [0,1)$ is an increasing differentiable function, then $\varphi(t) = \int_0^t g(x)dx$ satisfies property (A_1). It is clear that φ is increasing and $\varphi(0) = 0$. Since $\varphi'(t) = g(t)$ for all $t \geq 0$ and g is increasing, φ' is increasing. Hence, φ is convex. Since $1-g$ is continuous and $1-g > 0$, $\int(1-g) > 0$. Thus, $\varphi(t) < t$. Note that

$$\varphi(u+v) = \int_0^{u+v} g(x)dx = \int_0^{v} g(x)dx + \int_v^{u+v} g(x)dx$$

$$\leq \int_v^{u+v} 1dx + \int_0^{v} g(x)dx = u + \int_0^{v} g(x)dx = u + \varphi(v)$$

for all $u, v \geq 0$. Therefore, φ satisfies property (A_1). In particular, $\varphi(t) = t - \log(1+t)$ satisfies property (A_1).

Lemma 6.2

[21] Let $\{t_n\}$ be a sequence in $[0,1]$ such that $\sum_{n=1}^{\infty} t_n = \infty$, $\{c_n\}$ a sequence in $[0,\infty)$ such that $\sum_{n=1}^{\infty} c_n < \infty$. Also, suppose that $\{a_n\}$ and $\{b_n\}$ are two sequences in $[0,\infty)$ satisfying $b_n = o(t_n)$ and $a_{n+1} \leq (1-t_n)a_n + b_n + c_n$ for all $n \geq 1$. Then, $\lim_{n\to\infty} a_n = 0$.

Theorem 6.9

Suppose that X is a normed space, $\varphi : [0,\infty) \to [0,\infty)$ satisfies property (A_1) and $T : X \to X$ is a mapping satisfying $F(T) = \{q\}$ and

$$\|Tx - q\| \leq \varphi(\|x - q\|) \tag{6.4}$$

for all $x \in X$. Then Mann iteration is almost T-stable.

Proof: Let $\{u_n\}_{n\geq 1}$ denotes Mann iteration. Then, using (6.4) and the fact that $\varphi(t) \leq t$ for each $t \geq 0$,

$$\begin{aligned}
\|u_{n+1} - q\| &= \|(1-\alpha_n)u_n + \alpha_n T u_n - q\| \\
&\leq (1-\alpha_n)\|u_n - q\| + \alpha_n \|Tu_n - q\| \\
&\leq (1-\alpha_n)\|u_n - q\| + \alpha_n \varphi(\|u_n - q\|) \\
&\leq (1-\alpha_n)\|u_n - q\| + \alpha_n(\|u_n - q\|) = \|u_n - q\|.
\end{aligned}$$

Therefore $\{\|u_n - q\|\}_{n\geq 1}$ is a nonnegative decreasing sequence in X and is bounded. Assume that $\sum_{n=1}^{k} y_n$ converges, where

$$y_{n+1} = \|u_{n+1} - (1-\alpha_n)u_n - \alpha_n T u_n\|.$$

Take $M = \sup_{n\geq 1} \|u_n - q\|$. For each $\varepsilon > 0$, there exists a natural number p such that

$$\sum_{i=m}^{\infty} y_i \leq \frac{\varepsilon}{4}, \quad \varphi^m(M) < \frac{\varepsilon}{4},$$

for all $m \geq p$. By considering $d_n = \|u_n - q\|$ and $c_n = \|Tu_n - q\|$ we have

$$d_n \leq y_n + (1-\alpha_{n-1})d_{n-1} + \alpha_{n-1}\varphi d_{n-1}.$$

In fact

$$\begin{aligned}
d_n = \|u_n - q\| &\leq \|u_n - (1-\alpha_{n-1})u_{n-1} - \alpha_{n-1}Tu_{n-1}\| \\
&\quad + \|(1-\alpha_{n-1})u_{n-1} + \alpha_{n-1}Tu_{n-1} - q\| \\
&= \|u_n - (1-\alpha_{n-1})u_{n-1} - \alpha_{n-1}Tu_{n-1}\| \\
&\quad + \|(1-\alpha_{n-1})u_{n-1} + \alpha_{n-1}Tu_{n-1} - (1-\alpha_{n-1})q - \alpha_{n-1}q\| \\
&\leq y_n + (1-\alpha_{n-1})\|u_{n-1} - q\| + \alpha_{n-1}\|Tu_{n-1} - q\| \\
&\leq y_n + (1-\alpha_{n-1})\|u_{n-1} - q\| + \alpha_{n-1}\varphi(\|u_{n-1} - q\|) \\
&= y_n + (1-\alpha_{n-1})d_{n-1} + \alpha_{n-1}\varphi d_{n-1}.
\end{aligned}$$

Hence,

$$(1-\alpha_n)d_n \le (1-\alpha_n)y_n + (1-\alpha_n)(1-\alpha_{n-1})d_{n-1} + (1-\alpha_n)\alpha_{n-1}\varphi d_{n-1}.$$

On the other hand, since φ is increasing, we have

$$\varphi d_n \le \varphi(y_n + (1-\alpha_{n-1})d_{n-1} + \alpha_{n-1}\varphi d_{n-1}).$$

If $u = y_n$ and $v = (1-\alpha_{n-1})d_{n-1} + \alpha_{n-1}\varphi d_{n-1}$, then by using property (A_1) we have

$$\varphi d_n \le y_n + \varphi((1-\alpha_{n-1})d_{n-1} + \alpha_{n-1}\varphi d_{n-1}).$$

Since φ is a convex function, we obtain

$$\varphi d_n \le y_n + (1-\alpha_{n-1})\varphi d_{n-1} + \alpha_{n-1}\varphi^2 d_{n-1}.$$

These relations imply that

$$\|u_{n+1}-q\| \le y_{n+1} + (1-\alpha_n)d_n + \alpha_n\varphi d_n$$

$$\le y_{n+1} + (1-\alpha_n)y_n + \alpha_n y_n + (1-\alpha_n)(1-\alpha_{n-1})d_{n-1}$$
$$+(1-\alpha_n)\alpha_{n-1}\varphi d_{n-1} + \alpha_n(1-\alpha_{n-1})\varphi d_{n-1} + \alpha_n\alpha_{n-1}\varphi^2 d_{n-1}.$$

Therefore, by using similar methods, we have

$$\begin{aligned}
\|u_{n+1}-q\| &\le y_{n+1} + (1-\alpha_n)d_n + \alpha_n\varphi d_n \\
&\le y_{n+1} + (1-\alpha_n)y_n + \alpha_n y_n + (1-\alpha_n)(1-\alpha_{n-1})d_{n-1} \\
&+ (1-\alpha_n)\alpha_{n-1}\varphi d_{n-1} + \alpha_n(1-\alpha_{n-1})\varphi d_{n-1} + \alpha_n\alpha_{n-1}\varphi^2 d_{n-1} \\
&\le y_{n+1} + y_n + (1-\alpha_n)(1-\alpha_{n-1})y_{n-1} \\
&+ (1-\alpha_n)\alpha_{n-1}y_{n-1} + \alpha_n(1-\alpha_{n-1})y_{n-1} + \alpha_n\alpha_{n-1}y_{n-1} \\
&+ (1-\alpha_n)(1-\alpha_{n-1})(1-\alpha_{n-2})d_{n-2} \\
&+ (1-\alpha_n)(1-\alpha_{n-1})\alpha_{n-2}\varphi d_{n-2} + (1-\alpha_n)\alpha_{n-1}(1-\alpha_{n-2})\varphi d_{n-2} \\
&+ \alpha_n(1-\alpha_{n-1})(1-\alpha_{n-2})\varphi d_{n-2} + (1-\alpha_n)\alpha_{n-1}\alpha_{n-2}\varphi^2 d_{n-2} \\
&+ \alpha_n(1-\alpha_{n-1})\alpha_{n-2}\varphi^2 d_{n-2} \\
&+ \alpha_n\alpha_{n-1}(1-\alpha_{n-2})\varphi^2 d_{n-2} + \alpha_n\alpha_{n-1}\alpha_{n-2}\varphi^3 d_{n-2} \\
&= y_{n+1} + y_n + y_{n-1} + (1-\alpha_n)(1-\alpha_{n-1})(1-\alpha_{n-2})d_{n-2} \\
&+ (1-\alpha_n)(1-\alpha_{n-1})\alpha_{n-2}\varphi d_{n-2} \\
&+ (1-\alpha_n)\alpha_{n-1}(1-\alpha_{n-2})\varphi d_{n-2} + \alpha_n(1-\alpha_{n-1})(1-\alpha_{n-2})\varphi d_{n-2} \\
&+ (1-\alpha_n)\alpha_{n-1}\alpha_{n-2}\varphi^2 d_{n-2} + \alpha_n(1-\alpha_{n-1})\alpha_{n-2}\varphi^2 d_{n-2} \\
&+ \alpha_n\alpha_{n-1}(1-\alpha_{n-2})\varphi^2 d_{n-2} + \alpha_n\alpha_{n-1}\alpha_{n-2}\varphi^3 d_{n-2}.
\end{aligned}$$

By continuing these replacements, after a number of finite steps, we obtain

$$\|u_{n+1}-q\| \leq \sum_{i=p+1}^{n+1} y_i + \sum_{i=0}^{n-p} S_n^i \varphi^i d_p + (\prod_{i=p}^{n} \alpha_i)\varphi^{n-p+1} d_p,$$

where S_n^i are the coefficients of $\varphi^i d_p$. Note that

$$S_n^i = \sum_{\{r_1,r_2,\ldots,r_i\} \text{ is a subset of } \{p,p+1,\ldots,n\}} \alpha_{r_1}\alpha_{r_2}\ldots\alpha_{r_i} \prod_{k\in I_i}(1-\alpha_k),$$

where $I_i = \{p, p+1, \ldots, n\}\backslash\{r_1, r_2, \ldots, r_i\}$. We show that for each $k \geq 1$, $\sum_{i=0}^{k} S_k^i =$ 1. Clearly the equality holds when $k = 1$. Assume that the equality holds for all $k \leq n$. We will show that the equality holds for $n+1$. It is clear that for each $i \leq n$ we have

$$S_{n+1}^i = (1-\alpha_{n+1})S_n^i + \alpha_{n+1}S_n^{i-1}.$$

Hence,

$$\begin{aligned}
\sum_{i=0}^{n+1} S_{n+1}^i &= S_{n+1}^{n+1} + (1-\alpha_{n+1})\sum_{i=0}^{n} S_n^i + \alpha_{n+1}\sum_{i=0}^{n} S_n^{i-1} \\
&= \prod_{i=0}^{n+1}\alpha_i + (1-\alpha_{n+1} + \alpha_{n+1}(1-S_n^n) \\
&= \prod_{i=0}^{n+1}\alpha_i + (1-\alpha_{n+1}) + \alpha_{n+1} - \alpha_{n+1}\prod_{i=0}^{n}\alpha_i = 1.
\end{aligned}$$

We shall now prove that $\lim_{n\to\infty} S_n^i = 0$ for all $i \geq 1$. For $i = 1$,

$$S_n^1 = \alpha_p \prod_{\substack{i=p \\ i\neq p}}^{n}(1-\alpha_i) + \alpha_{p+1}\prod_{\substack{i=p \\ i\neq p+1}}^{n}(1-\alpha_i) + \cdots + \alpha_{n+1}\prod_{\substack{i=p \\ i\neq n}}^{n}(1-\alpha_i).$$

It is clear that $S_{n+1}^1 = (1-\alpha_{n+1})S_n^1 + \alpha_{n+1}\prod_{i=p}^{n}(1-\alpha_i)$. As we know, $1-x \leq e^{-x}$ for all $x > 0$. Since $\sum_{i=p}^{\infty}\alpha_i = \infty$, $\lim_{n\to\infty}\prod_{i=p}^{n}(1-\alpha_i) = 0$ and so by using Lemma 6.2, $\lim_{n\to\infty} S_n^1 = 0$. Assume that $\lim_{n\to\infty} S_n^{i-1} = 0$. Then

$$S_{n+1}^i = (1-\alpha_{n+1})S_n^i + \alpha_{n+1}S_n^{i-1}.$$

By using an argument similar to case $i = 1$ and Lemma 6.2, we get $\lim_{n\to\infty} S_n^i = 0$.

Let $n > 2p+1$. Since $\varphi t < t$ for all $t > 0$, we get

$$\begin{aligned}
\|u_{n+1}-q\| &\leq \sum_{i=p+1}^{n+1} y_i + (\sum_{i=0}^{p} S_n^i)M + (\sum_{i=p+1}^{n} S_n^i)\varphi^p M + (\prod_{i=p}^{n} \alpha_i)\varphi^p d_p \\
&\leq \sum_{i=p+1}^{n+1} y_i + (\sum_{i=0}^{p} S_n^i)M + \varphi^p M + \varphi^p M \\
&< \frac{\varepsilon}{4} + y_{n+1} + (\sum_{i=0}^{p} S_n^i)M + \frac{\varepsilon}{4} + \frac{\varepsilon}{4} \leq \varepsilon + y_{n+1} + (\sum_{i=0}^{p} S_n^i)M.
\end{aligned}$$

Thus, $\limsup_{n\to\infty} \|u_n - q\| \leq \varepsilon$ and so $\lim_{n\to\infty} \|u_n - q\| = 0$. This shows that the Mann iteration is almost T-stable.

♣

Example 89 *Define the function $T : R \to R$ by $T(x) = |x| - \log(1+|x|)$. Clearly 0 is unique fixed point of T. To show that Mann iteration is almost T-stable, by using Theorem 6.9, define $\varphi : [0,\infty) \to [0,\infty)$ by $\varphi(x) = x - \log(1+x)$.*

Corollary 6.2

Let X be a normed space, $\varphi : [0,\infty) \to [0,\infty)$ a map satisfying the property (A_1) and $T : X \to X$ a self-map satisfying

$$\|Tx - Ty\| \leq \varphi(\max\{\|x-y\|, \frac{1}{2}[\|x-Tx\| + \|y-Ty\|], \|x-Ty\|, \|y-Tx\|\}),$$

for all $x, y \in X$. Then Mann iteration is almost T-stable.

Proof: It is known that T has a unique fixed point q. Hence, it is sufficient we show that T satisfies (1). If $x \neq q$, then

$$\|Tx - q\| \leq \varphi(\max\{\|x-q\|, \frac{\|x-Tx\|}{2}, \|Tx-q\|\}).$$

If $\max\{\|x-q\|, \frac{\|x-Tx\|}{2}, \|Tx-q\|\} = \|Tx-q\|$, then $\|Tx-q\| = 0$. Thus, suppose that $\|Tx-q\| \leq \varphi(\max\{\|x-q\|, \frac{\|x-Tx\|}{2}\})$. If $\|x-q\| < \frac{\|x-Tx\|}{2}$, then

$$\begin{aligned}
\|Tx - q\| &\leq \varphi(\frac{\|x-Tx\|}{2}) \\
&\leq \varphi(\frac{1}{2}[\|x-q\| + \|q-Tx\|]) \\
&\leq \varphi(\max\{\|x-q\|, \|q-Tx\|\}).
\end{aligned}$$

Hence, (1) again holds.

♣

Putting $\varphi(t) = kt$ with $k \in [0,1)$, we obtain next result.

Corollary 6.3

Let X be a normed space, $k \in [0,1)$ and $T : X \to X$ a map satisfying

$$||Tx - Ty|| \leq k\max\{||x-y||, \frac{1}{2}[||x-Tx|| + ||y-Ty||], ||x-Ty||, ||y-Tx||\},$$

for all $x, y \in X$. Then Mann iteration is almost T-stable.

One can check that these results hold for Ishikawa iteration.

6.5 T-stability of Picard iteration for maps satisfy a contractive condition of integral type

In this section, we shall verify the T-stability of Picard iteration for mappings satisfying a contractive condition of integral type. Let R_+ be the set of nonnegative real numbers and

(*i*) $\psi : R_+ \to R_+$ is sub-additive, increasing and continuous from the right such that $\psi(t) < t$ for all $t > 0$;

(*ii*) $\varphi : R_+ \to R_+$ is a summable, Lebesgue-integrable and decreasing mapping on $(0,\infty)$ such that $\int_o^{\varepsilon} \varphi(t)dt > 0$ for each $\varepsilon > 0$.

Note that, there are many functions which satisfy (i). For example, suppose that $g : [0,\infty) \to [0,1]$ is a strictly decreasing map. Then, $\psi(t) = \int_o^t g(x)dx$ satisfies (*i*).

Let (X,d) be a complete metric space, $\psi, \varphi : R_+ \to R_+$ two maps satisfying the conditions (*i*) and (*ii*) respectively and $T : X \to X$ a map satisfying the following property (A_2):

$$\int_0^{d(Tx,Ty)} \varphi(t)dt \leq \psi\left(\int_0^{M(x,y)} \varphi(t)dt\right) \qquad (A_2)$$

for all $x, y \in X$, where

$$M(x,y) = \max\{d(x,y), d(x,Tx), d(y,Ty), d(x,Ty), d(y,Tx)\}.$$

By using Theorem 8 in [35], we conclude that, if there exists a bounded sequence $\{y_n\}_{n\geq 0}$ with $y_{n+1} = Ty_n$ for all $n \geq 0$, then T has a unique fixed point.

Theorem 6.10

Let (X,d) be a complete metric space and $T : X \to X$ a mapping satisfying property (A_2). If there exists a bounded sequence $\{y_n\}_{n\geq 0}$ with $\lim_{n\to\infty} d(y_{n+1}, Ty_n) = 0$, then $\{y_n\}_{n\geq 0}$ converges to the unique fixed point of T.

Proof: Let δ_n denote the diameter of the set $A_n = \{y_i\}_{i\geq n} \cup \{Ty_i\}_{i\geq n}$ for all $n \geq 0$. Put $\varepsilon_i = d(y_i, Ty_{i-1})$, for all $i \geq 1$. Then, for each $i, j \geq n$, we have

$$\begin{aligned} \int_0^{d(y_i,y_j)} \varphi(t)dt &\leq \int_0^{\varepsilon_i+\varepsilon_j+d(Ty_{i-1},Ty_{j-1})} \varphi(t)dt \\ &\leq \int_0^{\varepsilon_i+\varepsilon_j} \varphi(t)dt + \psi\left(\int_0^{\delta_{n-1}} \varphi(t)dt\right), \end{aligned} \tag{6.5}$$

$$\begin{aligned} \int_0^{d(y_i,Ty_j)} \varphi(t)dt &\leq \int_0^{\varepsilon_i+d(Ty_{i-1},Ty_j)} \varphi(t)dt \\ &\leq \int_0^{\varepsilon_i} \varphi(t)dt + \psi\left(\int_0^{\delta_{n-1}} \varphi(t)dt\right), \end{aligned} \tag{6.6}$$

$$\int_0^{d(Ty_i,Ty_j)} \varphi(t)dt \leq \psi\left(\int_0^{\delta_{n-1}} \varphi(t)dt\right). \tag{6.7}$$

Put $a_n = \sup_{i\geq n} 2\varepsilon_i$ for all $n \geq 1$. It is clear that $\lim_{n\to\infty} a_n = 0$. Note that, from (6.5), (6.6) and (6.7), we obtain

$$\int_0^{\delta_n} \varphi(t)dt \leq \int_0^{a_n} \varphi(t)dt + \psi\left(\int_0^{\delta_{n-1}} \varphi(t)dt\right)$$

for all $n \geq 1$. Now, assume that $M = \sup_{n\geq 1} \int_0^{\delta_n} \varphi(t)dt$ and $\varepsilon > 0$ is given. Choose a natural number k such that $\psi^k(M) < \frac{\varepsilon}{2}$. Then

$$\begin{aligned}\int_0^{\delta_{n+k}} \varphi(t)dt &\leq \int_0^{a_{n+k}} \varphi(t)dt + \psi\left(\int_0^{\delta_{n+k-1}} \varphi(t)dt\right) \\ &\leq \int_0^{a_{n+k}} \varphi(t)dt + \psi\left(\int_0^{a_{n+k-1}} \varphi(t)dt\right) + \psi^2\left(\int_0^{\delta_{n+k-2}} \varphi(t)dt\right) \\ &\leq \sum_{i=0}^{k-1} \psi^i\left(\int_0^{a_{n+k-i}} \varphi(t)dt\right) + \psi^k\left(\int_0^{\delta_n} \varphi(t)dt\right).\end{aligned}$$

It follows that $\lim\limits_{n\to\infty} \int_0^{\delta_n} \varphi(t)dt = 0$ and so $\lim\limits_{n\to\infty} \delta_n = 0$. Hence, $\{y_n\}_{n\geq 1}$ is a Cauchy sequence. Suppose that y_n converges q. Then, we have

$$\begin{aligned}\int_0^{d(q,Tq)} \varphi(t)dt &= \lim_{n\to\infty} \int_0^{d(Ty_n,Tq)} \varphi(t)dt \leq \limsup_{n\to\infty} \psi\left(\int_0^{M(y_n,q)} \varphi(t)dt\right) \\ &\leq \psi\left(\int_0^{d(q,Tq)} \varphi(t)dt\right).\end{aligned}$$

This implies $\int_0^{d(q,Tq)} \varphi(t)dt = 0$ and so $d(q,Tq) = 0$. By a method similar to the proof of Theorem 8 in [35], we can show that fixed point of T is unique.

♣

Corollary 6.4

Let (X,d) be a complete metric space and $T : X \to X$ a mapping satisfying property (A_2). If Picard iteration is convergent, then Picard iteration is T-stable.

Letting $\psi(t) = kt$ with $k \in [0,1)$ in (A_2), we obtain the following result.

Corollary 6.5

Let (X,d) be a complete metric space, $k \in [0,1)$ and $T : X \to X$ a map satisfying

$$\int_0^{d(Tx,Ty)} \varphi(t)dt \leq k \int_0^{M(x,y)} \varphi(t)dt$$

for all $x,y \in X$, where φ satisfies the property (ii). If there exist a bounded sequence $\{y_n\}_{n\geq 1}$ such that $\lim_{n\to\infty} d(y_{n+1}, Ty_n) = 0$, then the sequence $\{y_n\}_{n\geq 1}$ converges to the unique fixed point T.

By considering $\varphi(t) = 1$ and $\psi(t) = kt$ with $k \in [0,1)$ in (A_2), we obtain the following result.

Corollary 6.6

Let (X,d) be a complete metric space, $k \in [0,1)$ and the map $T : X \to X$ satisfying

$$d(Tx,Ty) \leq k\max\{d(x,y), d(x,Tx), d(y,Ty), d(x,Ty), d(y,Tx)\}$$

for all $x,y \in X$. Then Picard iteration is T-stable.

Chapter 7

Some false results on rate of convergence

7.1 Introduction

As you know, the Picard iteration has been extensively used in many works from different views. Let (X,d) be a metric space, $x_0 \in X$ and $T\colon X \to X$ a selfmap. The Picard iteration is defined by

$$x_{n+1} = Tx_n \tag{7.1}$$

for all $n \geq 0$. Let $\{\alpha_n\}_{n\geq 0}$, $\{\beta_n\}_{n\geq 0}$ and $\{\gamma_n\}_{n\geq 0}$ be sequences in $[0,1]$. Then, the Mann iteration method is defined by

$$x_{n+1} = \alpha_n x_n + (1-\alpha_n)Tx_n \tag{7.2}$$

for all $n \geq 0$ (for more information, see [71]). Also, the Ishikawa iteration method is defined by

$$\begin{aligned} x_{n+1} &= (1-\alpha_n)x_n + \alpha_n Ty_n, \\ y_n &= (1-\beta_n)x_n + \beta_n Tx_n \end{aligned} \tag{7.3}$$

for all $n \geq 0$ (for more information, see [52]). The Noor iteration method is defined by

$$\begin{aligned} x_{n+1} &= (1-\alpha_n)x_n + \alpha_n Ty_n, \\ y_n &= (1-\beta_n)x_n + \beta_n Tz_n, \\ z_n &= (1-\gamma_n)x_n + \gamma_n Tx_n \end{aligned} \tag{7.4}$$

for all $n \geq 0$ (for more information, see [75]). In 2007, Agarwal et al. defined their new iteration methods by

$$\begin{aligned} x_{n+1} &= (1-\alpha_n)Tx_n + \alpha_n Ty_n, \\ y_n &= (1-\beta_n)x_n + \beta_n Tx_n \end{aligned} \tag{7.5}$$

for all $n \geq 0$ (for more information, see [6]). In 2013, Abbas et al. defined their new iteration methods by

$$\begin{aligned} x_{n+1} &= (1-\alpha_n)Ty_n + \alpha_n Tz_n, \\ y_n &= (1-\beta_n)Tx_n + \beta_n Tz_n, \\ z_n &= (1-\gamma_n)x_n + \gamma_n Tx_n \end{aligned} \tag{7.6}$$

for all $n \geq 0$ (for more information, see [2]). In 2014, Thakur, Singh and Postolache defined their new iteration methods by

$$\begin{aligned} x_{n+1} &= (1-\alpha_n)Tx_n + \alpha_n Ty_n, \\ y_n &= (1-\beta_n)z_n + \beta_n Tz_n, \\ z_n &= (1-\gamma_n)x_n + \gamma_n Tx_n \end{aligned} \tag{7.7}$$

for all $n \geq 0$ (for more information, see [102]). Also, the Picard S-iteration defined by

$$\begin{aligned} x_{n+1} &= Ty_n, \\ y_n &= (1-\beta_n)Tx_n + \beta_n Tz_n, \\ z_n &= (1-\gamma_n)x_n + \gamma_n Tx_n \end{aligned} \tag{7.8}$$

for all $n \geq 0$ ([77]).

7.2 Rate of convergence

Assume that $(X, \|\cdot\|)$ is a normed space, $x_0 \in X$, $T\colon X \to X$ is a selfmap and $\{\alpha_n\}_{n\geq 0}$, $\{\beta_n\}_{n\geq 0}$ and $\{\gamma_n\}_{n\geq 0}$ are sequences in $(0,1)$. The Mann iteration is given by

$$x_{n+1} = (1-\alpha_n)x_n + \alpha_n Tx_n$$

for all $n \geq 0$. Note that, we can rewrite it as

$$x_{n+1} = \alpha_n x_n + (1-\alpha_n)Tx_n$$

for all $n \geq 0$. We call these cases by first and second forms of the Mann iteration method.

In next result we show that choosing type of the sequence $\{\alpha_n\}_{n\geq 0}$ in the Mann iteration has notable role in rate of convergence of the sequence $\{x_n\}_{n\geq 0}$.

Let $\{u_n\}_{n\geq 0}$ and $\{v_n\}_{n\geq 0}$ be two fixed point iteration procedures that converge to the same fixed point p and $\|u_n - p\| \leq a_n$ and $\|v_n - p\| \leq b_n$ for all $n \geq 0$. If the sequences $\{a_n\}_{n\geq 0}$ and $\{b_n\}_{n\geq 0}$ converge to a and b respectively and

$\lim_{n\to\infty} \frac{\|a_n - a\|}{\|b_n - b\|} = 0$, then we say that $\{u_n\}_{n\geq 0}$ converges faster than $\{v_n\}_{n\geq 0}$ to p (see [102]).

Proposition 7.1

Let C be a nonempty, closed and convex subset of a Banach space X, $x_1 \in C$, $T : C \to C$ a contraction with constant $k \in (0,1)$ and p a fixed point of T. Consider the first case for Mann iteration. If the coefficients of Tx_n are greater than the coefficients of x_n, that is $1 - \alpha_n < \alpha_n$ for some sufficient large n or equivalently $\{\alpha_n\}_{n\geq n_0}$ is a sequence in $(\frac{1}{2}, 1)$ for some natural number n_0, then this Mann iteration converges faster than the another Mann iteration in which the coefficients of x_n are greater than the coefficients of Tx_n.

Proof: Let $\{x_n\}$ be the sequence in the Mann iteration which the coefficients of Tx_n are greater than the coefficients of x_n, that is,

$$x_{n+1} = (1 - \alpha_n)x_n + \alpha_n T x_n \tag{7.9}$$

for all n. In this case, we have

$$\begin{aligned}\|x_{n+1} - p\| = \|(1-\alpha)x_n + \alpha_n T x_n - p\| &\leq (1-\alpha_n)\|x_n - p\| + \alpha_n \|Tx_n - p\| \\ &\leq (1 - \alpha_n(1-k))\,\|x_n - p\|\end{aligned}$$

for all $n \geq n_0$. Since $\alpha_n \in (\frac{1}{2}, 1)$ for all $n \geq n_0$, $1 - \alpha_n(1-k) < 1 - \frac{1}{2}(1-k)$. Put

$$a_n = (1 - \frac{1}{2}(1-k))^n \|x_1 - p\|$$

for all n. Let $\{x_n\}$ be the sequence in the Mann iteration in which the coefficients of x_n are greater than the coefficients of Tx_n. In this case, we have

$$\begin{aligned}\|x_{n+1} - p\| = \|\alpha x_n + (1-\alpha_n)Tx_n - p\| &\leq (\alpha_n \|x_n - p\| + (1-\alpha_n)\|Tx_n - p\| \\ &\leq (1 - (1-\alpha_n)(1-k))\,\|x_n - p\|\end{aligned}$$

for all $n \geq n_0$. Since $1 - \alpha_n < \alpha_n$ for all $n \geq n_0$, we get $1 - (1-\alpha_n)(1-k) < 1$ for all $n \geq n_0$. Put $b_n = \|x_1 - p\|$ for all n. Note that,

$$\lim_{n\to\infty} \frac{a_n}{b_n} = \lim_{n\to\infty} \frac{(1 - \frac{1}{2}(1-k))^n \|x_1 - p\|}{\|x_1 - p\|} = 0.$$

This completes the proof.

♣

One can use similar conditions instead the conditions which we use in our results.

As we know, we can consider four cases for writing the Ishikawa iteration method. In next result, we indicate each case by a different enumeration. Similar to the last result, we want to compare the Ishikawa iteration method with itself in the four possible cases. Again, we show that the coefficient sequences $\{\alpha_n\}_{n\geq 0}$ and $\{\beta_n\}_{n\geq 0}$ have effective role in rate of convergence of the sequence $\{x_n\}_{n\geq 0}$ in the Ishikawa iteration method.

Proposition 7.2

Let C be a nonempty, closed and convex subset of a Banach space X, $x_1 \in C$, $T: C \to C$ a contraction with constant $k \in (0,1)$ and p a fixed point of T. Consider the following cases of Ishikawa iteration method

$$\begin{cases} x_{n+1} = (1-\alpha_n)x_n + \alpha_n T y_n, \\ y_n = (1-\beta_n)x_n + \beta_n T x_n, \end{cases} \tag{7.10}$$

$$\begin{cases} x_{n+1} = \alpha_n x_n + (1-\alpha_n) T y_n, \\ y_n = \beta_n x_n + (1-\beta_n) T x_n, \end{cases} \tag{7.11}$$

$$\begin{cases} x_{n+1} = \alpha_n x_n + (1-\alpha_n) T y_n, \\ y_n = (1-\beta_n)x_n + \beta_n T x_n \end{cases} \tag{7.12}$$

and

$$\begin{cases} x_{n+1} = (1-\alpha_n)x_n + \alpha_n T y_n, \\ y_n = \beta_n x_n + (1-\beta_n) T x_n \end{cases} \tag{7.13}$$

for all $n \geq 0$. If $1-\alpha_n < \alpha_n$ and $1-\beta_n < \beta_n$ for all $n \geq n_0$ for some natural number n_0, then the case (7.10) converges faster than others. In fact, the Ishikawa iteration method is faster whenever the coefficients of Ty_n and Tx_n simultaneously greater than the related coefficients of x_n for all $n \geq n_0$.

Proof: Let $\{x_n\}_{n\geq 0}$ be the sequence in the case (7.10). Then, we have

$$\begin{aligned} \|y_n - p\| &= \|(1-\beta_n)x_n + \beta_n T x_n - p\| \\ &\leq (1-\beta_n)\|x_n - p\| + \beta_n \|T x_n - p\| \\ &\leq ((1-\beta_n) + \beta_n k)\|x_n - p\| \end{aligned}$$

and

$$\begin{aligned}
\|x_{n+1}-p\| &= \|(1-\alpha_n)x_n+\alpha_n Ty_n-p\| \\
&\leq (1-\alpha_n)\|x_n-p\|+\alpha_n\|Ty_n-p\| \\
&\leq (1-\alpha_n)\|x_n-p\|+k\alpha_n\|y_n-p\| \\
&\leq \left(1-\alpha_n)+k\alpha_n[(1-\beta_n)+\beta_n k\right)\|x_n-p\| \\
&\leq \left(1-\alpha_n+\alpha_n k-\alpha_n\beta_n k+\alpha_n\beta_n k^2\right)\|x_n-p\| \\
&\leq \left(1-\alpha_n(1-k)-\alpha_n\beta_n k(1-k)\right)\|x_n-p\|
\end{aligned}$$

for all $n \geq 0$. Since $\alpha_n, \beta_n \in (\frac{1}{2}, 1)$ for all $n \geq n_0$,

$$1-\alpha_n(1-k)-\alpha_n\beta_n k(1-k) < 1-\frac{1}{2}(1-k)-\frac{1}{4}k(1-k)$$

for all $n \geq 0$. Put $a_n = \left(1-\frac{1}{2}(1-k)-\frac{1}{4}k(1-k)\right)^n\|x_1-p\|$ for all n. If $\{x_n\}_{n\geq 0}$ is the sequence in the case (7.11), then we get

$$\begin{aligned}
\|y_n-p\| &= \|\beta_n x_n+(1-\beta_n)Tx_n-p\| \\
&\leq \beta_n\|x_n-p\|+(1-\beta_n)\|Tx_n-p\| \\
&\leq \left(1-(1-\beta_n)(1-k)\right)\|x_n-p\|
\end{aligned}$$

and

$$\begin{aligned}
\|x_{n+1}-p\| &= \|\alpha_n x_n+(1-\alpha_n)Ty_n-p\| \\
&\leq \alpha_n\|x_n-p\|+(1-\alpha_n)\|Ty_n-p\| \\
&\leq \alpha_n\|x_n-p\|+k(1-\alpha_n)\|y_n-p\| \\
&\leq \left(\alpha_n+k(1-\alpha_n)(1-(1-\beta_n)(1-k)\right)\|x_n-p\| \\
&= \left(\alpha_n+(1-\alpha_n)k-k(1-\alpha_n)(1-\beta_n)(1-k)\right)\|x_n-p\| \\
&= \left(1-(1-\alpha_n)(1-k)-(1-\alpha_n)(1-\beta_n)k(1-k)\right)\|x_n-p\|
\end{aligned}$$

for all $n \geq n_0$. Since $\alpha_n, \beta_n \in (\frac{1}{2}, 1)$ for all $n \geq n_0$,

$$1-(1-\alpha_n)(1-k)-(1-\alpha_n)(1-\beta_n)(1-k) < 1$$

for all $n \geq 0$. Put $b_n = \|x_1-p\|$ for all n. Since

$$1-\frac{1}{2}(1-k)-\frac{1}{4}k(k-1) < 1+\frac{1}{2}k(1-k),$$

we get $\lim_{n\to\infty}\frac{a_n}{b_n} = \lim_{n\to\infty}\frac{\left(1-\frac{1}{2}(1-k)-\frac{1}{4}k(1-k)\right)^n\|x_1-p\|}{\|x_1-p\|} = 0$ and so the iteration (7.10) converges faster than the case (7.11). Now, let $\{x_n\}_{n\geq 0}$ be the sequence in the case (7.12). Then,

$$\begin{aligned}\|y_n - p\| &= \|\beta x_n + (1-\beta_n)Tx_n - p\| \\ &\le \beta_n \|x_n - p\| + (1-\beta_n)\|Tx_n - p\| \\ &\le (\beta_n + k(1-\beta_n))\|x_n - p\| \\ &= (1-(1-\beta_n)(1-k))\|x_n - p\|\end{aligned}$$

and

$$\begin{aligned}\|x_{n+1} - p\| &= \|(1-\alpha_n)x_n + \alpha_n Ty_n - p\| \\ &\le (1-\alpha_n)\|x_n - p\| + \alpha_n \|Ty_n - p\| \\ &\le (1-\alpha_n + k\alpha_n[(1-(1-\beta_n)(1-k))])\|x_n - p\| \\ &= (1-\alpha_n + k\alpha_n - \alpha_n(1-\beta_n)k(1-k))\|x_n - p\| \\ &= (1-\alpha_n(1-k) - \alpha_n(1-\beta_n)k(1-k))\|x_n - p\|\end{aligned}$$

for all $n \ge n_0$. Since $\alpha_n, \beta_n \in (\frac{1}{2}, 1)$ for all $n \ge n_0$, $-(1-k) < -\alpha_n(1-k) < -\frac{1}{2}(1-k)$ and $\frac{-1}{2}k(1-k) < -\alpha_n(1-\beta_n)k(1-k) < 0$ for all $n \ge n_0$. Hence,

$$1-\alpha_n(1-k) - \alpha_n(1-\beta_n)k(1-k) < 1 - \frac{1}{2}(1-k)$$

for all $n \ge n_0$. Put $c_n = \left(1-\frac{1}{2}(1-k)\right)^n \|x_1 - p\|$ for all n. Thus, we obtain

$$\lim_{n\to\infty} \frac{a_n}{c_n} = \lim_{n\to\infty} \frac{\left(1-\frac{1}{2}(1-k) - \frac{1}{4}k(1-k)\right)^n \|x_1 - p\|}{\left(1-\frac{1}{2}(1-k)\right)^n \|x_1 - p\|} = 0$$

and so the iteration (7.10) converges faster than the case (7.12). Now, let $\{x_n\}_{n\ge 0}$ be the sequence in the case (7.13). Then, we have

$$\begin{aligned}\|y_n - p\| &= \|(1-\beta)x_n + \beta_n Tx_n - p\| \\ &\le (1-\beta_n)\|x_n - p\| + \beta_n \|Tx_n - p\| \\ &\le (1-\beta_n(1-k))\|x_n - p\|\end{aligned}$$

and

$$\begin{aligned}\|x_{n+1} - p\| &= \|\alpha_n x_n + (1-\alpha_n)Ty_n\| \\ &\le \alpha_n \|x_n - p\| + (1-\alpha_n)\|Ty_n - p\| \\ &\le \alpha_n \|x_n - p\| + k(1-\alpha_n)\|y_n - p\| \\ &\le (\alpha_n + k(1-\alpha_n)[1-\beta_n(1-k)])\|x_n - p\| \\ &\le (\alpha_n + k(1-\alpha_n) - (1-\alpha_n)\beta_n k(1-k))\|x_n - p\| \\ &\le (1-(1-\alpha_n) + k(1-\alpha_n) - (1-\alpha_n)\beta_n k(1-k))\|x_n - p\| \\ &\le (1-(1-\alpha_n)(1-k) - (1-\alpha_n)\beta_n k(1-k))\|x_n - p\|\end{aligned}$$

for all $n \geq n_0$. Since $\alpha_n, \beta_n \in (\frac{1}{2}, 1)$ for all $n \geq n_0$, $-(1-k^2) < -\alpha_n(1-k^2) < -\frac{1}{2}(1-k^2)$ and $-\frac{1}{2}k(1-k) < -(1-\alpha_n)\beta_n k(1-k) < 0$for all $n \geq n_0$ and so

$$1-\alpha_n(1-k)-(1-\alpha_n)\beta_n k(1-k) < 1-\frac{1}{2}(1-k)$$

for all $n \geq n_0$. Put $d_n = \left(1-\frac{1}{2}(1-k)\right)^n \|x_1 - p\|$ for all n. Then, we have

$$\lim_{n\to\infty} \frac{a_n}{d_n} = \lim_{n\to\infty} \frac{\left(1-\frac{1}{2}(1-k)-\frac{1}{4}k(1-k)\right)^n \|x_1-p\|}{\left(1-\frac{1}{2}(1-k)\right)^n \|x_1-p\|} = 0$$

and so the iteration (7.10) converges faster than the case (7.13).

♣

By using a similar condition, one can show that the iteration (7.13) is faster than the case (7.11).

Now consider eight cases for writing the Noor iteration method. By using a condition, we show that the coefficient sequences $\{\alpha_n\}_{n\geq 0}$, $\{\beta_n\}_{n\geq 0}$ and $\{\gamma_n\}_{n\geq 0}$ have effective role in rate of convergence of the sequence $\{x_n\}_{n\geq 0}$ in the Noor iteration method. We enumerate the cases of the Noor iteration method during the proof of our next result.

Theorem 7.1

Let C be a nonempty, closed and convex subset of a Banach space X, $x_1 \in C$, $T: C \to C$ a contraction with constant $k \in (0,1)$ and p a fixed point of T. Consider the case (7.4) of the Noor iteration method

$$\begin{cases} x_{n+1} = (1-\alpha_n)x_n + \alpha_n T y_n, \\ y_n = (1-\beta_n)x_n + \beta_n T z_n, \\ z_n = (1-\gamma_n)x_n + \gamma_n T x_n \end{cases}$$

for all $n \geq 0$. If $1-\alpha_n < \alpha_n$, $1-\beta_n < \beta_n$ and $1-\gamma_n < \gamma_n$ for all $n \geq n_0$ for some natural number n_0, then the iteration (7.4) is faster than other possible cases.

Proof: First, we compare the case (7.4) with the following Noor iteration case

$$\begin{cases} u_{n+1} = (1-\alpha_n)u_n + \alpha_n T v_n, \\ v_n = (1-\beta_n)u_n + \beta_n T w_n, \\ w_n = \gamma_n u_n + (1-\gamma_n)T u_n \end{cases} \tag{7.14}$$

for all $n \geq 0$. Note that,

$$\begin{aligned} \|z_n - p\| &= \|(1-\gamma_n)x_n + \gamma_n T x_n - p\| \\ &\leq (1-\gamma_n)\|x_n - p\| + k\gamma_n \|x_n - p\| \\ &= (1-(1-k)\gamma_n)\|x_n - p\| \end{aligned}$$

and

$$\begin{aligned} \|y_n - p\| &= \|(1-\beta_n)x_n + \beta_n T z_n - p\| \\ &\leq (1-\beta_n)\|x_n - p\| + k\beta_n \|z_n - p\| \\ &\leq (1-\beta_n) + k\beta_n((1-(1-k)\gamma_n))\|x_n - p\| \\ &\leq [1-\beta_n(1-k) - \beta_n\gamma_n k(1-k)]\|x_n - p\| \end{aligned}$$

for all $n \geq 0$. Also, we have

$$\begin{aligned} \|x_{n+1} - p\| &= \|(1-\alpha_n)x_n + \alpha_n T y_n - p\| \\ &\leq (1-\alpha_n)\|x_n - p\| + k\alpha_n \|y_n - p\| \\ &\leq (1-\alpha_n)\|x_n - p\| + k\alpha_n[1-\beta_n(1-k) - \beta_n\gamma_n k(1-k)]\|x_n - p\| \\ &\leq (1-\alpha_n + k\alpha_n(1-\beta_n(1-k) - \beta_n\gamma_n k(1-k)))\|x_n - p\| \\ &\leq (1-\alpha_n + k\alpha_n - k(1-k)\beta_n\alpha_n - \alpha_n\beta_n\gamma_n k^2(1-k))\|x_n - p\| \\ &\leq (1-(1-k)\alpha_n - k(1-k)\beta_n\alpha_n - \alpha_n\beta_n\gamma_n k^2(1-k))\|x_n - p\| \end{aligned}$$

for all $n \geq 0$. Since $\alpha_n, \beta_n, \gamma_n \in (\frac{1}{2}, 1)$ for all $n \geq n_0$,

$$-(1-k^2) < -\alpha_n(1-k^2) < -\frac{1}{2}(1-k^2),$$

$-k(1-k) < -\alpha_n\beta_n k(1-k) < -\frac{1}{4}k(1-k)$ and

$$-k^2(1-k) < -\alpha_n\beta_n\gamma_n k^2(1-k) < -\frac{1}{8}k^2(1-k)$$

for all $n \geq n_0$. This implies that

$$\begin{aligned} &1-(1-k)\alpha_n - k(1-k)\beta_n\alpha_n - \alpha_n\beta_n\gamma_n k^2(1-k) \\ &< 1 - \frac{1}{2}(1-k) - \frac{1}{4}k(1-k) - \frac{1}{8}k^2(1-k) \end{aligned}$$

for all $n \geq n_0$. Put $a_n = (1-\frac{1}{2}(1-k) - \frac{1}{8}k^2(1-k))^n \|x_1 - p\|$ for all n. Now for the sequences $\{u_n\}_{n\geq 0}$ with $u_1 = x_1$ and $\{v_n\}_{n\geq 0}$ in (7.14), we have

$$\begin{aligned} \|w_n - p\| &= \|\gamma_n u_n + (1-\gamma_n)T u_n - p\| \\ &\leq \gamma_n \|u_n - p\| + k(1-\gamma_n)\|u_n - p\| \\ &= (1-(1-\gamma_n)(1-k))\|u_n - p\| \end{aligned}$$

and

$$\begin{aligned}
\|v_n - p\| &= \|(1-\beta_n)u_n + \beta_n T w_n - p\| \\
&\le (1-\beta_n)\|u_n - p\| + k\beta_n \|w_n - p\| \\
&\le (1-\beta_n) + k\beta_n(1-(1-\gamma_n)(1-k))\|u_n - p\| \\
&\le (1-\beta_n + k\beta_n - \beta_n(1-\gamma_n)k(1-k))\|u_n - p\| \\
&\le (1-\beta_n(1-k) - \beta_n(1-\gamma_n)k(1-k))\|u_n - p\|
\end{aligned}$$

for all $n \ge 0$. Hence,

$$\begin{aligned}
\|u_{n+1} - p\| &= \|(1-\alpha_n)u_n + \alpha_n T v_n - p\| \\
&\le (1-\alpha_n)\|u_n - p\| + k\alpha_n \|v_n - p\| \\
&\le (1-\alpha_n)\|u_n - p\| \\
&+ k\alpha_n(1-\beta_n(1-k) - \beta_n(1-\gamma_n)k(1-k))\|u_n - p\| \\
&\le ((1-\alpha_n) + k\alpha_n \\
&- \alpha_n\beta_n k(1-k) - \alpha\beta_n(1-\gamma_n)K^2(1-k))\|u_n - p\| \\
&\le (1-\alpha_n(1-k) - \alpha_n\beta_n k(1-k) - \alpha\beta_n(1-\gamma_n)K^2(1-k))\|u_n - p\|
\end{aligned}$$

for all n. Since $\alpha_n, \beta_n, \gamma_n \in (\frac{1}{2}, 1)$ for all $n \ge n_0$,

$$-k(1-k) < -\alpha_n\beta_n k(1-k) < -\frac{1}{4}k(1-k)$$

and $\frac{1}{2}k^2(1-k) < -\alpha_n\beta_n(1-\gamma_n)k^2(1-k) < 0$ for all $n \ge n_0$. Hence,

$$1-\alpha_n(1-k) - \alpha_n\beta_n k(1-k) - \alpha\beta_n(1-\gamma_n)k^2(1-k) < 1 - \frac{1}{2}(1-k) - \frac{1}{4}k(1-k)$$

for all $n \ge n_0$. Put $b_n = (1 - \frac{1}{2}(1-k) - \frac{1}{4}k(1-k))^n \|u_1 - p\|$ for all n. Then, we have

$$\lim_{n\to\infty} \frac{a_n}{b_n} = \frac{(1-\frac{1}{2}(1-k) - \frac{1}{4}k(1-k) - \frac{1}{8}k^2(1-k))^n \|x_1 - p\|}{(1-\frac{1}{2}(1-k) - \frac{1}{4}k(1-k))^n \|u_1 - p\|} = 0.$$

Thus, $\{x_n\}_{n\ge 0}$ converges faster than the sequence $\{u_n\}_{n\ge 0}$. Now, we compare the case (7.4) with the following Noor iteration case

$$\begin{cases} u_{n+1} = (1-\alpha_n)u_n + \alpha_n T v_n \\ v_n = \beta_n u_n + (1-\beta_n) T w_n \\ w_n = (1-\gamma_n)u_n + \gamma_n T u_n \end{cases} \tag{7.15}$$

for all $n \ge 0$. Note that,

$$\begin{aligned}
\|w_n - p\| &= \|(1-\gamma_n)u_n + \gamma_n T u_n - p\| \\
&\le (1-\gamma_n)\|u_n - p\| + k\gamma_n \|u_n - p\| \\
&= (1-(1-k)\gamma_n)\|u_n - p\|
\end{aligned}$$

and

$$\begin{aligned}\|v_n - p\| &= \|\beta_n u_n + (1-\beta_n)Tw_n - p\| \\ &\leq \beta_n \|u_n - p\| + k(1-\beta_n)\|w_n - p\| \\ &\leq \beta_n + k(1-\beta_n) - \beta_n\gamma_n k(1-k)\|u_n - p\| \\ &\leq (1-(1-k)(1-\beta_n) - \beta_n\gamma_n k(1-k))\|u_n - p\|\end{aligned}$$

for all $n \geq 0$. Hence,

$$\begin{aligned}\|u_{n+1} - p\| &= \|(1-\alpha_n)u_n + \alpha_n Tv_n - p\| \\ &\leq (1-\alpha_n)\|u_n - p\| + k\alpha_n\|w_n - p\| \\ &\leq (1-\alpha_n)\|u_n - p\| \\ &+ k\alpha_n(1-(1-k)(1-\beta_n) - \beta_n\gamma_n k(1-k))\|u_n - p\| \\ &\leq ((1-\alpha_n) + k\alpha_n - k(1-k)\alpha_n(1-\beta_n) \\ &- \alpha_n\beta_n\gamma_n k^2(1-k))\|u_n - p\| \\ &\leq (1-(1-k)\alpha_n - \alpha_n(1-\beta_n)k(1-k) - \alpha_n\beta_n\gamma_n k^2(1-k))\|u_n - p\|\end{aligned}$$

for all $n \geq 0$. Since $\alpha_n, \beta_n, \gamma_n \in (\frac{1}{2}, 1)$ for all $n \geq n_0$,

$$-\frac{1}{2}k(1-k) < -\alpha_n(1-\beta_n)k(1-k) < 0$$

and $-k^2(1-k) < -\alpha_n\beta_n(1-\gamma_n)k^2(1-k) < -\frac{1}{8}k^2(1-k)$ and so

$$1-(1-k)\alpha_n - \alpha_n(1-\beta_n)k(1-k) - \alpha_n\beta_n\gamma_n k^2(1-k) < 1-\frac{1}{2}(1-k) - \frac{1}{8}k^2(1-k)$$

for all $n \geq n_0$. Put $c_n = (1-\frac{1}{2}(1-k) - \frac{1}{8}k^2(1-k))^n\|u_1 - p\|$ for all n. Then, we have

$$\lim_{n\to\infty}\frac{a_n}{c_n} = \frac{(1-\frac{1}{2}(1-k) - \frac{1}{4}k(1-k) - \frac{1}{8}k^2(1-k))^n\|x_1 - p\|}{(1-\frac{1}{2}(1-k) - \frac{1}{8}k^2(1-k))^n\|u_1 - p\|} = 0.$$

Thus, $\{x_n\}_{n\geq 0}$ converges faster than the sequence $\{u_n\}_{n\geq 0}$. Now, we compare the case (7.4) with the following Noor iteration case

$$\begin{cases} u_{n+1} = (1-\alpha_n)u_n + \alpha_n Tv_n \\ v_n = \beta_n u_n + (1-\beta_n)Tw_n \\ w_n = \gamma_n u_n + (1-\gamma_n)Tu_n \end{cases} \tag{7.16}$$

for all $n \geq 0$. Note that,

$$\begin{aligned}\|w_n - p\| &= \|\gamma_n u_n + (1-\gamma_n)Tu_n - p\| \\ &\leq \gamma_n\|u_n - p\| + k(1-\gamma_n)\|u_n - p\| \\ &= (1-(1-\gamma_n)(1-k))\|u_n - p\|\end{aligned}$$

and

$$\begin{aligned}\|v_n - p\| &= \|(1-\beta_n)u_n + \beta_n T w_n - p\| \\ &\leq (1-\beta_n)\|u_n - p\| + k\beta_n \|w_n - p\| \\ &\leq (1-\beta_n) + k\beta_n(1-(1-\gamma_n)(1-k))] \|u_n - p\| \\ &\leq (1-\beta_n + k\beta_n - \beta_n(1-\gamma_n)k(1-k)) \|u_n - p\| \\ &\leq (1-\beta_n(1-k) - \beta_n(1-\gamma_n)k(1-k)) \|u_n - p\|\end{aligned}$$

and so

$$\begin{aligned}\|u_{n+1} - p\| &= \|(1-\alpha_n)u_n + \alpha_n T v_n - p\| \\ &\leq (1-\alpha_n)\|u_n - p\| + k\alpha_n \|w_n - p\| \\ &\leq (1-\alpha_n)\|u_n - p\| \\ &+ k\alpha_n(1-\beta_n(1-k) - \beta_n(1-\gamma_n)k(1-k)) \|u_n - p\| \\ &\leq (1-\alpha_n + k\alpha_n - \alpha_n\beta_n k(1-k) - \alpha_n\beta_n(1-\gamma_n)k^2(1-k)) \|u_n - p\| \\ &\leq (1-(1-k)\alpha_n - \alpha_n\beta_n k(1-k) - \alpha_n\beta_n(1-\gamma_n)k^2(1-k)) \|u_n - p\|\end{aligned}$$

for all n. Since $\alpha_n, \beta_n, \gamma_n \in (\frac{1}{2}, 1)$ for all $n \geq n_0$,

$$-k(1-k) < -\alpha_n\beta_n k(1-k) < -\frac{1}{4}k(1-k)$$

and $-\frac{1}{2}k^2(1-k) < -\alpha_n\beta_n(1-\gamma_n)k^2(1-k) < 0$ for all $n \geq n_0$. This imply that

$$1-(1-k)\alpha_n - \alpha_n\beta_n k(1-k) - \alpha_n\beta_n(1-\gamma_n)k^2(1-k) < 1 - \frac{1}{2}(1-k) - \frac{1}{4}k(1-k)$$

for all $n \geq n_0$. Put $d_n = (1-\frac{1}{2}(1-k) - \frac{1}{4}k(1-k))^n \|u_1 - p\|$ for all n. Then, we get

$$\lim_{n\to\infty} \frac{a_n}{d_n} = \frac{(1-\frac{1}{2}(1-k) - \frac{1}{4}k(1-k) - \frac{1}{8}k^2(1-k))^n \|x_1 - p\|}{(1-\frac{1}{2}(1-k) - \frac{1}{4}k(1-k))^n \|u_1 - p\|} = 0$$

and so the sequence $\{x_n\}_{n\geq 0}$ converges faster than the sequence $\{u_n\}_{n\geq 0}$. By using similar proofs, one can show that the case (7.4) is faster than the following cases of the Noor iteration method

$$\begin{cases} u_{n+1} = \alpha_n u_n + (1-\alpha_n)T v_n \\ v_n = (1-\beta_n)u_n + \beta_n T w_n \\ w_n = (1-\gamma_n)u_n + \gamma_n T u_n, \end{cases} \tag{7.17}$$

$$\begin{cases} u_{n+1} = \alpha_n u_n + (1-\alpha_n)T v_n \\ v_n = (1-\beta_n)u_n + \beta_n T w_n \\ w_n = \gamma_n u_n + (1-\gamma_n)T u_n, \end{cases} \tag{7.18}$$

$$\begin{cases} u_{n+1} = \alpha_n u_n + (1-\alpha_n)Tv_n \\ v_n = \beta_n u_n + (1-\beta_n)Tw_n \\ w_n = (1-\gamma_n)u_n + \gamma_n Tu_n \end{cases} \tag{7.19}$$

and

$$\begin{cases} u_{n+1} = \alpha_n u_n + (1-\alpha_n)Tv_n \\ v_n = \beta_n u_n + (1-\beta_n)Tw_n \\ w_n = \gamma_n u_n + (1-\gamma_n)Tu_n \end{cases} \tag{7.20}$$

for all $n \geq 0$. This completes the proof.

♣

By using similar conditions, one can show that the case (7.15) converges faster than (7.16), (7.17) converges faster than (7.19), (7.19) converges faster than (7.18) and (7.18) converges faster than (7.20).

As we know, the Agarwal iteration method could be write in the following four cases

$$\begin{cases} x_{n+1} = (1-\alpha_n)Tx_n + \alpha_n Ty_n \\ y_n = (1-\beta_n)x_n + \beta_n Tx_n, \end{cases} \tag{7.21}$$

$$\begin{cases} x_{n+1} = \alpha_n Tx_n + (1-\alpha_n)Ty_n \\ y_n = \beta_n x_n + (1-\beta_n)Tx_n, \end{cases} \tag{7.22}$$

$$\begin{cases} x_{n+1} = \alpha_n Tx_n + (1-\alpha_n)Ty_n \\ y_n = (1-\beta_n)x_n + \beta_n Tx_n \end{cases} \tag{7.23}$$

and

$$\begin{cases} x_{n+1} = (1-\alpha_n)Tx_n + \alpha_n Ty_n \\ y_n = \beta_n x_n + (1-\beta_n)Tx_n \end{cases} \tag{7.24}$$

for all $n \geq 0$. One can easily show that the case (7.46) converges faster than other ones for contractive maps. We record it as next Lemma.

Lemma 7.1

Let C be a nonempty, closed and convex subset of a Banach space X, $x_1 \in C$, $T: C \to C$ a contraction with constant $k \in (0,1)$ and p a fixed point of T. If $1-\alpha_n < \alpha_n$ and $1-\beta_n < \beta_n$ for all $n \geq n_0$ for some natural number n_0, then the case (7.21) *converges faster than* (7.22), (7.23) *and* (7.24).

Also by using a similar condition, one can show that the case (7.24) converges faster than (7.22). Similar to Theorem 7.1, we can prove that for contractive maps, one case in the Abbas iteration method converges faster than other possible cases whenever elements of the sequences $\{\alpha_n\}_{n\geq 0}$, $\{\beta_n\}_{n\geq 0}$ and $\{\gamma_n\}_{n\geq 0}$ are in $(\frac{1}{2},1)$ for sufficiently large n. Also, one can show that for contractive maps, the case (7.7) of the Thakur-Thakur-Postolache iteration method converges faster than other possible cases whenever elements of the sequences $\{\alpha_n\}_{n\geq 0}$, $\{\beta_n\}_{n\geq 0}$ and $\{\gamma_n\}_{n\geq 0}$ are in $(\frac{1}{2},1)$ for sufficiently large n. For record, we provide these results as following.

Lemma 7.2

Let C be a nonempty, closed and convex subset of a Banach space X, $u_1 \in C$, $T: C \to C$ a contraction with constant $k \in (0,1)$ and p a fixed point of T. Consider the following case in the Abbas iteration method

$$\begin{cases} u_{n+1} = (1-\alpha_n)Tw_n + \alpha_n T v_n \\ v_n = (1-\beta_n)Tu_n + \beta_n T w_n \\ w_n = (1-\gamma_n)u_n + \gamma_n T u_n \end{cases} \tag{7.25}$$

for all n. If $1-\alpha_n < \alpha_n$, $1-\beta_n < \beta_n$ and $1-\gamma_n < \gamma_n$ for sufficiently large n, then the case (7.25) converges faster than other possible cases.

Also, by using a similar conditions in the Abbas iteration method, one can show that under similar conditions, the cases

$$\begin{cases} u_{n+1} = \alpha_n T v_n + (1-\alpha_n)Tw_n \\ v_n = \beta_n T u_n + (1-\beta_n)Tw_n \\ w_n = (1-\gamma_n)u_n + \gamma_n T u_n \end{cases} \tag{7.26}$$

and

$$\begin{cases} u_{n+1} = \alpha_n T v_n + (1-\alpha_n)Tw_n \\ v_n = (1-\beta_n)Tu_n + \beta_n T w_n \\ w_n = \gamma_n u_n + (1-\gamma_n)Tu_n \end{cases} \tag{7.27}$$

converge faster than the case

$$\begin{cases} u_{n+1} = \alpha_n T v_n + (1-\alpha_n)Tw_n \\ v_n = \beta_n T u_n + (1-\beta_n)Tw_n \\ w_n = \gamma_n u_n + (1-\gamma_n)Tu_n. \end{cases} \tag{7.28}$$

Also under similar conditions, the case

$$\begin{cases} u_{n+1} = (1-\alpha_n)Tv_n + \alpha_n T w_n \\ v_n = (1-\beta_n)Tu_n + \beta_n T w_n \\ w_n = (1-\gamma_n)u_n + \gamma_n T u_n \end{cases} \tag{7.29}$$

converge faster than the cases

$$\begin{cases} u_{n+1} = (1-\alpha_n)Tv_n + \alpha_n Tw_n \\ v_n = \beta_n Tu_n + (1-\beta_n)Tw_n \\ w_n = (1-\gamma_n)u_n + \gamma_n Tu_n, \end{cases} \tag{7.30}$$

and

$$\begin{cases} u_{n+1} = (1-\alpha_n)Tv_n + \alpha_n Tw_n \\ v_n = (1-\beta_n)Tu_n + \beta_n Tw_n \\ w_n = \gamma_n u_n + (1-\gamma_n)Tu_n \end{cases} \tag{7.31}$$

and

$$\begin{cases} u_{n+1} = (1-\alpha_n)Tv_n + \alpha_n Tw_n \\ v_n = \beta_n Tu_n + (1-\beta_n)Tw_n \\ w_n = \gamma_n u_n + (1-\gamma_n)Tu_n. \end{cases} \tag{7.32}$$

Lemma 7.3

Let C be a nonempty, closed and convex subset of a Banach space X, $u_1 \in C$, $T : C \to C$ a contraction with constant $k \in (0,1)$ and p a fixed point of T. If $1-\alpha_n < \alpha_n$, $1-\beta_n < \beta_n$ and $1-\gamma_n < \gamma_n$ for sufficiently large n, then the case (7.7) in the Thakur-Thakur-Postolache iteration method converges faster than other possible cases.

Also by using similar conditions, one can show that the cases

$$\begin{cases} u_{n+1} = (1-\alpha_n)Tu_n + \alpha_n Tv_n \\ v_n = \beta_n w_n + (1-\beta_n)Tw_n \\ w_n = (1-\gamma_n)u_n + \gamma_n Tu_n \end{cases} \tag{7.33}$$

and

$$\begin{cases} u_{n+1} = (1-\alpha_n)Tu_n + \alpha_n Tv_n \\ v_n = (1-\beta_n)w_n + \beta_n Tw_n \\ w_n = \gamma_n u_n + (1-\gamma_n)Tu_n \end{cases} \tag{7.34}$$

converge faster than the case

$$\begin{cases} u_{n+1} = (1-\alpha_n)Tu_n + \alpha_n Tv_n \\ v_n = \beta_n w_n + (1-\beta_n)Tw_n \\ w_n = \gamma_n u_n + (1-\gamma_n)Tu_n. \end{cases} \tag{7.35}$$

Also the case

$$\begin{cases} u_{n+1} = \alpha_n Tu_n + (1-\alpha_n)Tv_n \\ v_n = (1-\beta_n)w_n + \beta_n Tw_n \\ w_n = (1-\gamma_n)u_n + \gamma_n Tu_n \end{cases} \tag{7.36}$$

converge faster than the cases

$$\begin{cases} u_{n+1} = \alpha_n T u_n + (1-\alpha_n) T v_n \\ v_n = \beta_n w_n + (1-\beta_n) T w_n \\ w_n = (1-\gamma_n) u_n + \gamma_n T u_n \end{cases} \tag{7.37}$$

and

$$\begin{cases} u_{n+1} = \alpha_n T u_n + (1-\alpha_n) T v_n \\ v_n = (1-\beta_n) w_n + \beta_n T w_n \\ w_n = \gamma_n u_n + (1-\gamma_n) T u_n \end{cases} \tag{7.38}$$

and

$$\begin{cases} u_{n+1} = \alpha_n T u_n + (1-\alpha_n) T v_n \\ v_n = \beta_n w_n + (1-\beta_n) T w_n \\ w_n = \gamma_n u_n + (1-\gamma_n) T u_n. \end{cases} \tag{7.39}$$

Finally, we have a similar situation for the Picard S-iteration which we record it here.

Lemma 7.4

Let C be a nonempty, closed and convex subset of a Banach space X, $x_1 \in C$, $T : C \to C$ a contraction with constant $k \in (0,1)$ and p a fixed point of T. If $1-\alpha_n < \alpha_n$ and $1-\beta_n < \beta_n$ for sufficiently large n, then the case (7.8) *in the Picard S-iteration iteration method converges faster than other possible cases.*

For years we have been taught that one iterative method is faster than the others. Note that, the above and next results shatter all our initial assumptions about the rate of convergence. Young researchers could think to providing different results. It is recommended that readers of this book sometimes think about swimming against the current of the river.

7.2.1 Comparing of different iterations methods

Now, we compare the rate of convergence of some different iteration methods for contractive maps. Our goal is to show that the rate of convergence relates to the coefficients.

Theorem 7.2

Let C be a nonempty, closed and convex subset of a Banach space X, $u_1 \in C$, $T : C \to$

C a contraction with constant $k \in (0,1)$ and p a fixed point of T. Consider the case (7.6) in the Abbas iteration method

$$\begin{cases} u_{n+1} = (1-\alpha_n)Tv_n + \alpha_n Tw_n \\ v_n = (1-\beta_n)Tu_n + \beta_n Tw_n \\ w_n = (1-\gamma_n)u_n + \gamma_n Tu_n, \end{cases}$$

the case (7.25) in the Abbas iteration method

$$\begin{cases} u_{n+1} = \alpha_n Tv_n + (1-\alpha_n)Tw_n \\ v_n = (1-\beta_n)Tu_n + \beta_n Tw_n \\ w_n = (1-\gamma_n)u_n + \gamma_n Tu_n \end{cases}$$

and the case (7.7) in the Thakur-Thakur-Postolache iteration method

$$\begin{cases} u_{n+1} = (1-\alpha_n)Tu_n + \alpha_n Tv_n \\ v_n = (1-\beta_n)w_n + \beta_n Tw_n \\ w_n = (1-\gamma_n)u_n + \gamma_n Tu_n \end{cases}$$

for all $n \geq 0$. If $1-\alpha_n < \alpha_n$, $1-\beta_n < \beta_n$ and $1-\gamma_n < \gamma_n$ for sufficiently large n, then the case (7.25) in the Abbas iteration method converges faster than the case (7.7) in the Thakur-Thakur-Postolache iteration method. Also, the case (7.7) in the Thakur-Thakur-Postolache iteration method is faster than the case (7.6) in the Abbas iteration method.

Proof: Let $\{u_n\}_{n\geq 0}$ be the sequence in the case (7.25). Then, we have

$$\begin{aligned} \|w_n - p\| &= \|(1-\gamma_n)u_n + \gamma_n Tu_n - p\| \\ &\leq (1-\gamma_n)\|u_n - p\| + k\gamma_n \|u_n - p\| \\ &= (1-(1-k)\gamma_n)\|u_n - p\|, \\ \|v_n - p\| &= \|(1-\beta_n)Tu_n + \beta_n Tw_n - p\| \\ &\leq k(1-\beta_n)\|u_n - p\| + k\beta_n \|w_n - p\| \\ &\leq k[(1-\beta_n) + \beta_n(1-(1-k)\gamma_n)]\|u_n - p\| \\ &\leq k[1-\beta_n\gamma_n(1-k)]\|u_n - p\| \end{aligned}$$

and

$$\begin{aligned} \|u_{n+1} - p\| &= \|\alpha_n Tv_n + (1-\alpha_n)Tw_n - p\| \\ &\leq \alpha_n k\|v_n - p\| + k\alpha_n \|w_n - p\| \\ &\leq \alpha_n k^2(1-\beta_n\gamma_n(1-k))\|u_n - p\| \\ &\quad + k(1-\alpha_n)(1-(1-k)\gamma_n)\|u_n - p\| \\ &\leq k[k\alpha_n - \alpha_n\beta_n\gamma_n k(1-k) + (1-\alpha_n)(1-(1-k)\gamma_n)]\|u_n - p\| \\ &= k[k\alpha_n - \alpha_n\beta_n\gamma_n k(1-k) + 1 - \alpha_n - (1-\alpha_n)\gamma_n(1-k)]\|u_n - p\| \\ &= k[1-\alpha_n(1-k) - (1-\alpha_n)\gamma_n(1-k) - \alpha_n\beta_n\gamma_n k(1-k)]\|u_n - p\| \end{aligned}$$

for all n. Since $\alpha_n, \beta_n, \gamma_n \in (\frac{1}{2}, 1)$ for sufficiently large n, we have

$$-(1-k) < -\alpha_n(1-k) < -\frac{1}{2}(1-k),$$

$-\frac{1}{2}(1-k) < -\alpha_n\gamma_n(1-k) < 0$ and $-k(1-k) < -\alpha_n\beta_n\gamma_n k(1-k) < -\frac{1}{8}k(1-k)$ for sufficiently large n. Hence,

$$1-\alpha_n(1-k)-(1-\alpha_n)\gamma_n(1-k)-\alpha_n\beta_n\gamma_n k(1-k) < 1-\frac{1}{2}(1-k)-\frac{1}{8}k(1-k)$$

for sufficiently large n. Put $a_n = k^n(1-\frac{1}{2}(1-k)-\frac{1}{8}k(1-k))^n \|u_1 - p\|$ for all n. Now, let $\{u_n\}_{n\geq 0}$ be the sequence in the case (7.7). Then, we have

$$\begin{aligned}
\|w_n - p\| &= \|(1-\gamma_n)u_n + \gamma_n T u_n - p\| \\
&\leq (1-\gamma_n)\|u_n - p\| + k\gamma_n \|u_n - p\| \\
&= (1-(1-k)\gamma_n)\|u_n - p\|,
\end{aligned}$$

$$\begin{aligned}
\|v_n - p\| &= \|(1-\beta_n)w_n + \beta_n T w_n - p\| \\
&\leq (1-\beta_n)\|u_n - p\| + k\beta_n \|w_n - p\| \\
&\leq (1-\beta_n)(1-(1-k)\gamma_n) + k\beta_n((1-(1-k)\gamma_n))\|u_n - p\| \\
&\leq [1-\beta_n(1-k)][1-\gamma_n(1-k)]\|u_n - p\|
\end{aligned}$$

and

$$\begin{aligned}
\|u_{n+1} - p\| &= \|(1-\alpha_n)Tu_n + \alpha_n T v_n - p\| \\
&\leq (1-\alpha_n)k\|u_n - p\| + k\alpha_n \|v_n - p\| \\
&\leq k(1-\alpha_n)\|u_n - p\| + k\alpha_n[1-\beta_n(1-k)][1-\gamma_n(1-k)]\|u_n - p\| \\
&\leq k[1-\alpha_n + \alpha_n(1-\beta_n(1-k))(1-\gamma_n(1-k))\|u_n - p\| \\
&\leq k[1-\alpha_n + (\alpha_n - (1-k)\beta_n\alpha_n)((1-\gamma_n)+k\gamma_n)]\|u_n - p\| \\
&\leq k[1-\alpha_n + \alpha_n(1-\gamma_n) + \alpha_n\gamma_n k - \beta_n\alpha_n(1-\gamma_n)(1-k) \\
&- \alpha_n\beta_n\gamma_n k(1-k))]\|u_n - p\| \\
&\leq k\Big[1-\alpha_n\gamma_n(1-k) - \alpha_n\beta_n(1-\gamma_n)(1-k) \\
&- \alpha_n\beta_n\gamma_n k(1-k)\Big]\|u_n - p\|
\end{aligned}$$

for all n. Since $\alpha_n, \beta_n, \gamma_n \in (\frac{1}{2}, 1)$ for sufficiently large n, we have

$$-(1-k) < -\alpha_n\gamma_n(1-k) < -\frac{1}{4}(1-k),$$

$-\frac{1}{2}(1-k) < -\alpha_n\beta_n(1-\gamma_n)(1-k) < 0$ and $-k(1-k) < -\alpha_n\beta_n\gamma_n k(1-k) < -\frac{1}{8}k(1-k)$ for sufficiently large n. Hence,

$$1-\alpha_n\gamma_n(1-k)-\alpha_n\beta_n(1-\gamma_n)(1-k)-\alpha_n\beta_n\gamma_n k(1-k) < 1-\frac{1}{4}(1-k)-\frac{1}{8}k(1-k)$$

for sufficiently large n. Put $b_n = k^n(1-\frac{1}{4}(1-k)-\frac{1}{8}k(1-k))^n \|u_1-p\|$ for all n. Then,

$$\lim_{n\to\infty}\frac{a_n}{b_n}=\frac{k^n(1-\frac{1}{2}(1-k)-\frac{1}{8}k(1-k))^n\|u_1-p\|}{k^n((1-\frac{1}{4}(1-k)-\frac{1}{8}k(1-k))^n\|u_1-p\|}=0$$

Thus, the case (7.25) in the Abbas iteration method converges faster than the case (7.7) in the Thakur-Thakur-Postolache iteration method.

Now for the case (7.6), we have

$$\begin{aligned}\|w_n-p\| &= \|(1-\gamma_n u_n+\gamma_n Tu_n-p\| \\ &\leq (1-\gamma_n)\|u_n-p\|+k\gamma_n\|u_n-p\| \\ &= (1-(1-k)\gamma_n)\|u_n-p\|,\end{aligned}$$

$$\begin{aligned}\|v_n-p\| &= \|(1-\beta_n)Tu_n+\beta_n Tw_n-p\| \\ &\leq k(1-\beta_n)\|u_n-p\|+k\beta_n\|w_n-p\| \\ &\leq k[(1-\beta_n)+\beta_n(1-(1-k)\gamma_n)]\|u_n-p\| \\ &\leq k[1-\beta_n\gamma_n(1-k)]\|u_n-p\|\end{aligned}$$

and

$$\begin{aligned}\|u_{n+1}-p\| &= \|(1-\alpha_n)Tv_n+\alpha_n Tw_n-p\| \\ &\leq (1-\alpha_n)k\|v_n-p\|+k\alpha_n\|w_n-p\| \\ &\leq (1-\alpha_n)k^2(1-\beta_n\gamma_n(1-k))\|u_n-p\| \\ &\quad +k\alpha_n(1-(1-k)\gamma_n)\|u_n-p\| \\ &\leq k[(1-\alpha_n)k-(1-\alpha_n)\beta_n\gamma_n k(1-k)+\alpha_n-\alpha_n\gamma_n(1-k)]\|u_n-p\| \\ &\leq k\Big[1-(1-\alpha_n)(1-k)-\alpha_n\gamma_n(1-k) \\ &\quad -(1-\alpha_n)\beta_n\gamma_n k(1-k)\Big]\|u_n-p\|\end{aligned}$$

for all n. Since $\alpha_n,\beta_n,\gamma_n\in(\frac{1}{2},1)$ for sufficiently large n,

$$-\frac{1}{2}(1-k)<-(1-\alpha_n)(1-k)<0, -(1-k)<-\alpha_n\gamma_n(1-k)<-\frac{1}{4}(1-k)$$

and $-\frac{1}{2}k(1-k)<-(1-\alpha_n)\beta_n\gamma_n k(1-k)<0$ for sufficiently large n. Hence,

$$1-(1-\alpha_n)(1-k)-\alpha_n\gamma_n(1-k)-(1-\alpha_n)\beta_n\gamma_n k(1-k)<1-\frac{1}{4}(1-k)$$

for sufficiently large n. Put $c_n = k^n(1-\frac{1}{4}(1-k))^n\|x_1-p\|$ for all n. Then, we have

$$\lim_{n\to\infty}\frac{b_n}{c_n}=\frac{k^n(1-\frac{1}{4}(1-k)-\frac{1}{8}k(1-k))^n\|u_1-p\|}{k^n((1-\frac{1}{4}(1-k))^n\|u_1-p\|}=0$$

and so the case (7.7) in the Thakur-Thakur-Postolache iteration method is faster than the case (7.6) in the Abbas iteration method.

♣

By using a similar proof, we can compare the Thakur-Thakur-Postolache and the Agarwal iteration methods as following.

Theorem 7.3

Let C be a nonempty, closed and convex subset of a Banach space X, $x_1 \in C$, $T: C \to C$ a contraction with constant $k \in (0,1)$ and p a fixed point of T. If $1-\alpha_n < \alpha_n$, $1-\beta_n < \beta_n$ and $1-\gamma_n < \gamma_n$ for sufficiently large n, then the case (7.7) in the Thakur-Thakur-Postolache iteration method converges faster than the case (7.5) in the Agarwal iteration method and the case (7.5) in the Agarwal iteration method is faster than the cases (7.37) and (7.38) in the Thakur-Thakur-Postolache iteration method.

Also by using similar proofs, we can compare some other iteration methods. For the record, we list those as following.

Theorem 7.4

Let C be a nonempty, closed and convex subset of a Banach space X, $x_1 \in C$, $T: C \to C$ a contraction with constant $k \in (0,1)$ and p a fixed point of T. If $1-\alpha_n < \alpha_n$, $1-\beta_n < \beta_n$ and $1-\gamma_n < \gamma_n$ for sufficiently large n, then the case (7.4) in the Abbas iteration method converges faster than the case (7.3) in the Ishikawa iteration method and the case (7.3) in the Ishikawa iteration method is faster than the cases (7.19) and (7.20) in the Abbas iteration method.

It is notable that there are some cases in which the coefficients have no effective role in the rate of convergence. By using a similar proof one can check the next result. One can obtain some similar cases. This shows us that you should care more about probability of efficiency of coefficients in rate of convergence for iteration methods.

Theorem 7.5

Let C be a nonempty, closed and convex subset of a Banach space X, $x_1 \in C$, $T: C \to C$ a contraction with constant $k \in (0,1)$, p a fixed point of T and $\alpha_n, \beta_n, \gamma_n \in (0,1)$ for all $n \geq 0$. Then, the case (7.5) in the Agarwal iteration method is faster than the case (7.2) in the Mann iteration method, the case (7.6) in the Abbas iteration *method is faster than the case* (7.2) *in the Mann iteration method, the case* (7.7) *in the Thakur-Thakur-Postolache iteration method is faster than the case* (7.2) *in the Mann iteration method, the case* (7.5) *in the Agarwal iteration method is faster than the case* (7.3) *in the Ishikawa iteration method, the case* (7.6) *in the Abbas iteration method is faster than the case* (7.3) *in the Ishikawa iteration method and the case* (7.7) *in the Thakur-Thakur-Postolache iteration method is faster than the case* (7.3) *in the Ishikawa iteration method.*

7.2.2 Examples and figures

In this section, we provide some examples to illustrate our results.

Example 90 *Let* $X = \mathbb{R}$, $C = [1,60]$, $x_0 = 20$, $\alpha_n = 0.7$ *and* $\beta_n = 0.85$ *for all* $n \geq 0$. *Define the map* $T : C \to C$ *by the formula* $T(x) = (3x+18)^{\frac{1}{3}}$ *for all* $x \in C$. *It is easy to see that T is a contraction. In the following tables, we first compare two cases of the Mann iteration method and also four cases of the Ishikawa and Agarwal iteration methods separately. From a mathematical view, one can see that that Mann* (7.9) *is more than 2.82 times faster than Mann* (7.2)*, Ishikawa* (7.10) *is more than 1.07 times faster than Ishikawa* (7.12)*, Ishikawa* (7.10) *is more than 11.33 times faster than Ishikawa* (7.11)*, Ishikawa* (7.10) *is more than 11 times faster than Ishikawa* (7.13)*, Ishikawa* (7.12) *is more than 8.75 times faster than Ishikawa* (7.13)*, Agarwal* (7.21) *is 1.22 times faster than Agarwal* (7.22)*, Agarwal* (7.21) *is 1.11 times faster than Agarwal* (7.23)*, Agarwal* (7.21) *is 1.22 times faster than Agarwal* (7.24) *and so on. We add our first CPU time in these tables for each iteration method. Also, we provide a figure in last table by using at least 30 times calculating of CPU times for our faster cases in the methods. From computer-calculation view, we get a different answer. As you can see it in the CPU time table, we found that Agarwal* (7.21) *and Mann* (7.9) *are faster than Ishikawa* (7.10)*. This note emphasizes on the difference of mathematical and computer-calculation results which have appeared many times in the literature.*

Step	*Mann* (7.2)	*Mann* (7.9)
1	15.2817976045	8.9908610772
2	11.8962912491	5.186577882
3	9.4591508761	3.8138707904
4	7.6992520365	3.305644632
5	6.4247631019	3.1152016077
6	5.4994648986	3.0434826465
7	4.8262347919	3.0164213456
8	4.3355308466	3.0062028434
9	3.977352589	3.0023431856
10	3.7156123245	3.0008851876
11	3.5241766763	3.000334402
12	3.3840675849	3.0001263293
13	3.2814716521	3.0000477244
14	3.2063163994	3.0000180292
15	3.1512468009	3.000006811
16	3.11088634	3.0000025731
17	3.0813015724	3.000000972
18	3.0596130334	3.0000003672
19	3.0437118532	3.0000001387
20	3.0320530065	3.0000000524
21	3.0235042722	3.0000000198
22	3.0172357852	3.0000000075
23	3.0126392095	3.0000000028
24	3.0092685565	
25	3.0067968355	
26	3.004984289	
.	.	
.	.	
.	.	
63	3.0000000517	
64	3.0000000379	
65	3.0000000278	
66	3.0000000204	
cpu time	0.0010	0.0007

Step	*Ishikawa* (7.10)	*Ishikawa* (7.11)	*Ishikawa* (7.13)	*Ishikawa* (7.12)
1	*6.022745179*	*17.599516463*	*6.397259957*	*17.53342562*
2	*3.55504988*	*15.542488073*	*3.725044385*	*15.426710149*
3	*3.102829451*	*13.778956254*	*3.157958555*	*13.626959863*
4	*3.019085154*	*12.266356345*	*3.034584416*	*12.089125019*
5	*3.003543432*	*10.968408676*	*3.007580568*	*10.774826445*
6	*3.000657931*	*9.854176549*	*3.001661995*	*9.651358665*
7	*3.000122164*	*8.89726621*	*3.000364402*	*8.690843013*
8	*3.000022683*	*8.07514758*	*3.000079898*	*7.86950815*
9	*3.000004212*	*7.368577613*	*3.000017518*	*7.167078769*
10	*3.000000782*	*6.76111087*	*3.000003841*	*6.566256169*
11	*3.000000145*	*6.23868412*	*3.000000842*	*6.052276815*
12	*3.000000027*	*5.789263769*	*3.000000185*	*5.612537089*
13	*3.000000005*	*5.402546543*	*3.00000004*	*5.23627424*
14	*3.000000001*	*5.069705312*	*3.000000009*	*4.91429501*
15		*4.783173147*	*3.000000002*	*4.638744748*
16		*4.536459758*		*4.402910896*
17		*4.323995342*		*4.201055645*
18		*4.14099766*		*4.028273397*
19		*3.983358785*		*3.880369278*
20		*3.847548529*		*3.753755571*
21		*3.730532022*		*3.645363373*
22		*3.629699305*		*3.55256722*
23		*3.542805134*		*3.473120743*
24		*3.467917475*		*3.405101727*
25		*3.403373393*		*3.346865184*
26		*3.347741258*		*3.297003256*
27		*3.299788327*		*3.254310946*
28		*3.258452935*		*3.217756821*
29		*3.222820611*		*3.18645797*
30		*3.192103569*		*3.159658578*
31		*3.165623078*		*3.136711605*
32		*3.142794307*		*3.117063114*
33		*3.123113286*		*3.100238856*
34		*3.1061457*		*3.0858328*
35		*3.09151723*		*3.07349731*
CPU time	*0.00086*	*0.0035*	*0.0016*	*0.0085*

Step	*Agarwal* (7.21)	*Agarwal* (7.22)	*Agarwal* (7.24)	*Agarwal* (7.23)
1	*3.663643981*	*4.231276342*	*4.038158759*	*4.165185499*
2	*3.034148064*	*3.125898552*	*3.08652991*	*3.112771857*
3	*3.001785887*	*3.013368608*	*3.007415671*	*3.011314821*
4	*3.000093479*	*3.001425297*	*3.000637055*	*3.001139398*
5	*3.000004893*	*3.000152024*	*3.000054738*	*3.000114779*
6	*3.000000256*	*3.000016216*	*3.000004703*	*3.000011563*
7	*3.000000013*	*3.00000173*	*3.000000404*	*3.000001165*
8	*3.000000001*	*3.000000184*	*3.000000035*	*3.000000117*
9	*3*	*3.00000002*	*3.000000003*	*3.000000012*
10		*3.000000002*	*3*	*3.000000001*
11		*3*		*3*
CPU time	*0.00095*	*0.0034*	*0.0011*	*0.0011*

Finally, one can find the related CPU time of this Example as Figure 1 in [39].

The next example illustrates Lemma 7.2.

Example 91 Let $X = \mathbb{R}$, $C = [0, 2000]$, $x_0 = 1000$, $\alpha_n = 0.85$, $\beta_n = 0.65$ and $\gamma_n = 0.75$ for all $n \geq 0$. Define the map $T : C \to C$ by $T(x) = \sqrt[3]{x^2}$ for all $x \in C$. The following table shows us that Abbas (7.25) converges faster than other cases, Abbas (7.26) is 1.1 times faster than Abbas (7.28), Abbas (7.27) is 1.05 times faster than Abbas (7.28), Abbas (7.29) is 1.04 times faster than Abbas (7.30) and 1.3 times faster than Abbas (7.31) and Abbas (7.32). One can get similar results about the difference between the mathematical and computer-calculating views for this example.

Step	Abbas (7.25)	Abbas (7.26)	Abbas (7.27)	Abbas (7.28)	Abbas (7.29)	Abbas (7.30)	Abbas (7.31)	Abbas (7.32)
1	20.933947	23.074444	29.706456	30.294581	42.622758	43.000492	74.725586	74.829373
2	3.501533	3.915728	4.912771	5.052334	6.872931	6.975246	14.057893	14.097781
3	1.650123	1.789347	2.07514	2.127569	2.605814	2.644699	4.919453	4.938021
4	1.218545	1.278689	1.392374	1.417334	1.596596	1.615195	2.581994	2.592232
5	1.080705	1.109014	1.161005	1.174049	1.254442	1.264388	1.750749	1.757015
6	1.030883	1.044439	1.069469	1.076461	1.115609	1.121158	1.389425	1.39348
7	1.011982	1.018426	1.030642	1.034379	1.054109	1.057231	1.212285	1.214975
8	1.004673	1.007695	1.013649	1.015622	1.025684	1.02743	1.119022	1.120821
9	1.001827	1.003223	1.006106	1.007132	1.012273	1.013239	1.067815	1.069015
10	1.000715	1.001351	1.002737	1.003264	1.005884	1.006411	1.038999	1.039794
11	1.00028	1.000567	1.001228	1.001495	1.002825	1.00311	1.022548	1.02307
12	1.000109	1.000238	1.000551	1.000685	1.001357	1.00151	1.013078	1.013417
13	1.000043	1.0001	1.000247	1.000314	1.000653	1.000733	1.007599	1.007818
14	1.000017	1.000042	1.000111	1.000144	1.000314	1.000356	1.00442	1.00456
15	1.000007	1.000018	1.00005	1.000066	1.000151	1.000173	1.002572	1.002661
16	1.000003	1.000007	1.000022	1.00003	1.000073	1.000084	1.001498	1.001554
17	1.000001	1.000003	1.00001	1.000014	1.000035	1.000041	1.000872	1.000907
18		1.000001	1.000005	1.000006	1.000017	1.00002	1.000508	1.00053
19		1.000001	1.000002	1.000003	1.000008	1.00001	1.000296	1.00031
20		1	1.000001	1.000001	1.000004	1.000005	1.000172	1.000181
21		1	1	1.000001	1.000002	1.000002	1.0001	1.000106
22		1	1	1	1.000001	1.000001	1.000058	1.000062
23		1	1	1	1	1.000001	1.000034	1.000036
24			1	1	1	1	1.00002	1.000021
25				1	1	1	1.000012	1.000012
26				1	1	1	1.000007	1.000007
27					1	1	1.000004	1.000004
28					1	1	1.000002	1.000002
29					1	1	1.000001	1.000001
30							1.000001	1.000001
31							1	1
32							1	1
33							1	1
34							1	1
35							1	1
36							1	1
37							1	1

Next example illustrates Theorem 7.2.

Example 92 Let $X=\mathbb{R}, C=[1,60], x_0=40, \alpha_n=0.9, \beta_n=0.6$ and $\gamma_n=0.8$ for all $n\geq 0$. Define the map $T: C\to C$ by $T(x)=\sqrt{x^2-8x+40}$ for all $x\in C$ (see [102]). The following table shows Abbas (7.25) converges 1.09 times faster than the Thakur-Thakur-Postolache (7.7) and Thakur-Thakur-Postolache (7.7) is 1.16 times faster than Abbas (7.6) from mathematical view. Again, we get different results from the computer-calculating view by checking last presented table.

Step	Thakur (7.7)	Abbas (7.6)	Abbas (7.25)
1	31.77453587	33.18158852	31.22317681
2	23.81196041	26.52340588	22.75386567
3	16.33019829	20.11920431	14.88031305
4	9.89958703	14.1634562	8.4317634
5	5.97706669	9.11456867	5.36305686
6	5.07407177	5.96019967	5.01260299
7	5.00409402	5.0925653	5.00037245
8	5.00022019	5.00645474	5.00001094
9	5.00001182	5.00043527	5.00000032
10	5.00000063	5.00002928	5.00000001
11	5.00000003	5.00000197	5
12	5	5.00000013	
13	5	5.00000001	
14		5	
CPU time	0.0012	0.0012	0.0009

Finally, one can find the convergence behaviour of the iteration methods Thakur (7.7), Abbas (7.6) and Abbas (7.25) related to this Example as Figure 2 in [39]. Also, one can find the related CPU time of this Example as Figure 3 in [39].

Next example shows that choosing of the coefficients is very important in rate of convergence of an iteration method.

Example 93 *Let* $X=\mathbb{R}$, $C=[0,30]$ *and* $x_0=20$. *Define the map* $T:\mathbb{R}\to\mathbb{R}$ *by* $T(x)=\frac{x}{2}+1$ *for all* $x\in C$. *Consider the following coefficients separately in Thakur-Thakur-Postolache* (7.7).
(a) $\alpha_n=\beta_n=\gamma_n=1-\frac{1}{(n+1)^{10}}$,
(b) $\alpha_n=\beta_n=\gamma_n=1-\frac{1}{n+1}$,
(c) $\alpha_n=\beta_n=\gamma_n=1-\frac{1}{(n+1)^{\frac{1}{2}}}$,
(d) $\alpha_n=\beta_n=\gamma_n=1-\frac{1}{(n+1)^{\frac{1}{3}}}$
for all $n\geq 0$. *The following table shows that Thakur-Thakur-Postolache* (7.7) *with coefficients (a) is 1.25 times faster than Thakur-Thakur-Postolache* (7.7) *with coefficients (b), Thakur-Thakur-Postolache* (7.7) *with coefficients (a) is*

1.6 times faster than Thakur-Thakur-Postolache (7.7) *with coefficients* (*c*) *and Thakur-Thakur-Postolache* (7.7) *with coefficients* (*a*) *is 2.16 times faster than Thakur-Thakur-Postolache* (7.7) *with coefficients* (*d*). *This note satisfies for another iteration methods of course from mathematical view. Here, we find a little different computer-calculating result for CPU time table of this example which one can check it in last figure.*

Step	(*a*)	(*b*)	(*c*)	(*d*)
1	*4.2609841803*	*9.03125*	*10.2844561595*	*10.8540663001*
2	2.2826469537	4.2135416667	5.4804739263	6.2632682688
3	2.0353310377	2.6009419759	3.3595601275	4.0142167756
4	2.004416382	2.1466298421	2.5007642765	2.9360724936
5	2.0005520478	2.0330086855	2.1756764587	2.4287794141
6	2.000069006	2.0069770545	2.0591364356	2.1939030837
7	2.0000086257	2.0014018838	2.0192087915	2.0866824323
8	2.0000010782	2.0002701847	2.0060472121	2.0383477219
9	2.0000001348	2.0000502881	2.0018515929	2.0168034488
10	2.0000000168	2.0000090866	2.0005529869	2.0072985299
11	2.0000000021	2.0000016005	2.0001614712	2.0031443476
12	2.0000000003	2.0000002757	2.0000461907	2.0013443922
13		2.0000000466	2.0000129668	2.0005707329
14		2.0000000077	2.0000035774	2.0002406784
15		2.0000000013	2.0000009712	2.0001008564
16			2.0000002597	2.0000420126
17			2.0000000685	2.0000174019
18			2.0000000178	2.0000071693
19			2.0000000046	2.0000029385
20			2.0000000012	2.0000011985
21				2.0000004865
22				2.0000001966
23				2.0000000791
24				2.0000000317
25				2.0000000127
26				2.000000005
CPU time	*0.0013*	*0.0014*	*0.0015*	*0.0017*

Finally, one can find the related CPU time of this Example as Figure 4 in [39].

7.3 The iteration methods for quasi-contractive mappings

Let (X,d) be a metric space and T a self-map on X. We say that T is a quasi-contraction whenever there exists $k \in (0,1)$ such that

$$d(Tx,Ty) \leq k\max\{d(x,y),d(x,Tx),d(y,Ty),d(x,Ty),d(y,Tx)\}$$

for all $x,y \in X$. In this section, we deal with quasi-contraction mappings only. Let $(X,\|.\|)$ be a normed space, $x_0 \in X$, T a selfmap on X and $\{\alpha_n\}_{n\geq 0}$, $\{\beta_n\}_{n\geq 0}$ and $\{\gamma_n\}_{n\geq 0}$ three sequences in $(0,1)$. As we mentioned, the Mann iteration is given by

$$x_{n+1} = (1-\alpha_n)x_n + \alpha_n T x_n \tag{7.40}$$

for all $n \geq 0$. Note that, we can rewrite it as

$$x_{n+1} = \alpha_n x_n + (1-\alpha_n)Tx_n \tag{7.41}$$

for all $n \geq 0$. Let $\{u_n\}_{n\geq 0}$ and $\{v_n\}_{n\geq 0}$ be two fixed point iteration procedures that converge to the same fixed point p and $\|u_n - p\| \leq a_n$ and $\|v_n - p\| \leq b_n$ for all $n \geq 0$. If the non-negative sequences $\{a_n\}_{n\geq 0}$ and $\{b_n\}_{n\geq 0}$ converge to a and b respectively and $\lim_{n\to\infty} \frac{\|a_n - a\|}{\|b_n - b\|} = 0$, then as said before, we say that $\{u_n\}_{n\geq 0}$ converges faster than $\{v_n\}_{n\geq 0}$ to p.

Proposition 7.3

Let C be a non-empty, closed and convex subset of a Banach space X, $x_0 \in C$, $T : C \to C$ a quasi-contraction with constant $k \in (0,1)$ and p a fixed point of T. If there exists a natural number n_0 such that $1-\alpha_n < \alpha_n$ for all $n \geq n_0$ or equivalently $\{\alpha_n\}_{n\geq n_0}$ is a sequence in $(\frac{1}{2},1)$, then the iteration (7.40) is faster than the iteration (7.41).

Proof: Let $\{x_n\}_{n\geq 1}$ be the sequence in case (7.40) of the Mann iteration. Then, we have

$$\begin{aligned}
\|x_{n+1} - p\| &\leq (1-\alpha_n)\|x_n - p\| + \alpha_n \|Tx_n - p\| \\
&\leq (1-\alpha_n)\|x_n - p\| + \alpha_n k \max\{\|x_n - p\|, \|Tx_n - x_n\| \\
&\qquad \|Tp - p\|, \|Tx_n - p\|, \|Tp - x_n\|\} \\
&\leq (1-\alpha_n)\|x_n - p\| + \alpha_n k \max\{\|x_n - p\|, \|Tx_n - x_n\|\}
\end{aligned}$$

for all n. Hence, $\|x_{n+1} - p\| \leq (1-\alpha_n(1-k))\|x_n - p\|$ or

$$\|x_{n+1} - p\| \leq (1-\alpha_n(1-k))\|x_n - Tx_n\|$$

for infinitely many n. By following the method of [104], we know that the sequences $\{x_n\}$, $\{y_n\}$, $\{Tx_n\}$ and $\{Ty_n\}$ are bounded. For each n, put

$$r_n := \max\{\sup_{j\geq n} \|x_n - Tx_j\|, \sup_{j\geq n} \|x_n - Ty_j\|\}.$$

Then,

$$\|x_n - Tx_j\| \leq (1-(1-k)\alpha_{n-1})r_{n-1}$$

and

$$\|x_n - Ty_j\| \leq (1-(1-k)\alpha_{n-1})r_{n-1}.$$

By putting $j = n$, we obtain

$$\begin{aligned} \|x_n - Tx_n\| &\leq (1-(1-k)\alpha_{n-1})r_{n-1} \\ &= (1-(1-k)\alpha_{n-1})\max\{\|x_{n-1} - Tx_{n-1}\|, \|x_{n-1} - Tx_n\|, \\ &\quad \|x_{n-1} - Ty_{n-1}\|, \|x_{n-1} - Ty_n\|\} \end{aligned}$$

and so $r_n \leq (1-(1-k)\alpha_{n-1})r_{n-1}$ for all n. Put $H_n := \max\{r_n, \|x_n - p\|\}$ for all n. Hence, $r_n \leq H_n$ for all n. Since $\{x_n\}$ converges to p, $H_n \leq H_{n-1}$. By using above inequalities, we get $\|x_{n+1} - p\| \leq (1-\alpha_n(1-k))r_n \leq (1-\alpha_n(1-k))H_n$ for all n. Since $\alpha_n \in (\frac{1}{2}, 1)$ for all $n \geq n_0$, $-\alpha_n(1-k) \leq -\frac{1}{2}(1-k)$ and so $(1-\alpha_n(1-k)) \leq (1-\frac{1}{2}(1-k))$ for all $n \geq n_0$. Now, put $a_n = (1-\frac{1}{2}(1-k))^n H_0$ for all $n \geq n_0$. Let $\{x_n\}_{n\geq 1}$ be the sequence in case (7.41) of the Mann iteration. Then, we have

$$\begin{aligned} \|x_{n+1} - p\| &\leq \alpha_n \|x_n - p\| + (1-\alpha_n)\|Tx_n - p\| \\ &\leq \alpha_n \|x_n - p\| + (1-\alpha_n)k\max\{\|x_n - p\|, \|Tx_n - x_n\| \\ &\quad \|Tp - p\|, \|Tx_n - p\|, \|Tp - x_n\|\} \\ &\leq \alpha_n \|x_n - p\| + (1-\alpha_n)k\max\{\|x_n - p\|, \|Tx_n - x_n\|\} \end{aligned}$$

for all n. Again, put $r_n := \max\{\sup_{j\geq n}\|x_n - Tx_j\|, \sup_{j\geq n}\|x_n - Ty_j\|\}$ for all n. Similarly, for $j = n$ we get

$$\begin{aligned} \|x_n - Tx_n\| &\leq (1-(1-k)(1-\alpha_{n-1}))r_{n-1} \\ &= (1-(1-k)(1-\alpha_{n-1}))\max\{\|x_{n-1} - Tx_{n-1}\|, \|x_{n-1} - Tx_n\|, \\ &\quad \|x_{n-1} - Ty_{n-1}\|, \|x_{n-1} - Ty_n\|\} \end{aligned}$$

and so $r_n \leq (1-(1-k)(1-\alpha_{n-1}))r_{n-1}$ for all n. Put $H_n := \max\{r_n, \|x_n - p\|\}$ for all n. Then, $r_n \leq H_n$, $H_n \leq H_{n-1}$ and $\|x_{n+1} - p\| \leq (1-(1-\alpha_n)(1-k))H_n$ for all n. Since $\alpha_n \in (\frac{1}{2}, 1)$ for all $n \geq n_0$, $-(1-\alpha_n)(1-k) \leq 0$ and so $1-(1-\alpha_n)(1-k) \leq 1$ for all $n \geq n_0$. Now, put $b_n = H_0$ for all $n \geq n_0$. Since

$$\lim \frac{a_n}{b_n} = \lim \frac{(1-\frac{1}{2}(1-k))^n H0}{H0} = 0,$$

the iteration (7.40) is faster than the iteration (7.41).

♣

By using a similar proof and putting $a_n = (1 - \frac{1}{2}(1-k) - \frac{1}{4}k(1-k))^n H_0$, $b_n = d_n = H_0$, $c_n = (1-\frac{1}{2}(1-k))^n H_0$ for all $n \geq n_0$, we get $\lim \frac{a_n}{b_n} = \lim \frac{a_n}{c_n} = \lim \frac{a_n}{d_n} = 0$. Here, $H_0 := \max\{r_0, \|x_0 - p\|\}$. This implies next result.

Proposition 7.4

Let C be a non-empty, closed and convex subset of a Banach space X, $x_0 \in C$, $T : C \to C$ a quasi-contraction with constant $k \in (0,1)$ and p a fixed point of T. Numerate whole cases of the Ishikawa iterative method by

$$\begin{cases} x_{n+1} = (1-\alpha_n)x_n + \alpha_n T y_n \\ y_n = (1-\beta_n)x_n + \beta_n T x_n, \end{cases} \tag{7.42}$$

$$\begin{cases} x_{n+1} = \alpha_n x_n + (1-\alpha_n) T y_n \\ y_n = \beta_n x_n + (1-\beta_n) T x_n, \end{cases} \tag{7.43}$$

$$\begin{cases} x_{n+1} = \alpha_n x_n + (1-\alpha_n) T y_n \\ y_n = (1-\beta_n)x_n + \beta_n T x_n \end{cases} \tag{7.44}$$

and

$$\begin{cases} x_{n+1} = (1-\alpha_n)x_n + \alpha_n T y_n \\ y_n = \beta_n x_n + (1-\beta_n) T x_n \end{cases} \tag{7.45}$$

for all $n \geq 0$. If the exists a natural number n_0 such that $1-\alpha_n < \alpha_n$ and $1-\beta_n < \beta_n$ for all $n \geq n_0$, then the case (7.42) converges faster than others.

Here, we provide the next result without proof. In fact, its proof is similar to the proof of the last results.

Proposition 7.5

Let D be a non-empty, closed and convex subset of a Banach space E, $x_0 \in C$, $T : D \to D$ a quasi contraction with constant $k \in (0,1)$ and p a fixed point of T. Consider the case (7.4) of the Noor iterative method

$$\begin{cases} x_{n+1} = (1-\alpha_n)x_n + \alpha_n T y_n, \\ y_n = (1-\beta_n)x_n + \beta_n T z_n, \\ z_n = (1-\gamma_n)x_n + \gamma_n T x_n \end{cases}$$

for all $n \geq 0$. If $1-\alpha_n < \alpha_n$, $1-\beta_n < \beta_n$ and $1-\gamma_n < \gamma_n$ for all $n \geq n_0$, then the iteration (7.4) is faster than other possible cases.

As you know, the Agarwal iterative method could be written in the following four cases

$$\begin{cases} x_{n+1} = (1-\alpha_n)Tx_n + \alpha_n Ty_n \\ y_n = (1-\beta_n)x_n + \beta_n Tx_n, \end{cases} \tag{7.46}$$

$$\begin{cases} x_{n+1} = \alpha_n Tx_n + (1-\alpha_n)Ty_n \\ y_n = \beta_n x_n + (1-\beta_n)Tx_n, \end{cases} \tag{7.47}$$

$$\begin{cases} x_{n+1} = \alpha_n Tx_n + (1-\alpha_n)Ty_n \\ y_n = (1-\beta_n)x_n + \beta_n Tx_n \end{cases} \tag{7.48}$$

and

$$\begin{cases} x_{n+1} = (1-\alpha_n)Tx_n + \alpha_n Ty_n \\ y_n = \beta_n x_n + (1-\beta_n)Tx_n \end{cases} \tag{7.49}$$

for all $n \geq 0$. One can easily show that the case (7.46) converges faster than other ones for quasi-contractive maps. Similar to the above results, we can prove that for quasi-contractive maps one case in the Abbas iterative method converges faster than other possible cases whenever elements of the sequences $\{\alpha_n\}_{n\geq 0}$, $\{\beta_n\}_{n\geq 0}$ and $\{\gamma_n\}_{n\geq 0}$ belong to $(\frac{1}{2}, 1)$ for sufficiently large n. Also, one can show that for quasi-contractive maps the case (7.7) of the Thakur-Singh-Postolache iterative method converges faster than other possible cases whenever elements of the sequences $\{\alpha_n\}_{n\geq 0}$, $\{\beta_n\}_{n\geq 0}$ and $\{\gamma_n\}_{n\geq 0}$ are in $(\frac{1}{2}, 1)$ for sufficiently large n. Finally, one can show that the case (7.8) in the Picard S-iteration iterative method converges faster than other possible cases under similar conditions for quasi-contractive maps. Here we give an example.

Example 94 *Let $X = \mathbb{R}$, $C = [0, 100]$, $x_0 = 5$, $\alpha_n = \beta_n = \gamma_n = 0.7$, for all $n \geq 0$. Define the map $T: C \to C$ by $T(x) = \frac{3}{5}x - 1$ whenever $x \leq 0$, $T(x) = \frac{1}{8}x - 1$ whenever $0 < x < 1$ and $T(x) = -4x - 1$ whenever $x \geq 1$. It is easy to check that T is a quasi-contraction map. In next tables, we compare two cases of the Mann iterative method and four cases of the Ishikawa iterative method separately. The tables show us the difference of rate of convergence in all cases of these iteration methods.*

Step	Mann (7.40)	Mann (7.40)
1	-13.2	-2.8
2	-10.204	-2.764
3	-8.04688	-2.73232
4	-6.493754	-2.704442
5	-5.375503	-2.679909
.	.	
.	.	
.	.	
14	-2.649522	-2.556937
15	-2.607656	-2.550105
16	-2.577512	-2.544092
17	-2.555809	-2.538801
18	-2.540182	-2.534145
19	-2.528931	-2.530048
.	.	
.	.	
.	.	
49	-2.500002	-2.500649
50	-2.500001	-2.500571
51	-2.500001	-2.500503
52	-2.500001	-2.500442
53	-2.5	-2.500389
.	.	.
.	.	.
.	.	.
95		-2.500002
96		-2.500002
97		-2.500001
98		-2.500001
99		-2.500001
100		-2.500001
cpu time	0.016	0.01

Table1. Comparing Mann cases

Step	Ishikawa (7.42)	Ishikawa (7.43)	Ishikawa (7.44)	Ishikawa (7.45)
1	-4.744	2.696	0.824	-0.376
2	-3.8517856	1.29056	0.264512	-1.0777696
3	-3.31431565	0.4332416	-0.20433894	-1.54767452
4	-2.99054374	0.02125261	-0.59551959	-1.86232286
5	-2.79550355	-0.33568025	-0.92004305	-2.07301139
6	-2.67801134	-0.64214793	-1.18926771	-2.21408843
7	-2.60723403	-0.90521978	-1.4126165	-2.30855361
8	-2.56459778	-1.13104066	-1.59790664	-2.3718075
9	-2.5389137	-1.3248853	-1.75162335	-2.4141623
10	-2.52344161	-1.49128154	-1.87914673	-2.44252308
11	-2.51412123	-1.63411608	-1.98494013	-2.46151345
12	-2.50850663	-1.75672524	-2.07270633	-2.47422941
13	-2.50512439	-1.86197295	-2.14551717	-2.48274401
14	-2.50308693	-1.95231758	-2.20592105	-2.48844539
15	-2.50185957	-2.02986941	-2.2560321	-2.49226303
16	-2.5011202	-2.0964399	-2.29760423	-2.49481933
17	-2.50067481	-2.15358401	-2.33209247	-2.49653102
18	-2.50040651	-2.20263651	-2.36070391	-2.49767717
19	-2.50024488	-2.24474318	-2.38443997	-2.49844463
20	-2.50014752	-2.28088755	-2.4041314	-2.49895853
.	.	.	.	.
.	.	.	.	.
.	.	.	.	.
39	-2.50000001	-2.48795598	-2.49724461	-2.49999949
40		-2.48966141	-2.49771413	-2.49999966
41		-2.49112536	-2.49810364	-2.49999977
42		-2.49238201	-2.49842678	-2.49999985
43		-2.49346071	-2.49869486	-2.4999999
44		-2.49438668	-2.49891725	-2.49999993
45		-2.49518152	-2.49910175	-2.49999995
46		-2.49586382	-2.49925482	-2.49999997
47		-2.4964495	-2.4993818	-2.49999998
48		-2.49695225	-2.49948714	-2.49999999
49		-2.49738381	-2.49957453	
.	.	.	.	.
.	.	.	.	.
.	.	.	.	.
90		-2.499995	-2.4999998	
91		-2.49999571	-2.49999983	
92		-2.49999632	-2.49999986	
93		-2.49999684	-2.49999989	
94		-2.49999729	-2.4999999	
95		-2.49999767	-2.49999992	
96		-2.499998	-2.49999993	
97		-2.49999828	-2.49999995	
98		-2.49999853	-2.49999995	
99		-2.49999873		
100		-2.49999891		
CPU time	0.0008	0.0019	0.0011	0.0018

Table 2. Comparing Ishikawa cases

7.3.1 Comparing of different iterations methods

In this section, we compare the rate of convergence of some different iteration methods for quasi-contractions. Our goal is to show that the rate of convergence relates to the coefficients.

Proposition 7.6

Let D be a non-empty, closed and convex subset of a Banach space E, $p_0 \in D$, T :

$D \to D$ a quasi-contraction with contractive constant $k \in (0,1)$ and p a fixed point of T. If $\alpha_n, \beta_n, \gamma_n \in (\frac{1}{2}, 1)$ for sufficiently large n, then the Picard iteration converges faster than the Mann iteration (7.40).

Proof: By using Theorem 2.1 in [86], we know $\|T^{n+1}x_0 - p\| \le \frac{k^{n+1}}{(1-k)^2}\|x_0 - p\|$ for all n. Put $a_n = \frac{k^{n+1}}{(1-k)^2}\|x_0 - p\|$ for all n. Let $\{x_n\}_{n\ge 0}$ be the iterative sequence in the Mann iteration (7.40). Then, we have

$$\begin{aligned}\|x_{n+1} - p\| &\le (1-\alpha_n)\|x_n - p\| + \alpha_n \|Tx_n - p\| \\ &\le (1-\alpha_n)\|x_n - p\| + \alpha_n k \max\{\|x_n - p\|, \|Tx_n - x_n\| \\ &\quad + \|Tp - p\| + \|Tx_n - p\| + \|Tp - x_n\| \\ &\le (1-\alpha_n)\|x_n - p\| + \alpha_n k \max\{\|x_n - p\|, \|Tx_n - x_n\|\}\end{aligned}$$

for all n. If $\max\{\|x_n - p\|, \|Tx_n - x_n\|\} = \|x_n - p\|$, then

$$\|x_{n+1} - p\| \le (1 - \alpha_n(1-k))\|x_0 - p\|.$$

If $\max\{\|x_n - q\|, \|Tx_n - x_n\|\} = \|Tx_n - x_n\|$, then

$$\|x_{n+1} - q\| \le \frac{1-\alpha_n + k}{1-k}\|x_0 - p\|.$$

Note that, $1 - \alpha_n + \alpha_n k \le 1 - \alpha_n + k \le \frac{1-\alpha_n+k}{1-k}$ for sufficiently large n. Put

$$b_n = (1 - \alpha_n(1-k))^{n+1}\|x_0 - p\|$$

for all n. Then, $\lim_{n\to\infty} \frac{a_n}{b_n} = \frac{k^{n+1}}{(1-k)^2(1-\alpha_n(1-k))^{n+1}} = 0$ and so the Picard iteration converges faster than the Mann iteration (7.40).

♣

By using a similar method, one can show that the Ishikawa iteration (7.3) for quasi-contractions converges faster than the Mann iteration (7.40) under similar assumptions in the last result. We record it as follows.

Theorem 7.6

Let D be a non-empty, closed and convex subset of a Banach space E, $x_1 \in D$, $T : D \to D$ a quasi-contraction with contractive constant $k \in (0,1)$ and p a fixed point of T. If $\alpha_n, \beta_n, \gamma_n \in (\frac{1}{2}, 1)$ for sufficiently large n, then the Ishikawa iteration (7.3) *for quasi-contractions converges faster than the Mann iteration* (7.40).

Proof: By proof of proposition (7.3), put $a_n = (1 - \frac{1}{2}(1-k))^n H_0$ for all n. Let $\{x_n\}_{n\ge 0}$ be the sequence in the Ishikawa iteration. Then, we have

$$\begin{aligned}\|y_n - p\| &\le (1-\beta_n)\|x_n - p\| + \beta_n \|Tx_n - p\| \\ &\le (1-\beta_n)\|x_n - p\| + \beta_n k \max\{\|x_n - p\|, \|Tx_n - x_n\| \\ &\quad \|Tp - p\|, \|Tx_n - p\|, \|Tp - x_n\|\} \\ &\le (1-\beta_n)\|x_n - p\| + \beta_n k \max\{\|x_n - p\|, \|Tx_n - x_n\|\}\end{aligned}$$

and so

$$\|y_n - p\| \leq (1 - \beta_n(1-k))\,\|x_n - p\| \tag{7.50}$$

or

$$\|y_n - p\| \leq (1 - \beta_n(1-k))\,\|x_n - Tx_n\| \tag{7.51}$$

for all n. By using the related Lemma in [104], we get that the sequences $\{x_n\}$, $\{y_n\}$, $\{Tx_n\}$ and $\{Ty_n\}$ are bounded. Put, $r_n := \max\{\sup\|x_n - Tx_j\|, \sup\|x_n - Ty_j\|\}$ for all n. By using definition of $\{x_n\}$, we find

$$\|x_n - Tx_j\| \leq (1-(1-k)\beta_{n-1})r_{n-1}$$

and

$$\|x_n - Ty_j\| \leq (1-(1-k)\beta_{n-1})r_{n-1}$$

for all n because T is quasi-contraction. Hence,

$$\begin{aligned}\|x_n - Tx_n\| &\leq (1-(1-k)\beta_{n-1})r_{n-1}\\ &= (1-(1-k)\beta_{n-1})\max\{\|x_{n-1} - Tx_{n-1}\|, \|x_{n-1} - Tx_n\|,\\ &\quad \|x_{n-1} - Ty_{n-1}\|, \|x_{n-1} - Ty_n\|\}\end{aligned}$$

and so $r_n \leq (1-(1-k)\beta_{n-1})r_{n-1}$. Put $H_n := \max\{r_n, \|x_n - p\|\}$ for all n. Since $\{x_n\}$ converges to p, $H_n \leq H_{n-1}$ (see [104]). By using (7.50) and (7.51), we get

$$\|y_n - p\| \leq (1-\beta_n(1-k))H_n \tag{7.52}$$

for all n. Thus,

$$\begin{aligned}\|x_{n+1} - p\| &\leq (1-\alpha_n)\,\|x_n - p\| + \alpha_n\,\|Ty_n - p\|\\ &\leq (1-\alpha_n)\,\|x_n - p\| + \alpha_n\,\|Ty_n - p\|\\ &\leq (1-\alpha_n)\,\|x_n - p\| + \alpha_n k\max\{\|y_n - p\|, \|Ty_n - y_n\|\\ &\quad \|Tp - p\|, \|Tx_n - p\|, \|Tp - x_n\|\}\\ &\leq (1-\alpha_n)\,\|x_n - p\| + \alpha_n k\max\{\|y_n - p\|, \|Ty_n - y_n\|\}\\ &\leq (1-\alpha_n)\,\|x_n - p\| + \alpha_n k r_n\\ &\leq (1-\alpha_n H_n + k\alpha_n(1-\beta_n(1-k))r_{n-1}\\ &\leq (1-\alpha_n + k\alpha_n(1-\beta_n(1-k))H_{n-1}\\ &\leq (1-\alpha_n(1-k) - \alpha_n\beta_n k(1-k))H_{n-1}\end{aligned}$$

Since $\alpha_n, \beta_n \in (\frac{1}{2}, 1)$ for sufficiently large n, $\frac{1}{2}(1-k) \leq \alpha_n(1-k)$ and

$$\frac{1}{4}k(1-k) \leq \alpha_n\beta_n k(1-k).$$

Hence, $\frac{1}{2}(1-k) + \frac{1}{4}k(1-k) \leq \alpha_n(1-k) + \alpha_n\beta_n k(1-k)$ and so

$$-\alpha_n(1-k) - \alpha_n\beta_n k(1-k) \leq -\frac{1}{2}(1-k) - \frac{1}{4}k(1-k).$$

Thus, $1-\alpha_n k(1-k)-\alpha_n\beta_n k(1-k)\leq 1-\frac{1}{2}(1-k)-\frac{1}{4}k(1-k)$ for sufficiently large n. Put $b_n=(1-\frac{1}{2}(1-k)-\frac{1}{4}k(1-k))^n H_0$. Note that,

$$\lim_{n\to\infty}\frac{b_n}{a_n}=\frac{(1-\frac{1}{2}(1-k)-\frac{1}{4}k(1-k))^n H_0}{(1-\frac{1}{2}(1-k))^n H_0}=0$$

and so the Ishikawa iteration (7.3) for quasi-contractions converges faster than the Mann iteration (7.40).

♣

By using similar assumptions and a similar method, one can show that the Noor iteration (7.4) for quasi-contractions converges faster than the Ishikawa iteration (7.3) and the Agarwal iteration (7.5) for quasi-contractions converges faster than the Noor iteration (7.4). Also by using a similar proof of our results in the last section, one can prove that the Abbas iteration (7.6) for quasi-contractions converges faster than the Thakur, Singh and Postolache iteration (7.7) and the Thakur, Singh and Postolache iteration (7.7) for quasi-contractions converges faster than the Abbas iteration (7.6).

Example 95 *Let $X=\mathbb{R}$, $D=[0,1]$, $x_0=0.7$, $\alpha_n=0.8$, $\beta_n=0.65$ and $\gamma_n=0.7$ for all $n\geq 0$. Consider the map $T:D\to D$ defined by $Tx=\frac{1}{8}x+0.1$ whenever $0\leq x\leq\frac{1}{4}$, $Tx=\frac{1}{6}x+0.1$ whenever $\frac{1}{4}<x\leq\frac{1}{2}$, $Tx=\frac{1}{4}x+0.1$ whenever $\frac{1}{2}<x\leq\frac{3}{4}$ and $Tx=-x$ whenever $\frac{3}{4}<x\leq 1$. It is easy to check that T is a quasi-contraction. Next table and also Figure 1 in [41] show the difference of rate of convergence in these iteration methods. The Figure 1 in [41] includes first CPU time for each iterative method. In that figure, it has been used at least 30 times calculating of CPU times for our faster cases in the methods.*

Step	Picard (7.1)	Mann (7.40)	Ishikawa (7.3)	Noor (7.4)	Agarwal (7.5)	Thakur (7.7)
1	0.275	0.36	0.2765	0.26714722	0.1915	0.15994792
2	0.14583333	0.2	0.15447292	0.15075275	0.11954594	0.11619033
3	0.11822917	0.14	0.12405623	0.12297028	0.11464407	0.11436516
4	0.11477865	0.122	0.11666117	0.11635393	0.11431013	0.11428903
5	0.11434733	0.1166	0.11486325	0.11477826	0.11428738	0.11428585
6	0.11429342	0.11498	0.11442613	0.11440301	0.11428583	0.11428572
7	0.11428668	0.114494	0.11431985	0.11431365	0.11428572	0.11428571
8	0.11428583	0.1143482	0.11429401	0.11429237	0.11428571	
9	0.11428573	0.11430446	0.11428773	0.1142873		
10	0.11428572	0.11429134	0.1142862	0.11428609		
11	0.11428571	0.1142874	0.11428583	0.1142858		
12		0.11428622	0.11428574	0.11428574		
13		0.11428587	0.11428572	0.11428572		
14		0.11428576	0.11428572	0.11428572		
15		0.11428573				
16		0.11428572				
CPU time	0.00056	0.00075	0.00062	0.00073	0.00041	0.00045

Table3. Comparing some iteration methods

7.4 The iteration methods for Zamfirescu operators

A self-map T on a metric space (X,d) is called a Zamfirescu operator whenever there exist real numbers $a \in (0,1)$ and $b,c \in (0,\frac{1}{2})$ such that for each $x,y \in X$ at least one of the conditions $d(Tx,Ty) \leq b(d(x,Tx)+d(y,Ty))$, $d(Tx,Ty) \leq c(d(x,Ty)+d(y,Tx))$ and $d(Tx,Ty) \leq ad(x,y)$ holds. It has been proved that a self-map T is Zamfirescu operator if and only if there exists $k \in (0,1)$ such that

$$d(Tx,Ty) \leq k\max\{d(x,y), \frac{d(x,Tx)+d(y,Ty))}{2}, \frac{d(x,Ty)+d(y,Tx)}{2}\}$$

for all $x,y \in X$.

Proposition 7.7

Let C be a nonempty, closed and convex subset of a Banach space X, $x_1 \in C$, $T: C \to C$ a Zamfirescu with constant $k \in (0,1)$, p a fixed point of T and $\sum_{n=1}^{\infty} \alpha_n = \infty$. If $1-\alpha_n < \alpha_n$ for sufficiently large n, then the Mann iteration (7.40) *converges faster than the Mann iteration* (7.41).

Proof: Let $\{x_n\}$ be the sequence in the Mann iteration (7.40). In this case, we have

$$\begin{aligned}
\|x_{n+1}-p\| &\leq \|(1-\alpha)x_n+\alpha_n Tx_n - p\| \\
&\leq (1-\alpha_n)\|x_n-p\| + \alpha_n \|Tx_n-p\| \\
&\leq (1-\alpha_n)\|x_n-p\| + \alpha \max\Big\{ \|x_n-p\| \\
&\quad , \frac{\|x_n-Tx_n\|+\|p-Tp\|}{2}, \frac{\|x_n-Tp\|+\|p-Tx_n\|}{2} \Big\} \\
&\leq (1-\alpha_n(1-k))\|x_n-p\|
\end{aligned}$$

for all n. Since $1-\alpha_n < \alpha_n$ for sufficiently large n, we get $1-\alpha_n(1-k) < 1-\frac{1}{2}(1-k)$. Put $a_n = \left(1-\frac{1}{2}(1-k)\right)^n \|x_1-p\|$ for all n. Now, let $\{x_n\}$ be the sequence in the Mann iteration (7.41). In this case, we have

$$\begin{aligned}
\|x_{n+1}-p\| = \|\alpha x_n+(1-\alpha_n)Tx_n-p\| &\leq (\alpha_n\|x_n-p\| + (1-\alpha_n)\|Tx_n-p\| \\
&\leq (1-(1-\alpha_n)(1-k))\|x_n-p\|
\end{aligned}$$

for all n. Since $1-\alpha_n < \alpha_n$ for sufficiently large n, we get $1-(1-\alpha_n)(1-k) < 1$. If $b_n = \|x_1-p\|$ for all n, then $\lim \frac{a_n}{b_n} = \lim \frac{\left(1-\frac{1}{2}(1-k)\right)^n \|x_1-p\|}{\|x_1-p\|} = 0$. This shows that the Mann iteration (7.40) converges faster than the Mann iteration (7.41).

♣

Again, consider the following cases of the Ishikawa iteration method

$$\begin{cases} x_{n+1} = (1-\alpha_n)x_n + \alpha_n T y_n, \\ y_n = (1-\beta_n)x_n + \beta_n T x_n, \end{cases} \tag{7.53}$$

$$\begin{cases} x_{n+1} = \alpha_n x_n + (1-\alpha_n) T y_n, \\ y_n = \beta_n x_n + (1-\beta_n) T x_n, \end{cases} \tag{7.54}$$

$$\begin{cases} x_{n+1} = \alpha_n x_n + (1-\alpha_n) T y_n, \\ y_n = (1-\beta_n)x_n + \beta_n T x_n \end{cases} \tag{7.55}$$

and

$$\begin{cases} x_{n+1} = (1-\alpha_n)x_n + \alpha_n T y_n, \\ y_n = \beta_n x_n + (1-\beta_n) T x_n, \end{cases} \tag{7.56}$$

where $\{\alpha_n\}_{n\geq 0}$ $\{\beta_n\}_{n\geq 0}$ are two sequences in $[0,1]$ with $\sum_{n=1}^{\infty} \alpha_n = \infty$. By using a similar proof of last result, one can prove next proposition.

Proposition 7.8

Let C be a nonempty, closed and convex subset of a Banach space X, $x_0 \in C$, $T : C \to C$ a Zamfirescu map with constant $k \in (0,1)$ and p a fixed point of T. If $1-\alpha_n < \alpha_n$ and $1-\beta_n < \beta_n$ for sufficiently large n and $\sum_{n=1}^{\infty} \alpha_n = \infty$, then the Ishikawa iteration (7.53) *converges faster than others.*

Note that, the condition $\sum_{n=1}^{\infty} \alpha_n = \infty$ for the sequence $\{\alpha_n\}_{n\geq 0}$ be used for convergence of the sequence $\{x_n\}$ which we obtain from the iteration methods Mann, Ishikawa and Noor. Now, we give a similar result with the Noor iteration method.

Theorem 7.7

Let C be a nonempty, closed and convex subset of a Banach space X, $x_0 \in C$, $T : C \to C$ a Zamfirescu map with constant $k \in (0,1)$, p a fixed point of T and $\{\alpha_n\}$, $\{\beta_n\}$ and $\{\gamma_n\}$ three sequences in $[0,1]$ with $\sum_{n=1}^{\infty} \alpha_n(1-\alpha_n) = \infty$. If $1-\alpha_n < \alpha_n$, $1-\beta_n < \beta_n$ and $1-\gamma_n < \gamma_n$ for sufficiently large n, then the iteration (7.4) *is faster than other possible cases.*

Proof: We prove only one case because other cases have similar proofs. We show that the case (7.4) in the Noor iteration method is faster than the case

$$\begin{cases} u_{n+1} = (1-\alpha_n)u_n + \alpha_n T v_n, \\ v_n = (1-\beta_n)u_n + \beta_n T w_n, \\ w_n = \gamma_n u_n + (1-\gamma_n) T u_n \end{cases} \tag{7.57}$$

for all $n \geq 0$. Note that,

$$\begin{aligned} \|z_n - p\| &= \|(1-\gamma_n)x_n + \gamma_n T x_n - p\| \leq (1-\gamma_n)\|x_n - p\| \\ &\quad + k\gamma_n \max\{\|x_n - p\|, \frac{\|x_n - Tx_n\| + \|p - Tp\|}{2}, \frac{\|x_n - Tp\| + \|p - Tx_n\|}{2}\} \\ &= (1-(1-k)\gamma_n)\|x_n - p\|, \end{aligned}$$

and

$$\begin{aligned} \|y_n - p\| &= \|(1-\beta_n)x_n + \beta_n T z_n - p\| \leq (1-\beta_n)\|x_n - p\| \\ &\quad + k\beta_n \max\{\|z_n - p\|, \frac{\|z_n - Tz_n\| + \|p - Tp\|}{2}, \frac{\|z_n - Tp\| + \|p - Tz_n\|}{2}\} \\ &\leq (1-\beta_n) + k\beta_n((1-(1-k)\gamma_n))\|x_n - p\| \\ &\leq [1 - \beta_n(1-k) - \beta_n\gamma_n k(1-k)]\|x_n - p\| \end{aligned}$$

for all $n \geq 0$. Thus, we get

$$\begin{aligned} \|x_{n+1} - p\| &= \|(1-\alpha_n)x_n + \alpha_n T y_n - p\| \leq (1-\alpha_n)\|x_n - p\| + k\alpha_n \\ &\quad \times \max\{\|y_n - p\|, \frac{\|y_n - Ty_n\| + \|p - Tp\|}{2}, \frac{\|y_n - Tp\| + \|p - Ty_n\|}{2}\} \\ &\leq (1-\alpha_n)\|x_n - p\| + k\alpha_n[1 - \beta_n(1-k) - \beta_n\gamma_n k(1-k)]\|x_n - p\| \\ &\leq (1 - \alpha_n + k\alpha_n(1 - \beta_n(1-k) - \beta_n\gamma_n k(1-k)))\|x_n - p\| \\ &\leq (1 - \alpha_n + k\alpha_n - k(1-k)\beta_n\alpha_n - \alpha_n\beta_n\gamma_n k^2(1-k))\|x_n - p\| \\ &\leq (1 - (1-k)\alpha_n - k(1-k)\beta_n\alpha_n - \alpha_n\beta_n\gamma_n k^2(1-k))\|x_n - p\| \end{aligned}$$

for all n. Since $\alpha_n, \beta_n, \gamma_n \in (\frac{1}{2}, 1)$ for sufficiently large n, we get

$$-(1-k^2) < -\alpha_n(1-k^2) < -\frac{1}{2}(1-k^2),$$

$-k(1-k) < -\alpha_n\beta_n k(1-k) < -\frac{1}{4}k(1-k)$ and

$$-k^2(1-k) < -\alpha_n\beta_n\gamma_n k^2(1-k) < -\frac{1}{8}k^2(1-k)$$

for sufficiently large n. This implies that

$$1 - (1-k)\alpha_n - k(1-k)\beta_n\alpha_n - \alpha_n\beta_n\gamma_n k^2(1-k)$$

$$< 1 - \frac{1}{2}(1-k) - \frac{1}{4}k(1-k) - \frac{1}{8}k^2(1-k)$$

for sufficiently large n. Put $a_n = (1 - \frac{1}{2}(1-k) - \frac{1}{8}k^2(1-k))^n \|x_1 - p\|$ for all $n \geq 0$. Now for the sequences $\{u_n\}_{n\geq 0}$ and $\{v_n\}_{n\geq 0}$ in the iteration method (7.57), we have

$$\|w_n - p\| = \|\gamma_n u_n + (1-\gamma_n)Tu_n - p\| \leq \gamma_n \|u_n - p\|$$

$$+k(1-\gamma_n)\max\{\|u_n - p\|, \frac{\|u_n - Tu_n\| + \|p - Tp\|}{2}, \frac{\|u_n - Tp\| + \|p - Tu_n\|}{2}\}$$

$$= (1-(1-\gamma_n)(1-k))\|u_n - p\|$$

and

$$\|v_n - p\| = \|(1-\beta_n)u_n + \beta_n Tw_n - p\| \leq (1-\beta_n)\|u_n - p\|$$

$$+k\beta_n \max\{\|w_n - p\|, \frac{\|w_n - Ty_n\| + \|p - Tp\|}{2}, \frac{\|w_n - Tp\| + \|p - Tw_n\|}{2}\}$$

$$\leq (1-\beta_n) + k\beta_n(1-(1-\gamma_n)(1-k))\|u_n - p\|$$

$$\leq (1 - \beta_n + k\beta_n - \beta_n(1-\gamma_n)k(1-k))\|u_n - p\|$$

$$\leq (1 - \beta_n(1-k) - \beta_n(1-\gamma_n)k(1-k))\|u_n - p\|$$

for all $n \geq 0$. Hence, we get

$$\|u_{n+1} - p\| = \|(1-\alpha_n)u_n + \alpha_n Tv_n - p\| \leq (1-\alpha_n)\|u_n - p\|$$

$$+k\alpha_n \max\{\|v_n - p\|, \frac{\|v_n - Tv_n\| + \|p - Tp\|}{2}, \frac{\|v_n - Tp\| + \|p - Tv_n\|}{2}\}$$

$$\leq (1-\alpha_n)\|u_n - p\| + k\alpha_n(1 - \beta_n(1-k) - \beta_n(1-\gamma_n)k(1-k))\|u_n - p\|$$

$$\leq ((1-\alpha_n) + k\alpha_n - \alpha_n\beta_n k(1-k) - \alpha\beta_n(1-\gamma_n)K^2(1-k))\|u_n - p\|$$

$$\leq (1 - \alpha_n(1-k) - \alpha_n\beta_n k(1-k) - \alpha\beta_n(1-\gamma_n)K^2(1-k))\|u_n - p\|$$

for all n. Since $\alpha_n, \beta_n, \gamma_n \in (\frac{1}{2}, 1)$ for sufficiently large n,

$$-k(1-k) < -\alpha_n\beta_n k(1-k) < -\frac{1}{4}k(1-k)$$

and $\frac{1}{2}k^2(1-k) < -\alpha_n\beta_n(1-\gamma_n)k^2(1-k) < 0$ for sufficiently large n. Hence,

$$1 - \alpha_n(1-k) - \alpha_n\beta_n k(1-k) - \alpha\beta_n(1-\gamma_n)k^2(1-k) < 1 - \frac{1}{2}(1-k) - \frac{1}{4}k(1-k)$$

for sufficiently large n. Put

$$b_n = (1 - \frac{1}{2}(1-k) - \frac{1}{4}k(1-k))^n \|u_1 - p\|$$

for all $n \geq 0$. Then, we have

$$\lim_{n\to\infty} \frac{a_n}{b_n} = \frac{\left(1-\frac{1}{2}(1-k)-\frac{1}{4}k(1-k)-\frac{1}{8}k^2(1-k)\right)^n \|x_1-p\|}{\left(1-\frac{1}{2}(1-k)-\frac{1}{4}k(1-k)\right)^n \|u_1-p\|} = 0.$$

This shows that the sequence of the iteration (7.4) is faster than the iteration (7.57).

♣

Consider the following cases of the Ishikawa iteration method

$$\begin{cases} u_{n+1} = (1-\alpha_n)u_n + \alpha_n T v_n \\ v_n = \beta_n u_n + (1-\beta_n)T w_n \\ w_n = (1-\gamma_n)u_n + \gamma_n T u_n \end{cases} \tag{7.58}$$

$$\begin{cases} u_{n+1} = (1-\alpha_n)u_n + \alpha_n T v_n \\ v_n = \beta_n u_n + (1-\beta_n)T w_n \\ w_n = \gamma_n u_n + (1-\gamma_n)T u_n \end{cases} \tag{7.59}$$

$$\begin{cases} u_{n+1} = \alpha_n u_n + (1-\alpha_n)T v_n \\ v_n = (1-\beta_n)u_n + \beta_n T w_n \\ w_n = (1-\gamma_n)u_n + \gamma_n T u_n, \end{cases} \tag{7.60}$$

$$\begin{cases} u_{n+1} = \alpha_n u_n + (1-\alpha_n)T v_n \\ v_n = (1-\beta_n)u_n + \beta_n T w_n \\ w_n = \gamma_n u_n + (1-\gamma_n)T u_n, \end{cases} \tag{7.61}$$

$$\begin{cases} u_{n+1} = \alpha_n u_n + (1-\alpha_n)T v_n \\ v_n = \beta_n u_n + (1-\beta_n)T w_n \\ w_n = (1-\gamma_n)u_n + \gamma_n T u_n \end{cases} \tag{7.62}$$

and

$$\begin{cases} u_{n+1} = \alpha_n u_n + (1-\alpha_n)T v_n \\ v_n = \beta_n u_n + (1-\beta_n)T w_n \\ w_n = \gamma_n u_n + (1-\gamma_n)T u_n \end{cases} \tag{7.63}$$

for all $n \geq 0$. By using similar conditions, one can show that the case (7.58) converges faster than (7.59), (7.60) converges faster than (7.62), (7.62) converges faster than (7.61) and (7.61) converges faster than (7.63).

Now, we give an example to illustrate some of our results.

Example 96 *Define the map $T:[0,1]\to\mathbb{R}$ by $T(x)=\frac{1}{4}$ whenever $0\le x\le 2$ and $T(x)=\frac{1}{2}$ whenever $x>2$. One can check that T is a Zamfirescu map. Put $x_1=5$, $\alpha_n=0.85$ and $\beta_n=0.85$ for all n. We compare the two cases of the Mann and four cases of the Ishikawa methods separately in next tables. The tables show us the difference of rate of convergence in the iteration methods.*

Step	*Mann*(7.40)	*Mann*(7.41)
1	1.175	4.325
2	0.38875	3.75125
3	0.270813	3.263563
4	0.253122	2.849028
5	0.250468	2.496674
6	0.25007	2.197173
7	0.250011	1.942597
8	0.250002	1.688707
9	0.25	1.472901
10	0.25	1.289466
11	0.25	1.133546
12	0.25	1.001014
13	0.25	0.888362
14	0.25	0.792608
15	0.25	0.711217
16		0.642034
17		0.583229
18		0.533245
19		0.490758
20		0.454644
⋮	⋰	⋮
93		0.250001
94		0.250001
95		0.250001
96		0.250001
97		0.250001
98		0.250001
99		0.250001
100		0.25

Step	*Ishikawa*(7.53)	*Ishikawa*(7.54)	*Ishikawa*(7.55)	*Ishikawa*(7.56)
1	0.9625	4.325	1.175	4.2875
2	0.356875	3.75125	0.38875	3.681875
3	0.266031	3.263563	0.270813	3.167094
4	0.252405	2.849028	0.253122	2.72953
5	0.250361	2.496674	0.250468	2.3576
6	0.250054	2.197173	0.25007	2.04146
7	0.250008	1.905097	0.250011	1.772741
8	0.250001	1.656832	0.250002	1.54433
9	0.25	1.445808	0.25	1.35018
10	0.25	1.266436	0.25	1.185153
11	0.25	1.113971	0.25	1.04488
12	0.25	0.984375	0.25	0.925648
13		0.874219		0.824301
14		0.780586		0.738156
15		0.700998		0.664933
16		0.633348		0.602693
17		0.575846		0.549789
18		0.526969		0.50482
19		0.485424		0.466597
20		0.45011		0.434108
21		0.420094		0.406492
22		0.39458		0.383018
23		0.372893		0.363065
24		0.354459		0.346105
25		0.33879		0.33169
26		0.325472		0.319436
27		0.314151		0.309021
28		0.304528		0.300168
.	.	.	.	.
.	.	.	.	.
.	.	.	.	.
91		0.250002		0.250002
92		0.250002		0.250002
93		0.250001		0.250001
94		0.250001		0.250001
95		0.250001		0.250001
96		0.250001		0.250001
97		0.250001		0.250001
98		0.250001		0.250001
99		0.250001		0.25
100		0.25		0.25

Now, we compare rate of convergence for distinct iterative methods which were mentioned in the first section.

Proposition 7.9

Let C be a nonempty, closed and convex subset of a Banach space E, $p_0 \in C$, $T : C \to C$ a Zamfirescu map with constant $k \in (0,1)$ and p a fixed point of T. Then the Picard iteration converges faster than the Mann iteration.

Proof: Let $\{u_n\}_{n\geq 0}$ be the sequence in the Picard iteration. Then, we have

$$\|u_{n+1} - p\| = \|Tu_n - p\|$$

$$\leq k\max\{\|u_n - p\|, \frac{\|u_n - Tu_n\| + \|p - Tp\|}{2}, \frac{\|u_n - Tp\| + \|p - Txu_n\|}{2}\}$$

$$\leq k\|u_n - p\| \leq k^n \|u_1 - p\|$$

for all n. Put $a_n = k^n \|u_1 - p\|$ for all n. Now, let $\{x_n\}_{n\geq 0}$ be the sequence in the Mann iteration. Then, we have

$$\|y_{n+1} - p\| = \|(1-\alpha_n)x_n + \alpha_n Tx_n - p\| \leq (1-\alpha_n)\|x_n - p\| + \alpha_n \|Tx_n - p\|$$

$$\leq (1-\alpha_n)\|x_n - p\| + \alpha_n k\max\{\|x_n - p\|$$

$$, \frac{\|x_n - Tx_n\| + \|p - Tp\|}{2}, \frac{\|x_n - Tp\| + \|p - Tx_n\|}{2}\}$$

$$\leq ((1-\alpha_n) + \alpha_n k)\|x_n - p\| \leq ((1-\alpha_n(1-k))\|x_n - p\|$$

for all n. Put $b_n = (1-\alpha_n(1-k))^n$ for all n. Since $k < 1 - \alpha_n(1-k)$ for sufficiently large n, we get $\lim_{n\to\infty} \frac{a_n}{b_n} = \frac{k^n\|u_1 - p\|}{(1-\alpha_n(1-k))^n\|x_1 - p\|} = 0$. This implies that the Picard iteration is faster than the Mann iteration.

♣

Theorem 7.8

Let C be a nonempty, closed and convex subset of a Banach space X, $x_1 \in C$, $T : C \to C$ a Zamfirescu map with constant $k \in (0,1)$ and p a fixed point of T. If $1 - \alpha_n < \alpha_n$, $1 - \beta_n < \beta_n$ and $1 - \gamma_n < \gamma_n$ for sufficiently large n, then the case (7.4) *in the Noor iteration method converges faster than the case* (7.3) *in the Ishikawa iteration method and the case* (7.3) *in the Ishikawa iteration method is faster than the case* (7.40) *in the Mann iteration method.*

Proof: By using the proof of Theorem 7.7 for the case (7.4) in the Noor iteration method, we put $a_n = (1 - \frac{1}{2}(1-k) - \frac{1}{4}k(1-k) - \frac{1}{8}k^2(1-k))^n \|x_1 - p\|$ for all $n \geq 0$. Now for for the case (7.3) in the Ishikawa iteration method, we have

$$\|y_n - p\| = \|(1-\beta_n)x_n + \beta_n Tx_n - p\| \leq (1-\beta_n)\|x_n - p\|$$
$$+k\beta_n \max\{\|x_n - p\|, \frac{\|x_n - Tx_n\| + \|p - Tp\|}{2}, \frac{\|x_n - Tp\| + \|p - Tx_n\|}{2}\}$$
$$= (1-(1-k)\beta_n)\|x_n - p\|$$

and

$$\begin{aligned}\|x_{n+1} - p\| &= \|(1-\alpha_n)x_n + \alpha_n Ty_n - p\| \\ &\leq (1-\alpha_n)\|x_n - p\| + k\alpha_n \\ &\times \max\{\|y_n - p\|, \frac{\|y_n - Ty_n\| + \|p - Tp\|}{2}, \frac{\|y_n - Tp\| + \|p - Ty_n\|}{2}\} \\ &\leq (1-\alpha_n) + k\alpha_n((1-(1-k)\beta_n))\|x_n - p\| \\ &\leq [1 - \alpha_n(1-k) - \alpha_n\beta_n k(1-k)]\|x_n - p\|\end{aligned}$$

for all n. Since $\alpha_n, \beta_n \in (\frac{1}{2}, 1)$ for sufficiently large n, we get

$$-k(1-k) < -\alpha_n\beta_n k(1-k) < -\frac{1}{4}k(1-k)$$

for sufficiently large n. This implies that

$$1 - (1-k)\alpha_n - k(1-k)\beta_n\alpha_n < 1 - \frac{1}{2}(1-k) - \frac{1}{4}k(1-k)$$

for all n. Put $b_n = (1 - \frac{1}{2}(1-k) - \frac{1}{4}k(1-k))\|x_1 - p\|$ for all $n \geq 0$. Then, we have

$$\lim_{n\to\infty} \frac{a_n}{b_n} = \frac{(1 - \frac{1}{2}(1-k) - \frac{1}{4}k(1-k) - \frac{1}{8}k^2(1-k))^n \|x_1 - p\|}{(1 - \frac{1}{2}(1-k) - \frac{1}{4}k(1-k))^n \|x_1 - p\|} = 0$$

and so the case (7.4) in the Noor iteration method converges faster than the case (7.3) in the Ishikawa iteration method. By using the proof of Theorem 7.7 for the case (7.40) in the Mann iteration method, we put $c_n = (1 - \frac{1}{2}(1-k))^n \|x_1 - p\|$ for al n. Then, we have

$$\lim_{n\to\infty} \frac{b_n}{c_n} = \frac{(1 - \frac{1}{2}(1-k) - \frac{1}{4}k(1-k))^n \|x_1 - p\|}{(1 - \frac{1}{2}(1-k))^n \|x_1 - p\|} = 0$$

and so the case (7.3) in the Ishikawa iteration method converges faster than the case (7.40) in the Mann iteration method.

♣

Example 97 *Consider the map $T : [0,100] \to [0,100]$ defined by $T(x) = \frac{1}{2}x$ for all $x \in [0,1]$. Put $\alpha_n = \beta_n = \gamma_n = 0.9$ for all n and $x_1 = 10$. In fact, T is a contractive (and so Zamfirescu) map. The following table shows difference of rate of convergence in the mentioned iteration methods. Also, we provide cpu time of the iteration methods by using at least 30 times calculations.*

Step	*Picard*(7.1)	*Mann*(7.40)	*Ishikawa*(7.3)	*Noor*(7.4)
1	*5*	*5.5*	*3.475*	*2.56375*
2	*2.5*	*3.025*	*1.207563*	*0.657281*
3	*1.25*	*1.66375*	*0.419628*	*0.168511*
4	*0.625*	*0.915063*	*0.145821*	*0.043202*
5	*0.3125*	*0.503284*	*0.050673*	*0.011076*
6	*0.15625*	*0.276806*	*0.017609*	*0.00284*
7	*0.078125*	*0.152244*	*0.006119*	*0.000728*
8	*0.039063*	*0.083734*	*0.002126*	*0.000187*
9	*0.019531*	*0.046054*	*0.000739*	*0.000048*
10	*0.009766*	*0.02533*	*0.000257*	*0.000012*
11	*0.004883*	*0.013931*	*0.000089*	*0.000003*
12	*0.002441*	*0.007662*	*0.000031*	*0.000001*
13	*0.001221*	*0.004214*	*0.000011*	*0*
14	*0.00061*	*0.002318*	*0.000004*	*0*
15	*0.000305*	*0.001275*	*0.000001*	*0*
16	*0.000153*	*0.000701*	*0*	*0*
17	*0.000076*	*0.000386*	*0*	*0*
18	*0.000038*	*0.000212*	*0*	
19	*0.000019*	*0.000117*	*0*	
20	*0.00001*	*0.000064*	*0*	
21	*0.000005*	*0.000035*	*0*	
22	*0.000002*	*0.000019*		
23	*0.000001*	*0.000011*		
24	*0.000001*	*0.000006*		
25	*0*	*0.000003*		
26	*0*	*0.000002*		
27	*0*	*0.000001*		
28	*0*	*0.000001*		
29	*0*	*0*		
30	*0*	*0*		
31		*0*		
32		*0*		
33		*0*		
34		*0*		
35		*0*		
CPU time	*0.0044*	*0.0217*	*0.0192*	*0.0172*

Finally, one can find the related CPU time of this Example as Figure 1 in [40].

Chapter 8

Some false generalizations

8.1 Introduction

During last decades, various fixed point results have been proved. Many researchers have tried in this field, but it is natural that we find some false results in this area. The main problem of these type of results was that the related old work seemed to be generalized, but it was proved in 2011 that those are not real generalizations. The main goal of this chapter is that young researchers learn more about these type of false results and they will take more care.

8.2 Category of cone metric spaces

Let (E,τ) be a topological vector space, P a cone in E and (X,d) be a cone metric space based on P. By using Corollary 2.1 and Theorem 2.3 in section 5 of Chapter 2, τ_c is the Hausdorff cone topology on X. First, we provide next key result.

Lemma 8.1

Let (E,τ) be a topological vector space and P a cone in E. Then
(i) If $0 \ll c_1$ and $c_2 \in E$, then there exists a $0 \ll d$ such that $c_1 \ll d$ and $c_2 \ll d$.
(ii) If $0 \ll c_1$ and $0 \ll c_2$, then there exists a $0 \ll e$ such that $e \ll c_1$ and $e \ll c_2$.
(iii) If $0 \ll c_1$ and $c_2 \in E$, then there exists a natural number m such that $rc_2 \ll c_1$, for all $0 < r \leq \frac{1}{m}$.
(iv) If $x \in E$ and $0 \ll c$, then there exists a natural number m such that $-c \ll tx \ll c$, for all $|t| \leq \frac{1}{m}$.

Proof: (i) Choose a balanced absorbing neighborhood V of 0 such that $c_1 + V \subseteq intP$, then there is a $t > 0$ so that $-c_2 \in tV$. Hence, for each natural number $m > \max\{1,t\}$ we have $-c_2 \in mV$. Thus, $mc_1 - c_2 \in intP$ and so $c_2 \ll mc_1$. Therefore, by considering $d = mc_1$ we have $c_1 \ll d$ and $c_2 \ll d$.
(ii) In the proof of part (i), put $e = \frac{1}{m}c_2$.
(iii) By using part (i), there is a natural number $m > 1$ such that $c_2 \ll mc_1$. Hence $rc_2 \ll c_1$, for all $0 < r \leq \frac{1}{m}$.
(iv) Choose a balanced absorbing neighborhood V of 0 such that $c_1 + V \subseteq intP$, then there is a natural number $m \geq 1$ so that $x \in mV$. Thus, $tx \in V$ for all $|t| \leq \frac{1}{m}$. Hence, $c + tx, c - tx \in intP$ for all $|t| \leq \frac{1}{m}$.

♣

For each $x \in E$ and $0 \ll c$, define

$$N(x,c) = \{y \in E : -c \ll y - x \ll c\}.$$

First we show that

$$\mathcal{B} = \{N(x,c) : x \in E, 0 \ll c\}$$

is a basis for a topology on E.

Lemma 8.2

$\mathcal{B}$ is a basis for a topology on E.

Proof: First note that $x \in N(x,c)$, for all $x \in E$ and $0 \ll c$. Now, let

$$y \in N(x,c) \cap N(x',c').$$

Using Lemma 8.1, choose $0 \ll e$ so that

$$e \ll c - (y - x), e \ll c + (y - x), e \ll c' - (y - x') \text{ and } e \ll c' + (y - x').$$

If $z \in N(y,e)$, then $-c \ll z - x \ll c$ and $-c' \ll z - x' \ll c'$. Hence,

$$N(y,e) \subseteq N(x,c) \cap N(x',c').$$

♣

We denote this topology by τ_P.

Proposition 8.1

(E, τ_P) is a locally convex topological vector space.

Proof: First note that (E,τ_P) is a Hausdorff topological space. In fact, if $x,y \in E$ and $N(x,\frac{c}{2}) \cap N(y,\frac{c}{2}) \neq \emptyset$ for all $0 \ll c$, then $-c \ll x-y \ll c$. Thus, $\frac{-c}{n} \ll x-y \ll \frac{c}{n}$ for all $n \geq 1$. Since P is closed, $x-y \in P \cap (-P)$. Hence, $x=y$. Now, let $x_1,x_2 \in E$, $W \in \tau_P$ and $x_1+x_2 \in W$. Choose $0 \ll c$ so that $N(x_1+x_2,c) \subseteq W$. By using Lemma 8.1, choose $0 \ll e$ such that $0 \ll e \ll \frac{c}{2}$. If $t_1 \in N(x_1,e)$ and $t_2 \in N(x_2,e)$, then

$$-e \ll t_1 - x_1 \ll e \text{ and } -e \ll t_2 - x_2 \ll e.$$

Hence,

$$-c \ll -2e \ll (t_1+t_2)-(x_1+x_2) \ll 2e \ll c.$$

Thus, $N(x_1,e)+N(x_2,e) \subseteq W$ and so the addition operation is continuous. Now, suppose that $x \in E$, $\alpha \in \mathbb{R}$, $W \in \tau_P$ and $\alpha x \in W$.
case 1: $\alpha = 0$. Choose $0 \ll e$ such that $N(0,e) \subseteq W$. Put $V = N(0,e)$ and $r=1$. Let $|\beta| < r$ and $y \in V$. If $\beta > 0$, then $\beta(e-y), \beta(y+e) \in intP$. Hence, $-e \ll -\beta e \ll \beta y \ll \beta e \ll e$. If $\beta = 0$, then it is clear that $-e \ll \beta y \ll e$. If $\beta < 0$, then $-\beta(e-y), -\beta(y+e) \in intP$. Hence, $-e \ll \beta e \ll \beta y \ll -\beta e \ll e$. Thus, $\beta y \in W$ for all $y \in V$ and $|\beta| < r$.
case 2: $\alpha > 0$. Choose $0 \ll c$ and $0 \ll d$ such that $N(\alpha x,c) \subseteq W$ and $\alpha d \ll \frac{c}{3}$. By Lemma 8.1, there exists $0 < \delta < \alpha$ such that $\frac{-c}{3} \ll (\beta-\alpha)x \ll \frac{c}{3}$ for all $|\beta-\alpha| < \delta$. Let $y \in V = N(x,d)$. Then,

$$\frac{-c}{3} \ll -\delta d \ll (\beta-\alpha)(y-x) \ll \delta d \ll \frac{c}{3} \text{ and } \frac{-c}{3} \ll -\alpha d \ll \alpha(y-x) \ll \alpha d \ll \frac{c}{3}.$$

Since $\beta y - \alpha x = (\beta-\alpha)(y-x)+(\beta-\alpha)x+\alpha(y-x)$, so $-c \ll \beta y - \alpha x \ll c$ for all $y \in V$ and $|\beta-\alpha| < \delta$. Thus, $\beta y \in W$ for all $y \in V$ and $|\beta-\alpha| < \delta$.
case 3: $\alpha < 0$. Choose $0 \ll c$ and $0 \ll d$ such that $N(\alpha x,c) \subseteq W$ and $-\alpha d \ll \frac{c}{3}$. By Lemma 8.1, there exists $0 < \delta < -\alpha$ such that $\frac{-c}{3} \ll (\beta-\alpha)x \ll \frac{c}{3}$ for all $|\beta-\alpha| < \delta$. Let $y \in V = N(x,d)$. Then,

$$\frac{-c}{3} \ll -\delta d \ll (\beta-\alpha)(y-x) \ll \delta d \ll \frac{c}{3} \text{ and } \frac{-c}{3} \ll \alpha d \ll \alpha(y-x) \ll -\alpha d \ll \frac{c}{3}.$$

Since $\beta y - \alpha x = (\beta-\alpha)(y-x)+(\beta-\alpha)x+\alpha(y-x)$, so $-c \ll \beta y - \alpha x \ll c$ for all $y \in V$ and $|\beta-\alpha| < \delta$. Thus, $\beta y \in W$ for all $y \in V$ and $|\beta-\alpha| < \delta$.
Hence, the multiplication operation is continuous. Since for each $0 \ll c$ and $0 \ll e$ there is a natural number m such that $0 \ll \frac{c}{m} \ll e$, the family $\{N(0,c) : 0 \ll c\}$ is a local base for (E,τ_P). Finally note that every neighborhood $N(x,c)$ is convex and balanced. $N(x,c)$ is also absorbing by using Lemma 8.1. Thus, (E,τ_P) is a locally convex topological vector space.

♣

Lemma 8.3

The topological vector space (E,τ_P) is normable.

Proof: Let $0 \ll c$, $V = N(0,c)$, $W \in \tau_P$ and $0 \in W$. Choose $0 \ll d$ such that $N(0,d) \subseteq W$. By using Lemma 8.1, there is a natural number $m > 1$ such that $c \ll md$. Thus, $V \subseteq mW$. Hence, V is bounded. Thus, the topological vector space (E, τ_P) is normable and the norm is μ_V (see Theorem 1.39 in [89]). In fact, the Minkowski functional of V is defined by $\mu_V(x) = \inf\{t > 0 : t^{-1}x \in V\}$ for $x \in E$.

♣

Remark 8.1 The topological vector space (E, τ_P) is normal (respect to P and μ_V) and the normal constant is 1. ∎

Let $0 \ll c$, $V = N(0,c)$, $x, y \in E$, $0 \leq x \leq y$ and $\varepsilon > 0$. If $t > 0$ and $z \notin tV$, then $\mu_V(z) \geq t$. Thus, $y \in (\mu_V(y) + \varepsilon)V$ and so $-(\mu_V(y) + \varepsilon)c \ll y \ll (\mu_V(y) + \varepsilon)c$. Since $P + intP \subseteq intP$ and $y - x \in P$, we have $-(\mu_V(y) + \varepsilon)c \ll 0 \leq x \ll (\mu_V(y) + \varepsilon)c$. This implies $x \in (\mu_V(y) + \varepsilon)V$. Hence, $\mu_V(x) \leq \mu_V(y) + \varepsilon$. So, $\mu_V(x) \leq \mu_V(y)$. Now by using Lemma 2.1 in [85], the cone P is normal and the normal constant is $M = 1$.

Now by using Lemma 5 in [49], we can obtain the following result.

Corollary 8.1

Let (E, τ) be a topological vector space, P a cone in E and (X,d) a cone metric space. If $\{x_n\}$ and $\{y_n\}$ are two sequence in (X,d) such that $d(x_n, x) \to 0$ and $d(y_n, y) \to 0$, then $d(x_n, y_n)$ converges to $d(x,y)$ in (E, τ_P).

The following Lemma compares the topologies τ_P and τ.

Lemma 8.4

$\tau_P \subseteq \tau$.

Proof: Let $0 \ll c$, $V = N(0,c)$ and $y \in V$. Choose symmetric neighborhoods U_1, U_2 in τ such that $c - y + U_1 \subseteq intP$ and $c + y + U_2 \subseteq intP$. Thus, $-c \ll y + x \ll c$ for all $x \in U := U_1 \cap U_2$. Hence, $y + U \subseteq V$ and so $V \in \tau$.

♣

The following example shows that the converse is not true.

Example 98 *Let $E = C^2_{\mathbb{R}}([0,1])$ with the norm $\|f\| = \|f\|_\infty + \|f'\|_\infty$ and consider the cone $P = \{f \in E : f \geq 0\}$. One can check that P is a non-normal cone. Consider the sequence $\{\frac{x^n}{n}\}$ in E. Note that, $\|\frac{x^n}{n}\| = 1 + \frac{1}{n} \nrightarrow 0$. Since $0 \leq \frac{x^n}{n} \leq \frac{x}{n}$,*

$$0 \leq \mu_V(\frac{x^n}{n}) \leq \mu_V(\frac{x}{n}) = \frac{\mu_V(x)}{n}.$$

Thus, the sequence $\{\frac{x^n}{n}\}$ converges to 0 in (E, τ_P).

Now, we provide an answer to our question in first part of section 7 in chapter 1. By using Lemma 8.3, one should concludes that each category of metric and cone metric spaces are same. Thus, your results on cone metric spaces are not real generalizations for related results in metric spaces.

8.2.1 Some common fixed point results

In this section, we give some results about common fixed points of of two multifunctions on cone metric spaces. Again, we suppose that (E, τ) is a topological vector space, P is a (not necessary normal) cone in E with non-empty interior, (E, τ_P) is the normable topological vector space and by fixing a bounded, convex, balanced and absorbing neighborhood V of 0, μ_V is the norm on the space. Let us to use the notation $\|.\|$ instead of μ_V. If (X, d) is a cone metric space, A a compact set in (X, τ_c), $x \in X$,

$$f_x : (X, \tau_c) \to [0, \infty) \text{ and } g_A : (X, \tau_c) \to [0, \infty)$$

are defined by $f_x(y) = \|d(x,y)\|$ and $g_A(x) = \inf_{a \in A} \|d(x,a)\|$ respectively, then it is easy to see that the functions f_x and g_A are continuous. By using these functions, we have the following two results.

Lemma 8.5

Let (X,d) be a cone metric space and A a compact set in (X, τ_c). Then, for every $x \in X$ there exists $a_0 \in A$ such that $\|d(x, a_0)\| = \inf_{a \in A} \|d(x,a)\|$.

Lemma 8.6

Let (X,d) be a cone metric space and A, B two compact sets in (X, τ_c). Then, $\sup_{x \in B} d'(x,A) < \infty$, where $d'(x,A) = \inf_{a \in A} \|d(x,a)\|$.

Definition 8.1 Let (X,d) be a cone metric space, $\mathcal{H}_c(X)$ the set of all compact subsets of (X, τ_c) and $A \in \mathcal{H}_c(X)$. By using Lemma 8.6, we can define

$$h_A : \mathcal{H}_c(X) \to [0, \infty) \text{ and } d_H : \mathcal{H}_c(X) \times \mathcal{H}_c(X) \to [0, \infty)$$

by

$$h_A(B) = \sup_{x\in A} d'(x,B) \text{ and } d_H(A,B) = \max\{h_A(B), h_B(A)\},$$

respectively.

Remark 8.2 Let (X,d) be a cone metric space. Define $\rho : X \times X \to [0,\infty)$ by $\rho(x,y) = \|d(x,y)\|$. Then, (X,ρ) is a metric space. This implies that for each $A,B \in \mathcal{H}_c(X)$ and $x,y \in X$, we have the following relations
(i) $d'(x,A) \leq \|d(x,y)\| + d'(y,A)$,
(ii) $d'(x,A) \leq d'(x,B) + h_B(A)$,
(iii) $d'(x,A) \leq \|d(x,y)\| + d'(y,B) + h_B(A)$. ∎

Theorem 8.1

Let (X,d) be a complete cone metric space and $T_1, T_2 : X \to \mathcal{H}_c(X)$ two set-valued mappings satisfy the relation

$$d_H(T_1x, T_2y) \leq kd'(T_1x,x) + ld'(T_2y,y)$$

for all $x,y \in X$, where $k,l \in [0,1)$ with $k+l < 1$. Then, T_1 and T_2 have a common fixed point, that is, there exists $x \in X$ such that $x \in T_1x$ and $x \in T_2x$.

Proof: Let $x_0 \in X$ be given. By Lemma 8.5, there is $x_1 \in T_1x_0$ such that

$$d'(x_0, T_1x_0) = \|d(x_0,x_1)\|.$$

Also, there is $x_2 \in T_2x_1$ such that

$$d'(x_1, T_2x_1) = \|d(x_1,x_2)\|.$$

Thus, we obtain a sequence $\{x_n\}_{n\geq 1}$ in X such that

$$x_{2n-1} \in T_1x_{2n-2},\ x_{2n} \in T_2x_{2n-1},$$

$$d'(x_{2n-2}, T_1x_{2n-2}) = \|d(x_{2n-2},x_{2n-1})\|$$

and

$$d'(x_{2n-1}, T_2x_{2n-1}) = \|d(x_{2n-1},x_{2n})\|$$

for all $n \geq 1$. Thus, for all $n \geq 1$ we have

$$\|d(x_{2n},x_{2n+1})\| = d'(x_{2n}, T_1x_{2n}) \leq h_{T_2x_{2n-1}}(T_1x_{2n}) \leq d_H(T_1x_{2n}, T_2x_{2n-1})$$

$$\leq kd'(T_1x_{2n},x_{2n}) + ld'(T_2x_{2n-1},x_{2n-1}) = k\|d(x_{2n},x_{2n+1})\| + l\|d(x_{2n-1},x_{2n})\|.$$

Hence, $\|d(x_{2n},x_{2n+1})\| \leq \frac{l}{1-k}\|d(x_{2n-1},x_{2n})\|$ for all $n \geq 1$. Also,

$$\begin{aligned}\|d(x_{2n-1},x_{2n})\| &= d'(x_{2n-1},T_2x_{2n-1}) \\ &\leq h_{T_1x_{2n-2}}(T_2x_{2n-1}) \\ &\leq d_H(T_1x_{2n-2},T_2x_{2n-1}) \leq k\|d(x_{2n-2},x_{2n-1})\| + l \\ &\|d(x_{2n-1},x_{2n})\|,\end{aligned}$$

for all $n \geq 1$. Hence,

$$\|d(x_{2n-1},x_{2n})\| \leq \frac{k}{1-l}\|d(x_{2n-2},x_{2n-1})\|$$

for all $n \geq 1$. This implies that $\|d(x_m,x_{m+1})\| \leq s\|d(x_{m-1},x_m)\|$ for all $m \geq 1$, where $s = \max\{\frac{l}{1-k},\frac{k}{1-l}\}$¡1. Now, for $n > m$ we have

$$\|d(x_n,x_m)\| \leq \sum_{i=m+1}^{n}\|d(x_i,x_{i-1})\|$$

$$\leq (s^{n-1}+\cdots+s^m)\|d(x_0,x_1)\| \leq \frac{s^m}{1-s}\|d(x_0,x_1)\|.$$

This implies that $\lim_{m,n\to\infty}\|d(x_n,x_m)\| = 0$. By using Lemma 4 in [17], $\{x_n\}_{n\geq 1}$ is a Cauchy sequence in X. Thus, there exists $x^* \in X$ such that $x_n \to x^*$. Now by using Remark 8.2, we have

$$\begin{aligned}d'(x^*,T_1x^*) &\leq d'(x^*,T_2x_{2n-1}) + h_{T_2x_{2n-1}}(T_1x^*) \\ &\leq d'(x^*,T_2x_{2n-1}) + d_H(T_2x_{2n-1},T_1x^*) \\ &\leq \|d(x^*,x_{2n})\| + kd'(T_2x_{2n-1},x_{2n-1}) + ld'(T_1x^*,x^*)\end{aligned}$$

for all $n \geq 1$. Hence,

$$d'(x^*,T_1x^*) \leq \frac{k}{1-l}d'(T_2x_{2n-1},x_{2n-1}) + \frac{1}{1-l}\|d(x^*,x_{2n})\|$$

$$= \frac{k}{1-l}\|d(x_{2n-1},x_{2n})\| + \frac{1}{1-l}\|d(x^*,x_{2n})\|$$

for all $n \geq 1$. Therefore, $d'(x^*,T_1x^*) = 0$ and so $x^* \in T_1x^*$ being T_1x^* closed. Similarly,

$$\begin{aligned}d'(x^*,T_2x^*) &\leq d'(x^*,T_1x_{2n}) + h_{T_1x_{2n}}(T_2x^*) \\ &\leq d'(x^*,T_1x_{2n}) + d_H(T_1x_{2n},T_2x^*) \\ &\leq \|d(x^*,x_{2n+1})\| + kd'(T_1x_{2n},x_{2n}) + ld'(T_2x^*,x^*)\end{aligned}$$

for all $n \geq 1$. Hence,

$$d'(x^*, T_2x^*) \leq \frac{k}{1-l} d'(T_1x_{2n}, x_{2n}) + \frac{1}{1-l} \|d(x^*, x_{2n+1})\|$$

$$= \frac{k}{1-l} \|d(x_{2n+1}, x_{2n})\| + \frac{1}{1-l} \|d(x^*, x_{2n+1})\|$$

for all $n \geq 1$. Therefore, $d'(x^*, T_2x^*) = 0$ and so $x^* \in T_2x^*$ being T_2x^* closed. Thus, x^* is a common fixed point of T_1 and T_2.

♣

Corollary 8.2

Let (X, d) be a complete cone metric space and $T : X \to \mathcal{H}_c(X)$ a multifunction satisfy the relation

$$d_H(Tx, Ty) \leq kd'(Tx, x) + ld'(Ty, y)$$

for all $x, y \in X$, where $k, l \in [0, 1)$ with $k + l < 1$. Then, T has a fixed point.

Theorem 8.2

Let (X, d) be a complete cone metric space and $T_1, T_2 : X \to \mathcal{H}_c(X)$ two multifunctions satisfy the relation

$$d_H(T_1x, T_2y) \leq kd'(T_1x, y) + ld'(T_2y, x)$$

for all $x, y \in X$, where $k, l \in [0, \frac{1}{2})$ with $k + l < \frac{1}{2}$. Then, T_1 and T_2 have a common fixed point.

Proof: Let $x_0 \in X$ be given. By Lemma 8.5, there is $x_1 \in T_1x_0$ such that

$$d'(x_0, T_1x_0) = \|d(x_0, x_1)\|.$$

Also, there is $x_2 \in T_2x_1$ such that

$$d'(x_1, T_2x_1) = \|d(x_1, x_2)\|.$$

Thus, we obtain a sequence $\{x_n\}_{n \geq 1}$ in X such that

$$x_{2n-1} \in T_1x_{2n-2},\ x_{2n} \in T_2x_{2n-1},$$

$$d'(x_{2n-2}, T_1x_{2n-2}) = \|d(x_{2n-2}, x_{2n-1})\|$$

and

$$d'(x_{2n-1}, T_2x_{2n-1}) = \|d(x_{2n-1}, x_{2n})\|$$

for all $n \geq 1$. Thus, for all $n \geq 1$ we have

$$\begin{aligned}\|d(x_{2n},x_{2n+1})\| &= d'(x_{2n},T_1x_{2n}) \leq h_{T_2x_{2n-1}}(T_1x_{2n}) \leq d_H(T_1x_{2n},T_2x_{2n-1}) \\ &\leq kd'(T_1x_{2n},x_{2n-1}) + ld'(T_2x_{2n-1},x_{2n}) \\ &\leq k\|d(x_{2n},x_{2n-1})\| + kd'(T_1x_{2n},x_{2n}) + l\|d(x_{2n},x_{2n-1})\| + ld'(T_2x_{2n-1},x_{2n-1}) \\ &= k\|d(x_{2n},x_{2n-1})\| + k\|d(x_{2n},x_{2n+1})\| + l\|d(x_{2n},x_{2n-1})\| + l\|d(x_{2n-1},x_{2n})\|.\end{aligned}$$

Hence, $\|d(x_{2n},x_{2n+1})\| \leq \frac{k+2l}{1-k}\|d(x_{2n-1},x_{2n})\|$ for all $n \geq 1$. Also,

$$\begin{aligned}\|d(x_{2n-1},x_{2n})\| &= d'(x_{2n-1},T_2x_{2n-1}) \leq h_{T_1x_{2n-2}}(T_2x_{2n-1}) \\ &\leq d_H(T_1x_{2n-2},T_2x_{2n-1}) \\ &\leq kd'(T_1x_{2n-2},x_{2n-1}) + ld'(T_2x_{2n-1},x_{2n-2}) \\ &\leq k\|d(x_{2n-2},x_{2n-1})\| + kd'(T_1x_{2n-2},x_{2n-2}) \\ &\quad + l\|d(x_{2n-2},x_{2n-1})\| + ld'(T_2x_{2n-1},x_{2n-1}) \\ &\quad + k\|d(x_{2n-2},x_{2n-1})\| + k\|d(x_{2n-2},x_{2n-1})\| \\ &\quad + l\|d(x_{2n-2},x_{2n-1})\| + l\|d(x_{2n-1},x_{2n})\|,\end{aligned}$$

for all $n \geq 1$. Hence,

$$\|d(x_{2n-1},x_{2n})\| \leq \frac{l+2k}{1-l}\|d(x_{2n-2},x_{2n-1})\|$$

for all $n \geq 1$. This implies that $\|d(x_m,x_{m+1})\| \leq s\|d(x_{m-1},x_m)\|$ for all $m \geq 1$, where $s = \max\{\frac{k+2l}{1-k}, \frac{l+2k}{1-l}\}$¡1. Now, for $n > m$ we have

$$\|d(x_n,x_m)\| \leq \sum_{i=m+1}^{n}\|d(x_i,x_{i-1})\|$$

$$\leq (s^{n-1} + \cdots + s^m)\|d(x_0,x_1)\| \leq \frac{s^m}{1-s}\|d(x_0,x_1)\|.$$

This implies that $\lim_{m,n\to\infty}\|d(x_n,x_m)\| = 0$. By using Lemma 4 in [17], $\{x_n\}_{n\geq1}$ is a Cauchy sequence in X. Thus, there exists $x^* \in X$ such that $x_n \to x^*$. Now by using Remark 8.2, we have

$$\begin{aligned}d'(x^*,T_1x^*) &\leq d'(x^*,T_2x_{2n-1}) + h_{T_2x_{2n-1}}(T_1x^*) \\ &\leq d'(x^*,T_2x_{2n-1}) + d_H(T_2x_{2n-1},T_1x^*) \\ &\leq \|d(x^*,x_{2n})\| + kd'(T_2x_{2n-1},x^*) + ld'(T_1x^*,x_{2n-1}) \\ &\leq \|d(x^*,x_{2n})\| + k\|d(x^*,x_{2n})\| + ld'(T_1x^*,x^*) + l\|d(x^*,x_{2n-1})\|,\end{aligned}$$

for all $n \geq 1$. Hence,

$$d'(x^*,T_1x^*) \leq \frac{1+k}{1-l}\|d(x^*,x_{2n})\| + \frac{l}{1-l}\|d(x^*,x_{2n-1})\|$$

for all $n \geq 1$. Therefore, $d'(x^*, T_1x^*) = 0$ and so $x^* \in T_1x^*$ being T_1x^* closed. Similarly,

$$d'(x^*, T_2x^*) \leq d'(x^*, T_1x_{2n}) + h_{T_1x_{2n}}(T_2x^*) \leq d'(x^*, T_1x_{2n}) + d_H(T_1x_{2n}, T_2x^*)$$
$$\leq \|d(x^*, x_{2n+1})\| + kd'(T_1x_{2n}, x^*) + ld'(T_2x^*, x_{2n})$$
$$\leq \|d(x^*, x_{2n+1})\| + k\|d(x^*, x_{2n+1})\| + ld'(T_2x^*, x^*) + l\|d(x^*, x_{2n})\|$$

for all $n \geq 1$. Hence,

$$d'(x^*, T_2x^*) \leq \frac{1+k}{1-l}\|d(x^*, x_{2n+1})\| + \frac{l}{1-l}\|d(x^*, x_{2n})\|$$

for all $n \geq 1$. Therefore, $d'(x^*, T_2x^*) = 0$ and so $x^* \in T_2x^*$ being T_2x^* closed. Therefore, x^* is a common fixed point of T_1 and T_2.

♣

Corollary 8.3

Let (X,d) be a complete cone metric space and $T : X \to \mathcal{H}_c(X)$ a multifunction satisfy the relation

$$d_H(Tx, Ty) \leq kd'(Tx, y) + ld'(Ty, x)$$

for all $x, y \in X$, where $k, l \in [0, \frac{1}{2})$ with $k + l < \frac{1}{2}$. Then, T has a fixed point.

Here, we give an example which illustrates Corollary 8.3.

Example 99 *Let $X = \{a_1, a_2, a_3, \ldots\}$ be a countable set, $E = (l^2, \|.\|_2)$ and*

$$P = \{\{x_n\}_{n\geq 1} \in l^2 : x_n \geq 0\ (\forall n \geq 1)\}.$$

Put $x_i = \{\frac{5^i}{n}\}_{n\geq 1}$ for all $i \geq 1$ and note that $x_i \in l^2$ $(i \geq 1)$. Define the map $d : X \times X \to P$ by $d(a_i, a_j) = |x_i - x_j| = \{\frac{|5^i - 5^j|}{n}\}_{n\geq 1}$. It is easy to see that (X,d) is a cone metric space, the normal constant of P is $M = 1$ and there is no Cauchy sequence in (X,d). Hence, (X,d) is a complete cone metric space. Now, define the set-valued mapping

$$T : X \to \mathcal{H}_c(X)$$

by $Ta_1 = \{a_1\}$ and $Ta_i = \{a_1, a_2, \ldots, a_{i-1}\}$ for all $i \geq 2$. Then, we have $Ta_1 = Ta_2$ and for each $i \geq 3$

$$d_H(Ta_1, Ta_i) = \max\{d'(a_1, Ta_i), \sup_{b\in Ta_i} d'(Ta_1, b)\} = \|d(a_1, a_{i-1})\|_2$$

$$= \|x_1 - x_{i-1}\|_2 = \{\sum_{n=1}^{\infty} |\frac{5}{n} - \frac{5^{i-1}}{n}|^2\}^{\frac{1}{2}} = (5^{i-1} - 5)\sum_{n=1}^{\infty} \frac{1}{n^2}.$$

Also, we have $d'(a_1, Ta_i) = 0$ *and*

$$d'(a_i, Ta_1) = \|d(a_i, a_1)\|_2 = \|x_i - x_1\|_2 = (5^i - 5)\sum_{n=1}^{\infty} \frac{1}{n^2}.$$

Thus, $d_H(Ta_1, Ta_i) \le \frac{1}{5}(d'(a_1, Ta_i) + d'(a_i, Ta_1))$. *Now, suppose that* $j > i > 1$. *Then,* $Ta_i \subset Ta_j$ *and so* $\sup_{b \in Ta_i} d'(b, Ta_j) = 0$. *Hence,*

$$d_H(Ta_i, Ta_j) = \sup_{b \in Ta_j} d'(b, Ta_i) = \|d(a_{j-1}, a_{i-1})\|_2$$

$$= \|x_{i-1} - x_{j-1}\|_2 = (5^{j-1} - 5^{i-1})\sum_{n=1}^{\infty} \frac{1}{n^2}.$$

Also, we have $d'(a_i, Ta_j) = 0$ *and*

$$d'(a_j, Ta_i) = \|d(a_j, a_{i-1})\|_2 = (5^j - 5^{i-1})\sum_{n=1}^{\infty} \frac{1}{n^2}.$$

Thus, $d_H(Ta_i, Ta_j) \le \frac{1}{5}(d'(a_i, Ta_j) + d'(a_j, Ta_i))$ *and so* T *satisfies the assumptions of Corollary 8.3 and* a_1 *is the unique fixed point of* T.

8.3 Coupled fixed point results

In 1987, Guo and Lakshmikantham introduced the notion of coupled fixed points ([44]). Later, some researchers tried to provide different coupled fixed point results (see for examples, [26], [28], [46], [60] and [97]). In this section, our aim is to provide a simple method for obtaining some coupled fixed point results for α-ψ-contractive type mappings. In fact, we are going to show you that some fixed point results can be derived by using a simple method and it is not valuable for young researchers to try to review those.

Let $F : X \times X \to X$ be a mapping, where (X, d) is a metric space. We say that $(x_0, y_0) \in X \times X$ is a coupled fixed point of F whenever

$$F(x_0, y_0) = x_0 \text{ and } F(y_0, x_0) = y_0.$$

Define $T_F : X \times X \to X \times X$ by $T_F(x, y) = (F(x, y), F(y, x))$ for all $(x, y) \in X \times X$. Then, it is easy to check that (x_0, y_0) is a coupled fixed point of F if and only if (x_0, y_0) is a fixed point of T_f.

Now, define $\delta((x,y),(u,v)) = d(x,u) + d(y,v)$ and

$$m((x,y),(u,v)) = \max\{\delta((x,y),(u,v)),$$
$$\delta((x,y),(F(x,y),F(y,x))), \delta((u,v),(F(u,v),F(v,u))),$$
$$\frac{1}{2}[\delta((x,y),(F(u,v),F(v,u))) + \delta((u,v),(F(x,y),F(y,x)))]\}$$

for all $(x,y),(u,v) \in X \times X$.

Lemma 8.7

Let (X,d) be a complete metric space, $\alpha : X \times X \to [0,\infty)$ a function, $\psi \in \Psi$ and T a selfmap on X such that

$$\alpha(x,y)d(Tx,Ty) \leq \psi(\max\{d(x,y), d(x,Tx), d(y,Ty), \frac{1}{2}[d(x,Ty) + d(y,Tx)]\})$$

for all $x,y \in X$. Suppose that T is α-admissible and there exists $x_0 \in X$ such that $\alpha(x_0,Tx_0) \geq 1$. If X has the property (B), then T has a fixed point.

Proof: Take $x_0 \in X$ such that $\alpha(x_0,Tx_0) \geq 1$ and define the sequence $\{x_n\}$ in X by $x_{n+1} = Tx_n$ for all $n \geq 0$. If $x_n = x_{n+1}$ for some n, then $x^* = x_n$ is a fixed point of T. Assume that $x_n \neq x_{n+1}$ for all n. Since T is α-admissible, it is easy to check that $\alpha(x_n,x_{n+1}) \geq 1$ for all $n \geq 1$. Thus, for each natural number n we have

$$d(x_n,x_{n+1}) = d(Tx_{n-1},Tx_n) \leq \alpha(x_{n-1},x_n)d(Tx_{n-1},Tx_n)$$

$$\leq \psi(\max\{d(x_n,x_{n-1}), d(x_n,x_{n+1}), d(x_{n-1},x_n), \frac{1}{2}[d(x_n,x_n) + d(x_{n-1},x_{n+1})]\})$$
$$\leq \psi(\max\{d(x_n,x_{n-1}), d(x_n,x_{n+1}), \frac{1}{2}[d(x_n,x_{n-1}) + d(x_n,x_{n+1})]\})$$
$$= \psi(\max\{d(x_n,x_{n-1}), d(x_n,x_{n+1})\}).$$

If $\max\{d(x_n,x_{n-1}), d(x_n,x_{n+1})\} = d(x_n,x_{n+1})$, then

$$d(x_{n+1},x_n) \leq \psi(d(x_n,x_{n+1})) < d(x_{n+1},x_n)$$

which is contradiction. Thus, $\max\{d(x_n,x_{n-1}), d(x_n,x_{n+1})\} = d(x_n,x_{n-1})$ for all n. Hence, $d(x_{n+1},x_n) \leq \psi(d(x_n,x_{n-1}))$ and so $d(x_{n+1},x_n) \leq \psi^n(d(x_1,x_0))$ for all n. It is easy to check that $\{x_n\}$ is a Cauchy sequence. Thus, there exists $x^* \in X$ such that $x_n \to x^*$. By using the assumption, we have $\alpha(x_n,x^*) \geq 1$ for all n.

Thus,

$$\begin{aligned}
&d(Tx^*,x^*)\\
&\leq d(Tx^*,Tx_n)+d(x_{n+1},x^*)\leq \alpha(x_n,x^*)d(Tx^*,Tx_n)+d(x_{n+1},x^*)\\
&\leq \psi(\max\{d(x_n,x^*),d(x_n,x_{n+1}),d(x^*,Tx^*),\frac{1}{2}[d(x_n,Tx^*)+d(x^*,x_{n+1})]\})\\
&+d(x_{n+1},x^*)\\
&=\psi(\max\{d(x_n,x^*),d(x_n,x_{n+1})\})+d(x_{n+1},x^*),
\end{aligned}$$

for all n. Hence, $d(Tx^*,x^*)=0$ and so $Tx^*=x^*$.

♣

By using a similar argument we can prove next results.

Lemma 8.8

Let (X,d) be a complete metric space, $\alpha : X\times X\to[0,\infty)$ a function, $\psi\in\Psi$ and T a continuous selfmap on X such that

$$\alpha(x,y)d(Tx,Ty)\leq\psi(\max\{d(x,y),d(x,Tx),d(y,Ty),\frac{1}{2}[d(x,Ty)+d(y,Tx)]\})$$

for all $x,y\in X$. Suppose that T is α-admissible and there exists $x_0\in X$ such that $\alpha(x_0,Tx_0)\geq 1$. Then T has a fixed point.

Lemma 8.9

Let (X,d) be a complete metric space, $\alpha : X\times X\to[0,\infty)$ a function, $\psi\in\Psi$ and T a selfmap on X such that

$$\alpha(x,y)d(Tx,Ty)\leq\psi(d(x,u)+d(y,v))$$

for all $x,y\in X$. Suppose that T is α-admissible and there exists $x_0\in X$ such that $\alpha(x_0,Tx_0)\geq 1$. If X has the property (B), then T has a fixed point.

Lemma 8.10

Let (X,d) be a complete metric space, $\alpha : X\times X\to[0,\infty)$ a function, $\psi\in\Psi$ and T a continuous selfmap on X such that

$$\alpha(x,y)d(Tx,Ty)\leq\psi(d(x,u)+d(y,v))$$

for all $x,y\in X$. Suppose that T is α-admissible and there exists $x_0\in X$ such that $\alpha(x_0,Tx_0)\geq 1$. Then T has a fixed point.

Theorem 8.3

Let (X,d) be a complete metric space and $F : X \times X \to X$ a mapping. Suppose that there exist a mapping $\alpha : X^2 \times X^2 \to [0,+\infty)$ and $\psi \in \Psi$ such that $\alpha((x,y),(u,v))d(F(x,y),F(u,v)) \leq \frac{1}{2}\psi(m((x,y),(u,v)))$ for all $(x,y),(u,v) \in X \times X$, $\alpha((x,y),(u,v)) \geq 1$ implies

$$\alpha((F(x,y),F(y,x)),(F(u,v),F(v,u))) \geq 1$$

for all $(x,y),(u,v) \in X \times X$ and there exists $(x_0,y_0) \in X \times X$ such that

$$\alpha((x_0,y_0),(F(x_0,y_0),F(y_0,x_0))) \geq 1$$

and $\alpha((F(y_0,x_0),F(x_0,y_0)),(y_0,x_0)) \geq 1$.

Also, suppose that for each convergent sequences $\{x_n\}$ and $\{y_n\}$ with

$$x_n \to x, \quad y_n \to y,$$

$\alpha((x_n,y_n),(x_{n+1},y_{n+1})) \geq 1$ and $\alpha((y_{n+1},x_{n+1}),(y_n,x_n)) \geq 1$ for all n, we have

$$\alpha((x_n,y_n),(x,y)) \geq 1 \quad \text{and} \quad \alpha((y,x),(y_n,x_n)) \geq 1$$

for all n. Then F has a coupled fixed point.

Proof: It is easy to check that $m((x,y),(u,v)) = m((v,u),(y,x))$ for all $(x,y),(u,v)$ in $X \times X$, the metric space $(X \times X, \delta)$ is complete and $\beta(\xi,\eta)\delta(T_F\xi,T_F\eta) \leq \psi(m(\xi,\eta))$ for all $\xi = (\xi_1,\xi_2), \eta = (\eta_1,\eta_2) \in X \times X$, where

$$\beta((\xi_1,\xi_2),(\eta_1,\eta_2)) = \min\{\alpha((\xi_1,\xi_2),(\eta_1,\eta_2)),\alpha((\eta_2,\eta_1),(\xi_2,\xi_1))\}.$$

Thus, T_F is a β-ψ-contractive mapping. It is easy to check that T_F is β-admissible and $\beta((x_0,y_0),T_F(x_0,y_0)) \geq 1$. Since $(X \times X, \delta)$ has the property (B), T_F has a fixed point by using Lemma 8.7 and so F has a coupled fixed point.

♣

Theorem 8.4

Let (X,d) be a complete metric space and $F : X \times X \to X$ a continuous mapping. Suppose that there exist a mapping $\alpha : X^2 \times X^2 \to [0,+\infty)$ and $\psi \in \Psi$ such that $\alpha((x,y),(u,v))d(F(x,y),F(u,v)) \leq \frac{1}{2}\psi(m((x,y),(u,v)))$ for all $(x,y),(u,v) \in X \times X$, $\alpha((x,y),(u,v)) \geq 1$ implies

$$\alpha((F(x,y),F(y,x)),(F(u,v),F(v,u))) \geq 1$$

for all $(x,y),(u,v) \in X \times X$ and there exists $(x_0,y_0) \in X \times X$ such that

$$\alpha((x_0,y_0),(F(x_0,y_0),F(y_0,x_0))) \geq 1$$

and $\alpha((F(y_0,x_0),F(x_0,y_0)),(y_0,x_0)) \geq 1$. *Then* F *has a coupled fixed point.*

Proof: The metric space $(X \times X, \delta)$ is complete and $\beta(\xi,\eta)\delta(T_F\xi, T_F\eta) \leq \psi(m(\xi,\eta))$ for all $\xi = (\xi_1,\xi_2), \eta = (\eta_1,\eta_2) \in X \times X$, where

$$\beta((\xi_1,\xi_2),(\eta_1,\eta_2)) = \min\{\alpha((\xi_1,\xi_2),(\eta_1,\eta_2)), \alpha((\eta_2,\eta_1),(\xi_2,\xi_1))\}.$$

Thus, T_F is a continuous β-ψ-contractive mapping. Also, it is easy to check that T_F is β-admissible and $\beta((x_0,y_0),T_F(x_0,y_0)) \geq 1$. Now by using Lemma 8.8, T_F has a fixed point and so F has a coupled fixed point.

♣

Example 100 *Let* $X = [0,+\infty)$ *and* $d(x,y) = |x-y|$ *for all* $x,y \in X$. *Define the mapping* $F : X \times X \to X$ *by* $F(x,y) = \frac{x-y}{4}$ *whenever* $x \geq y$ *and* $F(x,y) = 0$ *whenever* $x < y$. *Then* F *is continuous mapping. Also, define* $\alpha : X^2 \times X^2 \to [0,+\infty)$ *by*

$$\alpha((x,y),(u,v)) = \begin{cases} 1 & x \geq y \text{ and } u \geq v, \\ 0 & \text{otherwise.} \end{cases}$$

Thus, for each $(x,y),(u,v) \in X \times X$ *we have*

$$\alpha((x,y),(u,v))d(F(x,y),F(u,v)) \leq |F(x,y) - F(u,v)| = |\frac{x-y}{4} - \frac{u-v}{4}|$$

$$= \frac{1}{4}|(x-u)+(v-y)| \leq \frac{1}{4}|(x-u)| + |(v-y)| = \frac{1}{4}(d(x,u)+d(y,v))$$

$$= \frac{1}{4}\delta((x,y),(u,v)) \leq \frac{1}{2}\psi(m((x,y),(u,v))),$$

where $\psi(t) = \frac{t}{2}$ *for all* $t \geq 0$. *Also, it is easy to check that* $\alpha((x,y),(u,v)) \geq 1$ *implies* $\alpha((F(x,y),F(y,x)),(F(u,v),F(v,u))) \geq 1$ *for all* $(x,y),(u,v) \in X \times X$. *Also,*

$$\alpha((x_0,y_0),(F(x_0,y_0),F(y_0,x_0))) \geq 1$$

and $\alpha((F(y_0,x_0),F(x_0,y_0)),(y_0,x_0)) \geq 1$, *where* $(x_0,y_0) = (1,1)$. *Thus by using Theorem 8.4,* F *has a coupled fixed point.*

Corollary 8.4

Let (X,d) *be a complete metric space,* $F : X \times X \to X$ *a mapping and* $\preceq$ *an order on* $X \times X$. *Suppose that there exist* $(x_0,y_0) \in X \times X$ *such that* (x_0,y_0) *and* $(F(x_0,y_0),F(y_0,x_0))$ *are comparable and also* (y_0,xy_0) *and* $(F(y_0,x_0),F(x_0,y_0))$ *are comparable, and a mapping* $\psi \in \Psi$ *such that*

$$d(F(x,y),F(u,v)) \leq \frac{1}{2}\psi(m((x,y),(u,v)))$$

for all comparable elements $(x,y),(u,v)$ in $X \times X$. Suppose that $(F(x,y),F(y,x))$ is comparable with $(F(u,v),F(v,u))$ whenever (x,y) is comparable with (u,v). Also, suppose that for each convergent sequences $\{x_n\}$ and $\{y_n\}$ with $x_n \to x$, $y_n \to y$, (x_n,y_n) is comparable with (x_{n+1},y_{n+1}) and (y_{n+1},x_{n+1}) is comparable with (y_n,x_n) for all n, we get that (x_n,y_n) is comparable with (x,y) and (y,x) is comparable with (y_n,x_n) for all n. Then F has a coupled fixed point.

Proof: Define the mapping $\alpha : X^2 \times X^2 \to [0,+\infty)$ by $\alpha((x,y),(u,v)) = 1$ whenever (x,y) and (u,v) are comparable and $\alpha((x,y),(u,v)) = 0$ otherwise. Then by using Theorem 8.3, F has a coupled fixed point.

♣

Corollary 8.5

Let (X,d) be a complete metric space, $F : X \times X \to X$ a continuous mapping, $(x^*,y^*) \in X \times X$ a fixed element and $\preceq$ an order on $X \times X$. Suppose that there exist $(x_0,y_0) \in X \times X$ such that

$$(x_0,y_0),\ (y_0,x_0),\ (F(x_0,y_0),F(y_0,x_0)) \text{ and } (F(y_0,x_0),F(x_0,y_0))$$

are comparable with (x^*,y^*), and a mapping $\psi \in \Psi$ such that $d(F(x,y),F(u,v)) \leq \frac{1}{2}\psi(m((x,y),(u,v)))$ for all (x,y) and (u,v) in $X \times X$ which are comparable with (x^*,y^*). Assume that $(F(x,y),F(y,x))$ and $(F(u,v),F(v,u))$ are comparable with (x^*,y^*) whenever (x,y) and (u,v) so are. Then F has a coupled fixed point.

Proof: Define the mapping $\alpha : X^2 \times X^2 \to [0,+\infty)$ by $\alpha((x,y),(u,v)) = 1$ whenever (x,y) and (u,v) are comparable with (x^*,y^*) and $\alpha((x,y),(u,v)) = 0$ otherwise. Then by using Theorem 8.4, F has a coupled fixed point.

♣

Theorem 8.5

Let (X,d) be a complete metric space and $F : X \times X \to X$ a mapping. Suppose that there exist a mapping $\alpha : X^2 \times X^2 \to [0,+\infty)$ and $\psi \in \Psi$ such that

$$\alpha((x,y),(u,v))d(F(x,y),F(u,v)) \leq \frac{1}{2}\psi(\delta((x,y),(u,v)))$$

for all $(x,y),(u,v) \in X \times X$, $\alpha((x,y),(u,v)) \geq 1$ implies

$$\alpha((F(x,y),F(y,x)),(F(u,v),F(v,u))) \geq 1$$

for all $(x,y),(u,v) \in X \times X$ *and there exists* $(x_0,y_0) \in X \times X$ *such that*

$$\alpha((x_0,y_0),(F(x_0,y_0),F(y_0,x_0))) \geq 1$$

and $\alpha((F(y_0,x_0),F(x_0,y_0)),(y_0,x_0)) \geq 1$.

Also, suppose that for each convergent sequences $\{x_n\}$ *and* $\{y_n\}$ *with*

$$x_n \to x,\ y_n \to y,$$

$\alpha((x_n,y_n),(x_{n+1},y_{n+1})) \geq 1$ *and* $\alpha((y_{n+1},x_{n+1}),(y_n,x_n)) \geq 1$ *for all n, we have*

$$\alpha((x_n,y_n),(x,y)) \geq 1 \text{ and } \alpha((y,x),(y_n,x_n)) \geq 1$$

for all n. Then F has a coupled fixed point.

Proof: The metric space $(X \times X, \delta)$ is complete and $\beta(\xi,\eta)\delta(T_F\xi,T_F\eta) \leq \psi(\delta(\xi,\eta))$ for all $\xi = (\xi_1,\xi_2), \eta = (\eta_1,\eta_2) \in X \times X$, where

$$\beta((\xi_1,\xi_2),(\eta_1,\eta_2)) = \min\{\alpha((\xi_1,\xi_2),(\eta_1,\eta_2)),\alpha((\eta_2,\eta_1),(\xi_2,\xi_1))\}.$$

Thus, $\beta(\xi,\eta)d(T_F\xi,T_F\eta) \leq \psi(\delta(\xi,\eta))$ for all $\xi,\eta \in X \times X$. Also, it is easy to check that T_F is β-admissible and $\beta((x_0,y_0),T_F(x_0,y_0)) \geq 1$. Since $(X \times X, \delta)$ has the property (B), T_F has a fixed point by using Lemma 8.9 and so F has a coupled fixed point.

♣

Theorem 8.6

Let (X,d) *be a complete metric space and* $F : X \times X \to X$ *a continuous mapping. Suppose that there exist a mapping* $\alpha : X^2 \times X^2 \to [0,+\infty)$ *and* $\psi \in \Psi$ *such that* $\alpha((x,y),(u,v))d(F(x,y),F(u,v)) \leq \frac{1}{2}\psi(\delta((x,y),(u,v)))$ *for all* $(x,y),(u,v) \in X \times X$, $\alpha((x,y),(u,v)) \geq 1$ *implies*

$$\alpha((F(x,y),F(y,x)),(F(u,v),F(v,u))) \geq 1$$

for all $(x,y),(u,v) \in X \times X$ *and there exists* $(x_0,y_0) \in X \times X$ *such that*

$$\alpha((x_0,y_0),(F(x_0,y_0),F(y_0,x_0))) \geq 1$$

and $\alpha((F(y_0,x_0),F(x_0,y_0)),(y_0,x_0)) \geq 1$. *Then F has a coupled fixed point.*

Proof: The metric space $(X \times X, \delta)$ is complete and $\beta(\xi,\eta)\delta(T_F\xi,T_F\eta) \leq \psi(\delta(\xi,\eta))$ for all $\xi = (\xi_1,\xi_2), \eta = (\eta_1,\eta_2) \in X \times X$, where

$$\beta((\xi_1,\xi_2),(\eta_1,\eta_2)) = \min\{\alpha((\xi_1,\xi_2),(\eta_1,\eta_2)),\alpha((\eta_2,\eta_1),(\xi_2,\xi_1))\}.$$

Thus, $\beta(\xi,\eta)d(T_F\xi,T_F\eta) \leq \psi(\delta(\xi,\eta))$ for all $\xi,\eta \in X \times X$. Also, it is easy to check that T_F is a continuous β-admissible mapping and $\beta((x_0,y_0),T_F(x_0,y_0)) \geq 1$. Now by using Lemma 8.10, T_F has a fixed point and so F has a coupled fixed point.

♣

Example 101 *Let $X = \mathbb{R}$ and $d(x,y) = |x-y|$ for all $x,y \in X$. Define the mapping $F : X \times X \to X$ by $F(x,y) = \frac{x-y}{3} + 1$ whenever $x \geq y$ and $F(x,y) = 0$ whenever $x < y$. Then F is a discontinuous mapping. Define $\alpha : X^2 \times X^2 \to [0,+\infty)$ by*

$$\alpha((x,y),(u,v)) = \begin{cases} 1 & x \geq y \text{ and } u \geq v, \\ 0 & \text{otherwise.} \end{cases}$$

Thus, for each $(x,y),(u,v) \in X \times X$ we have

$$\alpha((x,y),(u,v))d(F(x,y),F(u,v)) \leq |F(x,y) - F(u,v)| = |\frac{x-y}{3} - \frac{u-v}{3}|$$

$$= \frac{1}{3}|(x-u)+(v-y)| \leq \frac{1}{3}|(x-u)| + |(v-y)| = \frac{1}{3}(d(x,u)+d(y,v))$$

$$= \frac{1}{3}\delta((x,y),(u,v)) = \frac{1}{2}\psi(\delta((x,y),(u,v))),$$

where $\psi(t) = \frac{2t}{3}$ for all $t \geq 0$. Also, $\alpha((x,y),(u,v)) \geq 1$ implies

$$\alpha((F(x,y),F(y,x)),(F(u,v),F(v,u))) \geq 1$$

for all $(x,y),(u,v) \in X \times X$. Finally, $\alpha((x_0,y_0),(F(x_0,y_0),F(y_0,x_0))) \geq 1$ and

$$\alpha((F(y_0,x_0),F(x_0,y_0)),(y_0,x_0)) \geq 1,$$

where $(x_0,y_0) = (2,2)$. Thus by using Theorem 8.5, F has a coupled fixed point.

Corollary 8.6

Let (X,d) be a complete metric space, $F : X \times X \to X$ a mapping and $\preceq$ an order on $X \times X$. Suppose that there exists a mapping $\psi \in \Psi$ such that

$$d(F(x,y),F(u,v)) \leq \frac{1}{2}\psi(\delta((x,y),(u,v)))$$

for all comparable elements (x,y) and (u,v) in $X \times X$. Suppose that there exists $(x_0,y_0) \in X \times X$ such that $(F(x_0,y_0),F(y_0,x_0))$ and (x_0,y_0) are comparable and also $(F(y_0,x_0),F(x_0,y_0))$ and (y_0,x_0) are comparable, $(F(x,y),F(y,x))$ is comparable with $(F(u,v),F(v,u))$ whenever (x,y) is comparable with (u,v) and for each sequences $\{x_n\}$ and $\{y_n\}$ in X such that $x_n \to x$, $y_n \to y$, (x_n,y_n) and (x_{n+1},y_{n+1}) are *comparable elements of $X \times X$ for all n, (x_n, y_n) and (x, y) are comparable elements for all n. Then F has a coupled fixed point.*

Proof: Define the mapping $\alpha : X^2 \times X^2 \to [0, +\infty)$ by $\alpha((x,y),(u,v)) = 1$ whenever (x,y) and (u,v) are comparable and $\alpha((x,y),(u,v)) = 0$ otherwise. Then by using Theorem 8.5, F has a coupled fixed point.

♣

Corollary 8.7

Let (X,d) be a complete metric space, $F : X \times X \to X$ a continuous mapping, $(x^*, y^*) \in X \times X$ a fixed element and $\preceq$ an order on $X \times X$. Suppose that there exist $(x_0, y_0) \in X \times X$ such that

$$(x_0,y_0),\ (y_0,x_0),\ (F(x_0,y_0),F(y_0,x_0))\ \text{and}\ (F(y_0,x_0),F(x_0,y_0))$$

are comparable with (x^*, y^*), a mapping $\psi \in \Psi$ such that

$$d(F(x,y),F(u,v)) \le \frac{1}{2}\psi(\delta((x,y),(u,v)))$$

for all (x,y) and (u,v) in $X \times X$ which are comparable with (x^*, y^*). Assume that $(F(x,y),F(y,x))$ and $(F(u,v),F(v,u))$ are comparable with (x^*, y^*) whenever (x,y) and (u,v) so are. Then F has a coupled fixed point.

Proof: Define the mapping $\alpha : X^2 \times X^2 \to [0, +\infty)$ by $\alpha((x,y),(u,v)) = 1$ whenever (x,y) and (u,v) are comparable with (x^*, y^*) and $\alpha((x,y),(u,v)) = 0$ otherwise. Then by using Theorem 8.6, F has a coupled fixed point.

♣

8.4 Some false fixed point generalizations

Over last decades, many researchers tried to generalize fixed point results to common fixed point theorems. In this section, we show that some of these results are not real generalization, that is, we show that these common fixed point results could easily be obtained from the corresponding fixed point theorems. This shows that researchers should take care to obtain real generalizations.

First, we provide the next key result which is based on the axiom of choice.

Lemma 8.11

Let X be a nonempty set and $f : X \to X$ a function. Then there exists a subset $E \subseteq X$ such that $f(E) = f(X)$ and $f : E \to X$ is one-to-one.

Proof: Define a multifunction $F : f(X) \to 2^X$ by $F(y) = \{x \in X : f(x) = y\}$. By using the axiom of choice, F has a selector, that is, there is a function $g : f(X) \to X$ such that $g(y) \in F(y)$ for all $y \in f(X)$. Note that, $f(g(y)) = y$ for all $y \in f(X)$. Now, put $E = \{g(y) : y \in f(X)\}$. It is clear that f is one-to-one on E and $f(E) = f(X)$.

♣

8.4.1 Contractive maps

In 2007, Huang and Zhang proved the following result (Theorem 1 in [49]).

Theorem 8.7

Let (X,d) be a normal complete cone metric space and the mapping $f : X \to X$ satisfy the contractive condition

$$d(fx, fy) \leq kd(x,y),$$

for all $x,y \in X$, where $k \in [0,1)$ is a constant. Then, f has a unique fixed point in X. For each $x \in X$, the iterative sequence $\{f^n x\}_{n\geq 1}$ converges to the fixed point.

In 2008, Abbas and Jungck generalized it as following (Theorem 2.1 in [1]).

Theorem 8.8

Let (X,d) be a normal cone metric space and the self-mappings f and g on X satisfy the contractive condition

$$d(fx, fy) \leq kd(gx, gy),$$

for all $x,y \in X$, where $k \in [0,1)$ is a constant. If $f(X) \subseteq g(X)$ and $g(X)$ is a complete subset of X, then f and g have a unique point of coincidence in X. Moreover, if f and g are weakly compatible, then f and g have a unique common fixed point.

Theorem 8.9

Theorem 8.8 is a consequence of Theorem 8.7.

Proof: By using Lemma 8.11, there exists $E \subseteq X$ such that $g(E) = g(X)$ and $g : E \to X$ is one-to-one. Now, define a map $h : g(E) \to g(E)$ by $h(gx) = fx$. Since g is one-to-one on E, h is well-defined. Note that, $d(h(gx), h(gy)) \leq kd(gx, gy)$ for all $gx, gy \in g(E)$. Since $g(E) = g(X)$ is complete, by using Theorem 8.7, there exists $x_0 \in X$ such that $h(gx_0) = gx_0$. Hence, f and g have a point of, which is also unique coincidence. It is clear that f and g have a unique common fixed point whenever f and g are weakly compatible.

♣

Note that, not only Theorem 8.8 holds in non-normal cone metric spaces, but also by using a similar proof, it is a consequence of Theorem 2.3 of [85] in the non-normal case. Also, Abbas and Jungck proved the next result (Theorem 2.3 in [1]).

Theorem 8.10

Let (X,d) be a normal cone metric space and the self-mappings f and g on X satisfy the contractive condition

$$d(fx, fy) \leq k[d(fx, gx) + d(fy, gy)],$$

for all $x, y \in X$, where $k \in [0, \frac{1}{2})$ is a constant. If $f(X) \subseteq g(X)$ and $g(X)$ is a complete subset of X, then f and g have a unique point of coincidence in X. Moreover, if f and g are weakly compatible, then f and g have a unique common fixed point.

Also, Huang and Zhang proved the following result in 2007 (Theorem 3 in [49]).

Theorem 8.11

Let (X,d) be a normal complete cone metric space and the mapping $f : X \to X$ satisfy the contractive condition

$$d(fx, fy) \leq kd(x, fx) + kd(y, fy),$$

for all $x, y \in X$, where $k \in [0, \frac{1}{2})$ is a constant. Then, f has a unique fixed point in X. For each $x \in X$, the iterative sequence $\{f^n x\}_{n \geq 1}$ converges to the fixed point.

Theorem 8.12

Theorem 8.10 is a consequence of Theorem 8.11.

Proof: By Lemma 8.11, there exists $E \subseteq X$ such that $g(E) = g(X)$ and $g : E \to X$

is one-to-one. Now, define a map $h : g(E) \to g(E)$ by $h(gx) = fx$. Since g is one-to-one on E, h is well-defined. Note that, $d(h(gx),h(gy)) \le k[d(h(gx),gx) + d(h(gy),gy)]$ for all $gx, gy \in g(E)$. Since $g(E) = g(X)$ is complete, by using Theorem 3 in [49], there exists $x_0 \in X$ such that $h(gx_0) = gx_0$. Hence, f and g have a unique point of coincidence. It is clear that f and g have a unique common fixed point whenever f and g are weakly compatible.

♣

Again, not only Theorem 8.10 holds in non-normal cone metric spaces, but also by using a similar proof, it is a consequence of Theorem 2.6 of [85] in the non-normal case. By continuing this method, Theorem 2.4 of [1] is a consequence of Theorem 4 of [49] and is a consequence of Theorem 2.7 of [85] in the non-normal case.

The following example shows that completeness of $g(X)$ is necessary in Theorems 8.8 and 8.10.

Example 102 *Define the maps $f,g : [0,+\infty) \to [0,+\infty)$ by $fx = \frac{1}{x+1}$ and $gx = \frac{4}{x+1}$. Clearly, f and g satisfy the contractive conditions of Theorems 8.8 and 8.10, but f and g have not any coincidence point.*

In 2008, Ilic and Rakocevic proved the next result (Theorem 2.1 in [51]).

Theorem 8.4.1 *Let (X,d) be a normal complete cone metric space. Suppose that $f,g : X \to X$, f commutes with g, f or g is continuous, $g(X) \subseteq f(X)$ and g is f-quasicontraction, that is, there exists $\lambda \in (0,1)$ such that for each $x,y \in X$ there is*

$$u_{x,y} \in \{d(fx,fy), d(fx,gx), d(fx,gy), d(fy,gy), d(fy,gx)\}$$

so that $d(gx,gy) \le \lambda u_{x,y}$. Then f and g have a unique common fixed point.

Later, it proved without the assumption of normality (Theorem 2.1 in [84]). We recorded it before in Theorem 6.1. A selfmap $T : X \to X$ on a cone metric space (X,d) is called a quasi-contraction whenever there exists $\lambda \in (0,1)$ such that for each $x,y \in X$ there is $u \in \{d(x,y), d(x,Tx), d(y,Ty), d(x,Ty), d(y,Tx)\}$ so that $d(Tx,Ty) \le \lambda u$. Let us to restate Theorem 2.1 in [84].

Theorem 8.13

Let (X,d) be a cone metric space and $T : X \to X$ a quasi-contraction and $x \in X$. Then, the following statements hold.
(i) The orbit $O(x;\infty)$ is bounded,
(ii) $\{T^n x\}$ is Cauchy,

(iii) If X is complete, $\lim_{n\to\infty} T^n x = \bar{x}$ *exists and* $T(\bar{x}) = \bar{x}$*,*
(iv) $\bar{x}$ *is a unique fixed point of T.*

Theorem 8.14

Theorem 8.4.1 without the assumption of normality is a consequence of Theorem 8.13.

Proof: By Lemma 8.11, there exists $E \subseteq X$ such that $f(E) = f(X)$ and $f : E \to X$ is one-to-one. Now, define a map $h : f(E) \to f(E)$ by $h(fx) = gx$. Since f is one-to-one on E, h is well-defined. Note that, for each $x, y \in X$ there is an element

$$u_{x,y} \in \{d(fx, fy), d(fx, h(fx)), d(fx, h(fy)), d(fy, h(fy)), d(fy, h(fx))\}$$

such that $d(h(fx), h(fy)) \leq \lambda u_{x,y}$. Let $x_0 \in E$ and $y_0 = f(x_0)$. Choose $x_1 \in E$ such that $y_1 := f(x_1) = g(x_0) = h(y_0)$. By continuing this process, we obtain a sequence $\{y_n\}$ in $f(E)$ such that $y_{n+1} := f(x_{n+1}) = g(x_n) = h(y_n)$ for all $n \geq 0$. Now by using Theorem 8.13, $\{y_n\}$ is a Cauchy sequence in X because h is a quasi-contraction. Note that in the proof, we do not use the assumptions of completeness or normality, and main part of the proof of Lemma 2.1 in [51] relates to proving that the sequence $\{y_n\}$ is Cauchy in a normal cone metric space. Since X is complete, there exists $y \in X$ such that $y_n \to y$. Thus, $\lim_{n\to\infty} f(x_n) = \lim_{n\to\infty} g(x_n) = y$. Now, suppose that f is continuous. Then, for each $n \geq 1$ there exists

$$u_{x_n,y} \in \{d(f^2x_n, fy), d(f^2x_n, gfx_n), d(fy, gy), d(fy, gfx_n), d(f^2x_n, gy)\}$$

such that $d(fgx_n, gy) = d(gfx_n, gy) \leq \lambda u_{x_n,y}$. If $n \to \infty$, then we get $fy = gy$. Now, choose

$$u_{gy,y} \in \{d(fgy, fy), d(fgy, ggy), d(fgy, gy), d(fy, gy), d(fy, ggy)\}$$

such that $d(ggy, gy) = d(gfx_n, gy) \leq \lambda u_{gy,y}$. Hence, we get $ggy = gy$. Thus,

$$ggy = gfy = fgy = gy.$$

Hence, gy is a common fixed point of f and g. Now, suppose that g is continuous. Then, $fgx_n = gfx_n \to gy$. Thus, for each $n \geq 1$ there exists

$$u_{gx_n,x_n} \in \{d(fgx_n, fx_n), d(fgx_n, ggx_n), d(fgx_n, gx_n), d(fx_n, gx_n), d(fx_n, ggx_n)\}$$

such that $d(ggx_n, gx_n) \leq \lambda u_{gx_n,x_n}$. If $n \to \infty$, then we obtain $gy = y$. Since $g(X) \subseteq f(X)$, there exists $y' \in X$ such that $y = gy = fy'$. Now, for each $n \geq 1$ there exists

$$u_{gx_n,y'} \in \{d(fgx_n, fy'), d(fgx_n, ggx_n), d(fgx_n, gy'), d(fy', gy'), d(fy', ggx_n)\}$$

such that $d(ggx_n, gy') \leq \lambda u_{gx_n,y'}$. Hence, $gy' = y$ and so $fy = fgy' = gfy'gy = y$. Thus, y is a common fixed point of f and g. It is easy to see that f and g have a unique common fixed point.

♣

8.4.2 Contractive multifunctions

In 2007, Berinde et al. proved the following results (Theorems 3 and 4 in [20]). Let (X,d) be a metric space and $T : X \to \mathcal{P}(X)$ a set-valued operator. We say that T is a weak contraction or a (θ, L)-weak contractive set-valued map whenever there exist two constants $\theta \in (0,1)$ and $L \geq 0$ such that

$$H(Tx,Ty) \leq \theta d(x,y) + LD(y,Tx)$$

for all $x, y \in X$, where H is the Hausdorff metric and $D(x,Ty) = \inf_{t \in Ty} d(x,t)$.

Theorem 8.15

Let (X,d) be a complete metric space and $T : X \to \mathcal{P}(X)$ a multivalued (θ, L)-weak contraction. Then T has a fixed point and for any $x_0 \in X$ there exists an orbit $\{x_n\}$ of T at the point x_0 that converges to a fixed point u of T, for which $d(x_n, u) \leq \frac{h^n}{1-h} d(x_0, x_1)$ and $d(x_n, u) \leq \frac{h}{1-h} d(x_n, x_{n-1})$ for all $n \geq 1$, for a certain constant $h < 1$.

Theorem 8.16

Let (X,d) be a complete metric space and $T : X \to \mathcal{CB}(X)$ a generalized multivalued (α, L)-weak contraction, that is, there exists a function $\alpha : [0,\infty) \to [0,1)$ satisfying $\limsup_{r \to t^+} \alpha(r) < 1$ for every $t \in [0,\infty)$, such that

$$H(Tx,Ty) \leq \alpha(d(x,y))d(x,y) + LD(y,Tx)$$

for all $x, y \in X$. Then T has a fixed point.

Immediately, Kamran generalized the notions and the results in 2007 as follows (Theorems 2.9 and 3.5 in [63]). Let (X,d) be a metric space, $f : X \to X$ a mapping and $T : X \to \mathcal{CB}(X)$ a multivalued operator. T is said to be a f-weak contraction or a multivalued (f, θ, L)-weak contraction whenever there exist two constants $\theta \in (0,1)$ and $L \geq 0$ such that

$$H(Tx,Ty) \leq \theta d(fx,fy) + LD(fy,Tx)$$

for all $x, y \in X$.

Theorem 8.17

Let (X,d) be a metric space, $f : X \to X$ a map and $T : X \to \mathcal{CB}(X)$ a multivalued (f,θ,L)-weak contraction such that $TX \subseteq fX$. Suppose that fX is complete. Then $C(f,T) = \{u \in X : fu \in Tu\} \neq \emptyset$ and for any $x_0 \in X$ there exists an orbit $O_f(x_0) = \{fx_n : n \geq 1\}$ of T at x_0 such that $fx_n \to fu$, where u is a coincidence point of f and T, for which $d(fx_n, fu) \leq \frac{h^n}{1-h} d(fx_0, fx_1)$ and $d(fx_n, fu) \leq \frac{h}{1-h} d(fx_n, fx_{n-1})$ for all $n \geq 1$, for a certain constant $h < 1$. Further, if f is T-weakly commuting at u, that is, $ffu \in Tfu$, and $ffu = fu$, then f and T have a common fixed point.

Theorem 8.18

Let (X,d) be a metric space, $f : X \to X$ a map and $T : X \to \mathcal{CB}(X)$ a generalized (f,α,L)-weak contraction. Suppose that fX is complete. Then $C(f,T) = \{u \in X : fu \in Tu\} \neq \emptyset$. Further, if f is T-weakly commuting at u and $ffu = fu$, then f and T have a common fixed point.

Theorem 8.19

Theorem 8.17 is a consequence of Theorem 8.15.

Proof: By Lemma 8.11, there exists $E \subseteq X$ such that $f(E) = f(X)$ and $f : E \to X$ is one-to-one. Now, define a map $G : f(E) \to \mathcal{CB}(f(E))$ by $G(fx) = Tx$. Since f is one-to-one on E, G is well-defined. Note that,

$$H(G(fx), G(fy)) \leq \theta d(fx, fy) + LD(fy, G(fx))$$

for all $fx, fy \in f(E)$. By using Theorem 8.15, there exists $fx_0 \in f(E)$ such that $fx_0 \in H(fx_0) = Tx_0$. Hence, $C(f,T) = \{u \in X : fu \in Tu\} \neq \emptyset$. Now, we can complete the proof by following second part of the proof of Theorem 2.9 in [63].

♣

Finally, by using a similar proof we can prove the next result.

Theorem 8.20

Theorem 8.18 is a consequence of Theorem 8.16.

8.4.3 Nonexpansive maps

The following result of Suzuki ([99]) generalizes the well-known fixed point theorems of Edelestein (see [37]).

Theorem 8.21

Let (X,d) be a compact metric space and $S : X \to X$. Assume that

$$\frac{1}{2}d(x,Sx) < d(x,y) \text{ implies } d(Sx,Sy) < d(x,y)$$

for $x,y \in X$. Then S has a unique fixed point.

Singh and Mishra generalized Theorem 8.21 as follows (see Theorem 2.1 in [96]).

Theorem 8.22

Let $S,T : Y \to X$ be such that $S(Y) \subseteq T(Y)$ and $T(Y)$ is a compact subspace of X. Assume that for $x,y \in Y$,

$$\frac{1}{2}d(Tx,Sx) < d(x,y) \text{ implies } d(Sx,Sy) < d(Tx,Ty),$$

and $Tx = Ty$ implies $Sx = Sy$. Then S and T have a coincidence, that is, there exist $z \in Y$ such that $Sz = Tz$. Further, if $Y = X$, then S and T have a unique common fixed point provided that S and T commute at z.

Theorem 8.23

Theorem 8.22 is a consequence of Theorem 8.21.

Proof: By Lemma 8.11, there exists $E \subseteq X$ such that $T(E) = T(X)$ and $T : E \to X$ is one-to-one. Now, define a map $h : T(E) \to T(E)$ by $h(Tx) = Sx$. Since T is one to one, h is well-defined. Also, for any $Tx, Ty \in T(E)$, we have

$$\frac{1}{2}d(Tx,h(Tx)) < d(x,y) \text{ implies } d(h(Tx),h(Ty)) < d(Tx,Ty).$$

By Theorem 8.21, h has a fixed point in $T(E)$. Thus, S and T have a coincidence point. Further, if $Y = X$, clearly, S and T have a unique common fixed point provided that S and T commute at their coincidence point.

♣

In 2008, Suzuki generalized the result of Browder (in [23] and [24]) and Gohde (in [43]) as follows.

Theorem 8.24

Let C be a convex subset of a Banach space E and $S : C \to C$. Assume that

$$\frac{1}{2}\|x - Sx\| \leq \|x - y\| \text{ implies } \|Sx - Sy\| \leq \|x - y\|$$

for all $x,y \in C$. Assume further that either of the following holds:

- *C is compact;*
- *C is weakly compact and E has the Opial property;*
- *C is weakly compact and E is uniformly convex in every direction (UCED).*

Then S has a fixed point.

Again, Singh and Mishra generalized 8.24 as follows (see Theorem 3.2 in [96]).

Theorem 8.25

Let E be a Banach space and $S,T : Y \to E$ such that $S(Y) \subseteq T(Y)$ and $T(Y)$ is a convex subset of E. Assume that for $x,y \in Y$,

$$\tfrac{1}{2}\|Tx - Sx\| \leq \|x-y\| \text{ implies } \|Sx - Sy\| \leq \|Tx - Ty\|$$

and $Tx = Ty$ implies $Sx = Sy$. Assume further that either of the following holds:

- *$T(Y)$ is compact;*
- *$T(Y)$ is weakly compact and E has the Opial property;*
- *$T(Y)$ is weakly compact and E is (UCED).*

Then S and T have a coincidence.

By using a similar proof, one can prove the next result.

Theorem 8.26

Theorem 8.25 is the consequence of Theorem 8.24.

8.5 Partial metrics

In 1992, Matthews introduced partial metric spaces ([72]). In a partial metric space, the distance of a point from itself may not be zero. After the definition of partial metric spaces, Matthews proved the partial metric version of the Banach fixed point theorem. Later, many researchers studied partial metric spaces and related topological properties. Also, some researchers provided some fixed point generalizations on partial metric spaces. But, there is a big misunderstanding in partial metric fixed point results and researchers should know about it. In this section, we show that fixed point generalizations on partial metric spaces are not always real generalizations. We consider some cases to demonstrate this fact. A partial metric is a function $p : X \times X \to [0,\infty)$ satisfying the following conditions

(a) $p(x,y) = p(y,x)$,

(b) If $p(x,x) = p(x,y) = p(y,y)$, then $x = y$,

(c) $p(x,x) \leq p(x,y)$,

(d) $p(x,z) + p(y,y) \leq p(x,y) + p(y,z)$,

for all $x,y,z \in X$. Then (X,p) is called a partial metric space. Also, each partial metric p on X generates a T_0 topology τ_p on X with a base of the family of open p-balls $\{B_p(x,\varepsilon) : x \in X,\ \varepsilon > 0\}$, where $B_p(x,\varepsilon) = \{y \in X : p(x,y) < p(x,x) + \varepsilon\}$ ([72] and [87]).

If p is a partial metric p on X, then $p^s : X \times X \to [0,\infty)$ defined by

$$p^s(x,y) = 2p(x,y) - p(x,x) - p(y,y)$$

is a metric on X ([72]). In fact, $\lim_{n\to\infty} p^s(x_n,x) = 0$ if and only if

$$p(x,x) = \lim_{n\to\infty} p(x_n,x) = \lim_{n,m\to\infty} p(x_n,x_m).$$

Let (X,p) be a partial metric space. A sequence $\{x_n\}_{n\geq 1}$ converges to $x \in X$ with respect to τ_p if and only if $\lim_{n\to\infty} p(x_n,x) = p(x,x)$. Also, $\{x_n\}_{n\geq 1}$ is called a Cauchy sequence whenever $\lim_{n,m\to\infty} p(x_n,x_m)$ exists and finite. A partial metric space (X,p) is said to be complete whenever every Cauchy sequence $\{x_n\}_{n\geq 1}$ in X converges to a point $x \in X$ with respect to τ_p, that is, $\lim_{n,m\to\infty} p(x_n,x_m) = p(x,x)$.

Definition 8.2 ([50]) A sequence $\{x_n\}_{n\geq 1}$ in a partial metric space (X,p) is called 0-Cauchy if $\lim_{n,m\to\infty} p(x_n,x_m) = 0$. We say that (X,p) is 0-complete if each 0-Cauchy sequence in X converges to a point $x \in X$ such that $p(x,x) = 0$.

Note that, each 0-Cauchy sequence in (X,p) is Cauchy in (X,p^s) and every complete partial metric space is 0-complete.

Lemma 8.12

([72]) Let (X,p) be a partial metric space. Then
(i) $\{x_n\}_{n\geq 1}$ is a Cauchy sequence in (X,p) if and only if $\{x_n\}_{n\geq 1}$ is a Cauchy sequence in the metric space (X,p^s),
(ii) (X,p) is complete if and only if the metric space (X,p^s) is complete.

Here, we provide the next key result which has been proved by Hitzler and Seda in [47]. Since it is important for our aim, we reprove it here. In fact, it is the mixing of Propositions 4.8.8 and 4.8.10 in [47].

Proposition 8.2

Let (X,p) be a partial metric space. Then the function $d : X \times X \to [0,+\infty)$ defined by $d(x,y) = 0$ whenever $x = y$ and $d(x,y) = p(x,y)$ whenever $x \neq y$, is a metric on X such that $\tau_{p^s} \subseteq \tau_d$. Moreover, (X,d) is complete if and only if (X,p) is 0-complete.

Proof: It is clear that $d(x,y) = 0$ if and only if $x = y$ and $d(x,y) = d(y,x)$ for all $x,y \in X$. Let $x,y,z \in X$. Note that, $d(x,y) \leq p(x,y) \leq p(x,z) + p(y,z) - p(z,z)$. If $x \neq y$ and $x = z$, then $d(x,y) \leq p(z,z) + d(z,y) - p(z,z) = d(z,y) = d(x,y)$. If $x \neq y$ and $y = z$, then $d(x,y) \leq d(x,z) + p(y,y) - p(y,y) = d(x,y)$. If $x = y$, then

$$d(x,y) = 0 \leq d(x,z) + d(z,y).$$

Thus, (X,d) is a metric space. Now, suppose that (X,p) is 0-complete and $\{x_n\}_{n\geq 1}$ is a Cauchy sequence in (X,d). Without lose of the generality, we can assume that $x_n \neq x_m$ for all $n \neq m$. Hence, $d(x_n,x_m) = p(x_n,x_m)$ for all $n,m \geq 1$. Thus, $\{x_n\}_{n\geq 1}$ is a Cauchy sequence in (X,p). Since (X,p) is 0-complete, $\lim_{n\to\infty} p(x_n,x^*) = 0$ for some $x^* \in X$. Note that, $x^* \neq x_n$ for all n. Therefore, $\lim_{n\to\infty} d(x_n,x^*) = 0$ and so (X,d) is a complete metric space. Now, suppose that (X,d) is complete and $\{x_n\}_{n\geq 1}$ is a 0-Cauchy sequence in (X,p). Without lose of the generality, assume that $x_n \neq x_m$, for all $n \neq m$. Then, $p(x_n,x_m) = d(x_n,x_m)$ for all $n,m \geq 1$. Hence, $\{x_n\}_{n\geq 1}$ is a Cauchy sequence in (X,d). Since (X,d) is complete, there exists $x^* \in X$ such that $\lim_{n\to\infty} d(x_n,x^*) = 0$. Thus, $\lim_{n\to\infty} p(x_n,x^*) = 0$ and so (X,p) is 0-complete.

♣

Now, we show that there are partial fixed point results which can be obtained from the corresponding results in metric spaces. Thus, you should be careful in obtaining partial metric fixed point results.

Let (X,p) a partial metric space, T a selfmap on X, d the constructed metric in Proposition 8.2 and $x,y \in X$. Define

$$M_d(x,y) = \max\{d(x,y), d(x,Tx), d(y,Ty), \frac{1}{2}[d(x,Ty) + d(y,Tx)]\}$$

and

$$M_p(x,y) = \max\{p(x,y), p(x,Tx), p(y,Ty), \frac{1}{2}[p(x,Ty) + p(y,Tx)]\}.$$

Lemma 8.13

$M_d(x,y) = M_p(x,y)$ for all $x,y \in X$ with $x \neq y$.

Proof: First, we prove $M_p(x,y) \leq M_d(x,y)$ for all $x,y \in X$ with $x \neq y$. Let $x,y \in X$ with $x \neq y$. If $M_p(x,y) = p(x,y)$, then $M_p(x,y) = d(x,y) \leq M_d(x,y)$. If $M_p(x,y) = p(x,Tx)$, then $M_p(x,y) = p(x,x) \leq d(x,y) \leq M_d(x,y)$ whenever $x = Tx$ and

$$M_p(x,y) = d(x,Tx) \leq M_d(x,y)$$

whenever $x \neq Tx$. Similarly, If $M_p(x,y) = p(y,Ty)$, then $M_p(x,y) \leq M_d(x,y)$. Finally, let $M_p(x,y) = \frac{1}{2}[p(x,Ty) + p(y,Ty)]$. If $x = Ty$ and $y = Tx$, then

$$M_p(x,y) \leq d(x,y) \leq M_d(x,y).$$

If $x = Ty$ and $y \neq Tx$, then $M_p(x,y) \leq \frac{1}{2}[d(x,y) + d(y,Tx)] \leq M_d(x,y)$.

Similarly, $M_p(x,y) \leq M_d(x,y)$ whenever $y = Tx$. If $x \neq Ty$ and $y \neq Tx$, then

$$M_p(x,y) = \frac{1}{2}[d(x,Ty) + d(y,Tx)] \leq M_d(x,y).$$

Now, we show that $M_d(x,y) \leq M_p(x,y)$ for all $x,y \in X$ with $x \neq y$. Let $x,y \in X$ with $x \neq y$. If $M_d(x,y) = d(x,y)$, then $M_d(x,y) = p(x,y) \leq M_p(x,y)$. If $M_d(x,y) = d(x,Tx)$, then $M_d(x,y) = 0 \leq M_p(x,y)$ whenever $x = Tx$ and

$$M_d(x,y) = p(x,Tx) \leq M_p(x,y)$$

whenever $x \neq Tx$. Similarly, If $M_d(x,y) = d(y,Ty)$, then $M_d(x,y) \leq M_p(x,y)$. Let $M_d(x,y) = \frac{1}{2}[d(x,Ty) + d(y,Ty)]$. If $x = Ty$ and $y = Tx$, then $M_d(x,y) = 0 \leq M_p(x,y)$. If $x = Ty$ and $y \neq Tx$, then

$$M_d(x,y) = \frac{1}{2}p(y,Tx) \leq \frac{1}{2}[p(x,Ty) + p(y,Tx)] \leq M_p(x,y).$$

Similarly, $M_d(x,y) \leq M_p(x,y)$ whenever $y = Tx$. If $x \neq Ty$ and $y \neq Tx$, then $M_d(x,y) = \frac{1}{2}[p(x,Ty) + p(y,Tx)] \leq M_p(x,y)$. Thus,

$$M_p(x,y) = M_d(x,y)$$

for all $x,y \in X$ with $x \neq y$.

♣

The next result shows that Theorem 1 in [13] is a consequence of Theorem 2 in [22].

Theorem 8.27

Let (X,p) be a complete partial metric space, $\phi : [0,+\infty) \to [0,+\infty)$ a continuous and nondecreasing function such that $\phi(t) < t$ for all $t > 0$ and T a selfmap on X satisfying $p(Tx,Ty) \leq \phi(M_p(x,y))$ for all $x,y \in X$. Then T has a unique fixed point.

Proof: By using Lemma 8.13, $M_p(x,y) = M_d(x,y)$ for all $x,y \in X$ with $x \neq y$, where d is the constructed metric in Proposition 8.2. Thus,

$$d(Tx,Ty) \leq p(Tx,Ty) \leq \phi(M_p(x,y)) = \phi(M_d(x,y))$$

for all $x,y \in X$ with $x \neq y$. But, the inequality holds trivially for $x = y$. Hence,

$$d(Tx,Ty) \leq p(Tx,Ty) \leq \phi(M_p(x,y)) = \phi(M_d(x,y))$$

for all $x,y \in X$. By using Proposition 8.2, (X,d) is a complete metric space. Now by using special case of Theorem 2 in [22], T has a unique fixed point.

♣

Also, the next result shows that Theorem 6 in [5] is a consequence of the main result of [36].

Theorem 8.28

Let $\psi, \varphi : [0,+\infty) \to [0,+\infty)$ continuous and nondecreasing functions such that $\phi(t) = \varphi(t) = 0$ if and only if $t = 0$, (X,p) be a complete partial metric space and T a selfmap on X satisfying $\psi(p(Tx,Ty)) \leq \psi(p(x,y)) - \varphi(p(x,y))$ for all $x,y \in X$ with $x \neq y$. Then T has a unique fixed point.

Proof: By using Proposition 8.2, (X,d) is a complete metric space, where d is the constructed metric. Thus,

$$\psi(d(Tx,Ty)) \leq \psi(p(Tx,Ty)) \leq \psi(p(x,y)) - \varphi(p(x,y)) = \psi(d(x,y)) - \varphi(d(x,y))$$

for all $x,y \in X$ with $x \neq y$. Again, the inequality holds trivially for $x = y$. Hence,

$$\psi(d(Tx,Ty)) \leq \psi(p(Tx,Ty)) \leq \psi(p(x,y)) - \varphi(p(x,y)) = \psi(d(x,y)) - \varphi(d(x,y))$$

for all $x,y \in X$. Now by using the main result of [36], T has a unique fixed point.

♣

The next result shows that Theorem 5 in [4] is a consequence of Theorem 1.3 of [8].

Theorem 8.29

Let (X,p) be a complete partial metric space, $\phi : [0,+\infty) \to [0,+\infty)$ a continuous and nondecreasing function such that $\phi(t) > 0$ for all $t > 0$ and $\phi(0) = 0$. Suppose that T and S are selfmaps on X such that

$$p(Sx,Ty) \leq M_p(x,y,S,T) - \phi(M_p(x,y,S,T))$$

for all $x,y \in X$, where

$$M_p(x,y,S,T) = \max\{p(x,y), p(Sx,x), p(y,Ty), \frac{1}{2}[p(Sx,y) + p(x,Ty)]\}.$$

Then T and S have a unique common fixed point.

Proof: If $x = y = Sx = Ty$, then we have nothing to prove. Otherwise, by using a similar argument in proof of Lemma 8.13, one can show that

$$M_p(x,y,S,T) = M_d(x,y,S,T).$$

Thus, we obtain $d(Sx,Ty) \leq M_d(x,y,S,T) - \phi(M_d(x,y,S,T))$ for all $x,y \in X$, where

$$M_d(x,y,S,T) = \max\{d(x,y), d(Sx,x), d(y,Ty), \frac{1}{2}[d(Sx,y) + d(x,Ty)]\}.$$

Now by using Theorem 1.3 in [8], S and T have a unique common fixed point.

♣

The following result shows that Theorem 2 in [13] is a consequence of the main result of [30].

Theorem 8.30

Let (X,p) be a complete partial metric space, a,b,c,d,e nonnegative numbers such that $a+b+c+d+e < 1$ when $d \geq e$ and $a+b+c+d+2e < 1$ when $d < e$ and T a selfmap on X satisfying

$$p(Tx,Ty) \leq ap(x,y) + bp(x,Tx) + cp(y,Ty) + dp(x,Ty) + ep(y,Tx)$$

for all $x,y \in X$. Then T has a unique fixed point.

Proof: If $\lambda = a+b+c+d+e$, then $\lambda < 1$. By using a similar argument in proof of Lemma 8.13, one can show that

$$\max\{p(x,y), p(x,Tx), p(y,Ty), p(x,Ty), p(y,Tx)\}$$

$$= \max\{d(x,y), d(x,Tx), d(y,Ty), d(x,Ty), d(y,Tx)\}$$

for all $x, y \in X$ with $x \neq y$. Thus, we get

$$d(Tx,Ty) \leq p(Tx,Ty) \leq \lambda \max\{p(x,y), p(x,Tx), p(y,Ty), p(x,Ty), p(y,Tx)\}$$

$$= \lambda \max\{d(x,y), d(x,Tx), d(y,Ty), d(x,Ty), d(y,Tx)\}$$

for all $x, y \in X$ with $x \neq y$. But, the inequality holds trivially for $x = y$. Hence,

$$d(Tx,Ty) \leq \lambda \max\{d(x,y), d(x,Tx), d(y,Ty), d(x,Ty), d(y,Tx)]\}$$

for all $x, y \in X$. Since (X,d) is a complete metric space, by using the main result of [30], T has a unique fixed point.

♣

Note that the cone version of the main result of [30] has been proved by the authors in [84]. Next result shows that Theorem 8 in [3] is a consequence of Theorem 3.1 in [29].

Theorem 8.31

Let $\psi, \varphi : [0,+\infty) \to [0,+\infty)$ continuous and nondecreasing functions such that $\phi(t) = \varphi(t) = 0$ if and only if $t = 0$, (X,p) be a complete partial metric space and T a selfmap on X satisfying

$$\psi(p(Tx,Ty)) \leq \psi(M_p(x,y)) - \phi(\max\{p(x,y), p(y,Ty)\})$$

for all $x, y \in X$ with $x \neq y$. Then T has a unique fixed point.

Proof: By using Proposition 8.2, (X,d) is a complete metric space, where d is the constructed metric. Again by using a similar argument in proof of Lemma 8.13, one can show that $\max\{p(x,y), p(y,Ty)\} = \max\{d(x,y), d(y,Ty)\}$ for all $x, y \in X$ with $x \neq y$. Thus, by using Lemma 8.13 we obtain

$$\psi(d(Tx,Ty)) \leq \psi(p(Tx,Ty)) \leq \psi(M_d(x,y)) - \phi(\max\{d(x,y), d(y,Ty)\})$$

for all $x, y \in X$ with $x \neq y$. If we review the proof of Theorem 3.1 in [29] carefully, then we get the result holds again when one use the main inequality for all $x, y \in X$ with $x \neq y$ instead for all $x, y \in X$. Thus, by using this version of Theorem 3.1 in [29], T has a unique fixed point.

♣

Finally, our next result shows that Theorem 2.1 in [31] is a consequence of Theorem 2.4 of [8].

Theorem 8.32

Let (X,p) be a complete partial metric space, $\phi : [0,+\infty) \to [0,+\infty)$ a continuous and nondecreasing function such that $\phi(t) < t$ for all $t > 0$, $\phi(t) = 0$ if and only if $t = 0$ and A, B, S and T four selfmaps on X such that $AX \subseteq TX$, $BX \subseteq SX$ and

$$p(Ax,By) \le \phi(\max\{p(Sx,Ty),p(Ax,Sx),p(By,Ty),\frac{1}{2}[p(Sx,By)+p(Ax,Ty)]\})$$

for all $x,y \in X$. If one of the ranges AX, BX, TX and SX is a closed subset of X, then A and S have a coincidence point and A and S have a coincidence point. Moreover, if the pairs $\{A,S\}$ and $\{B,T\}$ are weakly compatible, then A, B, T and S have a unique common fixed point.

Proof: By using a similar argument in proof of Lemma 8.13, one can show that

$$\max\{p(Sx,Ty),p(Ax,Sx),p(By,Ty),\frac{1}{2}[p(Sx,By)+p(Ax,Ty)]\})$$

$$= \max\{d(Sx,Ty),d(Sx,Ax),d(By,Ty),\frac{1}{2}[d(Sx,By)+d(Ax,Ty)]\}$$

for all $x,y \in X$ with $x \neq y$ and also

$$\max\{p(Sx,Tx),p(Ax,Sx),p(Bx,Tx),\frac{1}{2}[p(Sx,Bx)+p(Ax,Tx)]\})$$

$$= \max\{d(Sx,Tx),d(Sx,Ax),d(Bx,Tx),\frac{1}{2}[d(Sx,Bx)+d(Ax,Tx)]\}.$$

Thus, we obtain

$$d(Ax,By) \le \phi(\max\{d(Sx,Ty),d(Sx,Ax),d(By,Ty),\frac{1}{2}[d(Sx,By)+d(Ax,Ty)]\})$$

for all $x,y \in X$. Now by considering $I - \phi$ instead of ϕ in Theorem 2.4 of [8], one can obtain the result.

♣

Now it is up to you whether you want to present results that are questionable or not? This does happen. Just be a little more careful.

Our whole goal in this book was to raise your awareness in this field and we hope that by reading this book you will be able to present more beautiful results.

It may have been clear to many at first that this was not a textbook, and that its purpose was to raise awareness among young researchers.

References

[1] M. Abbas and G. Jungck. Common fixed point results for noncommuting mappings without continuity in cone metric spaces. *Journal of Mathematical Analysis and Applications*, 341(1):416–420, 2008.

[2] M. Abbas and T. Nazir. A new faster iteration process applied to constrained minimization and feasibility problems. *Mat. Vesnik*, 66(2):223–234, 2014.

[3] T. Abdeljawad. Fixed points for generalized weakly contractive mappings in partial metric spaces. *Mathematical and Computer Modelling*, 54(11-12):2923–2927, 2011.

[4] T. Abdeljawad, E. Karapinar, and K. Tas. Existence and uniqueness of a common fixed point on partial metric spaces. *Applied Mathematics Letters*, 24(11):1900–1904, 2011.

[5] T. Abdeljawad, E. Karapinar, and K. Tas. A generalized contraction principle with control functions on partial metric spaces. *Computers and Mathematics with Applications*, 63(3):716–719, 2012.

[6] R. P. Agarwal, D. O'Regan, and D. R. Sahu. Iterative construction of fixed points of nearly asymptotically nonexpansive mappings. *Nonlinear Convex Analysis*, 8(1):61–79, 2007.

[7] R. P. Agarwal, D. O'Regan, and D. R. Sahu. *Fixed point theory for Lipschitzian-type mappings with applications*. Springer-Verlag, 2009.

[8] M. Akkouchi. Common fixed points of four maps using generalized weak contractivity and well-posedness. *International Journal of Nonlinear Analysis and Applications*, 2(1):73–81, 2011.

[9] M. A. Al-Thagafi and N. Shahzad. Convergence and existence results for best proximity points. *Nonlinear Analysis: Theory, Methods and Applications*, 70(10):3665–3671, 2009.

[10] S. M. A. Aleomraninejad, Sh. Rezapour, and N. Shahzad. Some fixed point results on a metric space with a graph. *Topology and its Applications*, 159:659–663, 2012.

[11] M. A. Alghamdi, Sh. H. Alnafei, S. Radenovic, and N. Shahzad. Fixed point theorems for convex contraction mappings on cone metric spaces. *Mathematical and Computer Modelling*, 54:2020–2026, 2011.

[12] I. Altun and A. Erduran. Fixed point theorems for monotone mappings on partial metric spaces. *Fixed Point Theory and Applications*, 2011. Article ID 508730.

[13] I. Altun, F. Sola, and H. Simsek. Generalized contractions on partial metric spaces. *Topology and its Applications*, 157(18):2778–2785, 2010.

[14] A. Amini-Harandi. Endpoints of set-valued contractions in metric spaces. *Nonlinear Analysis: Theory, Methods and Applications*, 72(1):132–134, 2010.

[15] A. Amini-Harandi. Fixed point theory for set-valued quasi-contraction maps in metric spaces. *Applied Mathematics Letters*, 24(11):1791–1794, 2011.

[16] D. Ariza-Ruiza, A. Jimenez-Melado, and G. Lopez-acedo. A fixed point theorem for weakly zamfirescu mappings. *Nonlinear Analysis: Theory, Methods and Applications*, 74:1628–1640, 2011.

[17] A. Azam, M. Arshad, and I. Beg. Common fixed points of two maps in cone metric spaces. *Rendiconti del Circolo Matematico di Palermo*, 57:433–441, 2008.

[18] J. S. Bae and M. S. Park. Fixed point theorems for multivalued mappings in banach spaces. *Chungcheong Math. Soc*, 3:659–663, 1990.

[19] M. Berind. Approximate fixed point theorems. *Stud. Univ. Babes-Bolyai Math.*, 51:11–25, 2006.

[20] M. Berinde and V. Berinde. On general class of multivalued weakly picard mappings. *Journal of Mathematical Analysis and Applications*, 326(2):772–782, 2007.

[21] V. Berinde. *Iterative approximation of fixed points*. Springer-Verlag, 2007.

[22] V. Berinde. A common fixed point theorem for compatible quasi contractiv self mappings in metric spaces. *Applied Mathematics and Computation*, 213(2):348–354, 2009.

[23] F. E. Browder. Nonexpansive nonlinear operators in a banach spaces. *Proceedings of the National Academy of Sciences*, 54(4):1041–1044, 1965.

[24] F. E. Browder. Nonexpansive nonlinear operators in a banach spaces. *Proceedings of the National Academy of Sciences*, 53(6):1272–1276, 1965.

[25] A. Bucur, L. Guran, and A. Petrusel. Fixed points for multivalued operators on a set endowed with vector-valued metrics and applications. *Fixed Point Theory*, 10:19–34, 2009.

[26] S. S. Chang and Y. H. Ma. Coupled fixed points for mixed monotone condensing operators and an existence theorem of the solutions for a class of functional equations arising in dynamic programming. *Journal of Mathematical Analysis and Applications*, 160(2):468–479, 1991.

[27] S. K. Chatterjea. Fixed-point theorems. *C. R. Acad. Bulgare Sci.*, 25:727–730, 1972.

[28] Y. Z. Chen. Existence theorems of coupled fixed points. *Journal of Mathematical Analysis and Applications*, 154(1):142–150, 1991.

[29] B. S. Choudhury, P. Konar, B. E. Rhoades, and N. Metiya. Fixed point theorems for generalized weakly contractive mappings. *Nonlinear Analysis: Theory, Methods and Applications*, 74(6):2116–2126, 2011.

[30] L. B. Ciric. A generalization of banachs contraction principle. *Proceedings of the American Mathematical society*, 45(2):267–273, 1974.

[31] L. B. Ciric, B. Samet, H. Aydi, and C. Vetro. Common fixed points of generalized contractions on partial metric spaces and an application. *Applied Mathematics and Computation*, 218(6):2398–2406, 2011.

[32] A. Constantin. A random fixed point theorem for multifunctions. *Stochastic Analysis and Applications*, 12:65–73, 1994.

[33] S. Czerwik. Nonlinear set-valued contraction mappings in b-metric spaces. *Atti Sem. Mat. Fis. Univ. Modena*, 46:263–276, 1998.

[34] S. Dhompongsa and H. Yingtaweesittikul. Fixed point for multivalued mappings and the metric completeness. *Fixed Point Theory and Applications*, 2009. Article ID 972395.

[35] A. Djoudi and F. Merghadi. Common fixed point theorems for maps under contractive condition of integral type. *Journal of Mathematical Analysis and Applications*, 341(2):953–960, 2008.

[36] N. P. Dutta and B. S. Choudhury. A generalizations of contractiion principle in metric spaces. *Fixed Point Theory and Applications*, pages 1–8, 2008. Article ID 406368.

[37] M. Edelstein. On fixed and periodic points under contractive mappings. *Journal of the London Mathematical Society*, 1(1):74–79, 1962.

[38] A. A. Eldered and P. Veeramani. Existence and convergence of best proximity points. *Journal of Mathematical Analysis and Applications*, 323(2):1001–1006, 2006.

[39] Sh. Fathollahi, A. Ghiura, M. Postolache, and Sh. Rezapour. A comparative study on the convergence rate of some iteration methods involving contractive mappings. *Fixed Point Theory and Applications*, 2015(234):1–24, 2015.

[40] Sh. Fathollahi and Sh. Rezapour. On the rate of convergence of some iteration methods for zamfirescu maps. *J. Adv. Math. Stud.*, 10(1):84–94, 2017.

[41] Sh. Fathollahi and Sh. Rezapour. Efficacy of coefficients on rate of convergence of some iteration methods for quasi-contractions. *Iranian Journal of Science and Technology, Transactions A: Science*, 42:1517–1523, 2018.

[42] V. Ghorbanian, Sh. Rezapour, and N. Shahzad. Some ordered fixed point results and the property (p). *Computers and Mathematics with Applications*, 63:1361–1368, 2012.

[43] D. Gohde. Zum prinzip der kontractiven abbildung. *Mathematische Nachrichten*, 30(3-4):251–258, 1965.

[44] D. Guo and V. Lakshmikantham. Coupled fixed points of nonlinear operators with applications. *Nonlinear Analysis: Theory, Methods and Applications*, 11(5):623–632, 1987.

[45] R. H. Haghi, Sh. Rezapour, and N. Shahzad. Some fixed point generalizations are not real generalizations. *Nonlinear Analysis: Theory, Methods and Applications*, 74(5):1799–1803, 2011.

[46] S. Heikkila and V. Lakshmikantham. A unified theory for first-order discontinuous scalar differential equations. *Nonlinear Analysis: Theory, Methods and Applications*, 26(4):785–797, 1996.

[47] P. Hitzler and A. Seda. *Mathematical Aspects of Logic Programming Semantics*. Taylor and Francis, 2011.

[48] L. G. Huang and X. Zhang. Cone metric spaces and fixed point theorems of contractive mappings. *J. Math. Anal. Appl*, 332(2):1468–1476, 2007.

[49] L. G. Huang and X. Zhang. Cone metric spaces and fixed point theorems of contractive mappings. *J. Math. Anal. Appl.*, 332(2):1468–1476, 2007.

[50] D. Ilic, V. Pavlovic, and V. Rakoceviv. Some new extensions of banach's contraction principle to partial metric space. *Applied Mathematics Letters*, 24(8):1326–1330, 2011.

[51] D. Ilic and V. Rakocevic. Common fixed points for maps on cone metric space. *Journal of Mathematical Analysis and Applications*, 341(2):876–882, 2008.

[52] S. Ishikawa. Fixed points by a new iteration method. *Proceedings of the American Mathematical Society*, 44(1):147–150, 1974.

[53] V. I. Istratescu. Some fixed point theorems for convex contraction mappings and mappings with convex diminishing diameters-i. *Annali di Matematica Pura ed Applicata*, 130:89– 104, 1982.

[54] J. Jachymski. The contraction principle for mappings on a metric space with a graph. *Proceedings of the American Mathematical Society*, 136:1359–1373, 2008.

[55] J. Jachymski. Equivalent conditions for generalized contractions on (ordered) metric spaces. *Nonlinear Analysis: Theory, Methods and Applications*, 74(3):768–774, 2011.

[56] T. D. Jeffres. *Kahler-Einstein cone metrics*. State University of New York at Stony Brook, 1996.

[57] T. D. Jeffres. Schwarz lemma for kahler cone metrics. *International Mathematics Research Notices*, 2000(7):371–382, 2000.

[58] T. D. Jeffres. Uniqueness of kahler-einstein cone metrics. *Publicacions matematiques*, 44(2):437–448, 2000.

[59] S. G. Jeong and B. E. Rhoades. *Maps for which $F(T) = F(T^n)$*, volume 6. Nova Sci. Publ., New York, 2007.

[60] S. Jingxian and L. Lishan. Iterative method for coupled quasi-solutions of mixed monotone operator equations. *Applied Mathematics and Computation*, 52(2-3):301–308, 1992.

[61] M. Jleli, E. Karapınar, and B. Samet. Further generalizations of the banach contraction principle. *Journal of Inequalities and Applications*, 439, 2014.

[62] Z. Kadelburg, S. Radenovic, and V. Rakocevic. Remarks on "quasi-contracion on a cone metric space". *Applied Mathematics Letters*, 22:1674–1679, 2009.

[63] T. Kamran. Multivalued f-weakly picard mappings. *Nonlinear Analysis: Theory, Methods and Applications*, 67(7):2289–2296, 2007.

[64] R. Kannan. Some results on fixed points. *Bull. Calcutta Math. Soc.*, 60:71–76, 1968.

[65] R. P. Kelisky and T. J. Rivlin. Iterates of bernstein polynomials. *Pacific Journal of Mathematics*, 21:511–520, 1967.

[66] M. A. Khamsi. Remarks on cone metric spaces and fixed point theorems of contractive mappings. *Fixed Point Theory and Applications*, 2010. Article ID 315398.

[67] M. Kikkawa and T. Suzuki. Some similarity between contractions and kannan mappings. *Bull. Kyushu Inst. Technol. Pure Appl. Math*, 55:1–13, 2008.

[68] W. A. Kirk, P. S. Srinavasan, and P. Veeramani. Fixed points for mappings satisfying cyclical contractive conditions. *Fixed Point Theory*, 4:79–89, 2003.

[69] M. A. Krasnoseljski and P. P. Zabrejko. *Geometrical Methods in Nonlinear Analysis*. Springer-Verlag, 1984.

[70] M. Lassonde. Fixed points of kakutani factorizable multifunctions. *Journal of Mathematical Analysis and Applications*, 152:46–60, 1990.

[71] W. R. Mann. Mean value methods in iteration. *Proceedings of the American Mathematical Society*, 4(3):506–510, 1953.

[72] S. G. Matthews. Partial metric topology. *Annals of the New York Academy of Sciences*, 728(1):183–197, 1994.

[73] G. Mot and A. Petrusel. Fixed point theory for a new type of contractive multivalued operators. *Nonlinear Analysis: Theory, Methods and Applications*, 70:3371–3377, 2009.

[74] S. B. Nadler and B. Sam. Multi-valued contraction mappings. *Pacific J. Math.*, 39:475–488, 1969.

[75] M. A. Noor. New approximation schemes for general variational inequalities. *Journal of Mathematical Analysis and applications*, 251(1):217–229, 2000.

[76] D. O'Regan and A. Petrusel. Fixed point theorems for generalized contractions in ordered metric spaces. *Journal of Mathematical Analysis and Applications*, 341:1241–1252, 2008.

[77] F. Ozturk Celikler. Convergence analysis for a modified sp iterative method. *The Scientific World Journal*, 2014. Article ID 840504.

[78] J. R. Parker. Harmonic maps into surfaces with two-dimensional cone metrics. *VIII School on Differential Geometry (Campinas, 1992) Mat. Contemp*, 4:139–151, 1993.

[79] J. R. Parker. Cone metrics on the sphere and livné's lattices. *Acta Math.*, 196(1):1–64, 2006.

[80] H. K. Pathak and N. Shahzad. Fixed point results for generalized quasi-contraction mappings in abstract metric spaces. *Nonlinear Analysis: Theory, Methods and Applications*, 71(12):6068–6076, 2009.

[81] G. Petrusel. Cyclic representations and periodic points. *Studia Univ. Babes-Bolyai. Math.*, 50:107–112, 2005.

[82] A. C. M. Ran and M. C. B. Reurings. A fixed point theorem in partially ordered sets and some applications to matrix equations. *proceedings of the American Mathematical Society*, 132:1435–1443, 2003.

[83] S. Reich and A. J. Zaslavski. Convergence of inexact iterative schemes for nonexpansive set-valued mappings. *Fixed Point Theory and Applications*, pages 1–10, 2010. Article ID 518243.

[84] Sh. Rezapour, R. H. Haghi, and N. Shahzad. Some notes on fixed points of quasi-contraction maps. *Applied Mathematics Letters*, 23(4):498–502, 2010.

[85] Sh. Rezapour and R. Hamlbarani Haghi. Some notes on the paper "cone metric spaces and fixed point theorems of contractive mappings". *Journal of Mathematical Analysis and Applications*, 345(2):719–724, 2008.

[86] B. E. Rhoades and Z. Xue. Comparison of the rate of convergence among picard, mann, ishikawa, and noor iterations applied to quasicontractive maps. *Fixed Point Theory and Applications*, pages 1–12, 2010. Article ID 169062.

[87] S. Romaguera. A kirk type characterization of completeness for partial metric spaces. *Fixed Point Theory and Applications*, 2010. Article ID 493298.

[88] H. L. Royden. *Real analysis*. Macmillan publishing company, third edition, 1988.

[89] W. Rudin. *Functional Analysis*. McGraw-Hill, Second edition, 1991.

[90] I. A. Rus. *Principles and applications of fixed point theory*. Editura Dacia, Cluj-Napoca, 1979.

[91] I. A. Rus. Iterates of bernstein operators, via contraction principle. *Journal of Mathematical Analysis and Applications*, 292:259–261, 2004.

[92] I. A. Rus, A. Petrusel, and A. Santamarian. Data dependence of the fixed point set of multivalued weakly picard operators. *Nonlinear Analysis: Theory, Methods and Applications*, 52:1947–1959, 2003.

[93] F. Khojasteh S. Moradi. Endpoints of multi-valued generalized weak contraction mappings. *Nonlinear Analysis: Theory, Methods and Applications*, 74(6):2170–2174, 2011.

[94] B. Samet, C. Vetro, and P. Vetro. Fixed point theorems for α-ψ-contractive type mappings. *Nonlinear analysis: theory, methods and applications*, 75(4):2154–2165, 2012.

[95] G. Schumacher and S. Trapani. Variation of cone metrics on riemann surfaces. *Journal of mathematical analysis and applications*, 311(1):218–230, 2010.

[96] S. L. Singh and S. N. Mishra. Remarks on recent fixed point theorems. *Fixed Point Theory and Applications*, 2010. Article ID 452905.

[97] Y. Sun. A fixed point theorem for mixed monotone operators with applications. *Journal of mathematical analysis and applications*, 156(1):240–252, 1991.

[98] T. Suzuki. A generalized banach contraction principle that characterizes metric completeness. *Proceedings of the American mathematical Society*, 136:1861–1869, 2008.

[99] T. Suzuki. A new type of fixed point theorem in metric spaces. *Nonlinear Analysis: Theory, Methods and Applications*, 71(11):5313–5317, 2009.

[100] T. Suzuki. Basic inequality on ab-metric space and its applications. *Journal of inequalities and applications*, 2017(1):256, 2017.

[101] T. Suzuki. A generalization of the banach contraction principle in non-complete metric spaces. *Filomat*, 31:3357–3363, 2017.

[102] D. Thakur, B. S. Thakur, and M. Postolache. New iteration scheme for numerical reckoning fixed points of nonexpansive mappings. *Journal of Inequalities and Applications*, 2014(328):1–15, 2014.

[103] J. S. Vandergraft. Newton's method for convex operators in partially ordered spaces. *SIAM Journal on Numerical Analysis*, 4(3):406–432, 1967.

[104] H. Xu. A note on the Ishikawa iteration scheme. *Journal of Mathematical Analysis and Applications*, 167(2):582–587, 1992.

[105] K. Yanagi. On some fixed point theorems for multivalued mappings. *Pacific J. Math.*, 87(1):233–240, 1980.

[106] P. P. Zabrejko. K-metric and k-normed linear spaces: survey. *Collectanea Mathematica*, 48(4):825–859, 1997.

[107] T. Zamfirescu. Fixed point theorems in metric spaces. *Archiv der Mathematik*, 23:292–298, 1972.

Index